AGRICULTURAL SYSTEM MODELS
in Field Research and Technology Transfer

AGRICULTURAL SYSTEM MODELS
in Field Research and Technology Transfer

Lajpat R. Ahuja
Liwang Ma
Terry A. Howell

LEWIS PUBLISHERS

A CRC Press Company
Boca Raton London New York Washington, D.C.

Library of Congress Cataloging-in-Publication Data

Agricultural system models in field research and technology transfer / [edited by] Lajpat
R. Ahuja, Liwang Ma, Terry A. Howell
    p. cm.
    Includes bibliographical references (p.).
    ISBN 1-56670-563-0
    1. Agricultural systems—Computer simulation. 2. Agriculture--Research--Computer
simulation. 3. Agriculture--Technology transfer--Computer simulation. I. Ahuja, L.
(Lajpat) II. Ma, Liwang. III. Howell, Terry A.

S494.5.D3 A4313 2002
630'.1'13—dc21
      2002016077
      CIP

Visit the CRC Press Web site at www.crcpress.com

© 2002 by CRC Press LLC
Lewis Publishers is an imprint of CRC Press LLC

No claim to original U.S. Government works
International Standard Book Number 1-56670-563-0
Library of Congress Card Number 20022016077
Printed in the United States of America 1 2 3 4 5 6 7 8 9 0
Printed on acid-free paper

Foreword

I am pleased to provide this foreword on such an exciting topic. Research on agricultural systems models represents both a new frontier and a return to a holistic look at the world.

When the first scientists began to examine the world from an analytical instead of a mythological viewpoint, they had few tools other than their powers of observation. The only way to understand the natural world was to dissect it. As time went on, specialization allowed researchers to understand amazing details about the world, including all aspects of agriculture. We can never overestimate the importance of these discoveries, such as uncovering the mysteries of how plants use light to grow or how genes govern the structure and function of living organisms. At the same time, we have always known that no specific organism or process acts in isolation.

Although a myriad of things still need to be learned and understood, we are now at the point where we can start to put the world back together. We also have the technology in computers that allows us to store and organize the millions of individual pieces of information that make up a system. User-friendly interfaces let producers and other users obtain meaningful results while inputting only a small amount of site-specific information. This systems approach is not just an intellectual exercise or academic luxury — it is a necessity.

Agricultural producers face a much more complex world than their ancestors. More people need to be fed with fewer natural resources (land, water) as well as more competing uses for these resources. Quality of our natural resources is a public concern. Many small family farms have given way to national and international conglomerates. Traditional crops have lost market value, creating the need for nontraditional farming approaches and reduced input costs. Today's farmers not only have to worry about the health, success, and marketing of their product — enough of a challenge in itself — they must also ensure that, while providing high value, low-cost food for consumers, they maintain a healthy environment. Because farms often lie next to urban areas, producers must also consider social demands to reduce odor and noise. In short, today's farmer is a total resource manager. Producers need agricultural systems models to help them make appropriate decisions amid an ever-changing environment.

Even apparently minor adjustments to agricultural practices and treatments could have major impacts. In this information age, through tools like the Internet and remote sensing, producers have access to a variety of real-time data about crops, range conditions, and weather. But they need quantitative tools to make sense of these data. The models can also give the producers objective ground for presenting and supporting their decisions to neighbors, legislators, and other interested parties who may question management choices.

In the 21st century, agricultural systems models and decision support systems with ancillary information and databases will increasingly play a vital role in transferring knowledge and technology such that it becomes useful in addressing society's needs. The Agricultural Research Service (ARS) has been a leader in developing and promoting applications of systems models, and its role in this area will greatly increase in the coming decades. These developments were often made in partnership with our university and international collaborators.

In this book, we will see demonstrations of some of these first sophisticated models. We will also see discussion of existing problems, knowledge gaps, challenges and ways those challenges can be met. This book should produce greater understanding of the science issues involved and some guidance on how to address these issues.

I commend ARS scientists for taking a leadership role in this endeavor, and I look forward to watching this field grow and to fostering the important work presented here.

Floyd P. Horn
Administrator, Agricultural Research Service
U.S. Department of Agriculture

Preface

The purpose of this book is to present the state-of-science of applications of agricultural system models, and tremendous benefits to be derived from the use of these computer models in agricultural research and technology transfer in the 21st Century. Leading international agricultural system scientists present their experiences and provide guidance on how the models can be used to enhance the quality of field research, transfer of research information and technology to farmers, and decision support for agricultural management. They also present expert review of the existing problems and possible solutions to improve these applications in the future. An international modular modeling computer framework is proposed to build problem-specific models in the future. Future research needs to fill major knowledge gaps are identified. The presentations cover modeling of natural resources, crop production, grazing lands, and animal production systems.

The first chapter summarizes the current status of whole-system integration and modeling in agriculture, existing problems, and future vision for their highly useful applications in research and technology transfer. The second chapter outlines the approaches taken by the CSIRO Plant Industry in Australia to develop models and decision support tools for managing grazing enterprises, and presents examples of their applications in sheep and cattle grazing industries. Chapter 3 presents experiences with the use of a cotton simulation model/decision support system, GOSSYM-COMAX, for management of water, nitrogen, herbicide, and growth regulator applications in cotton crop on farmers' fields, problem and policy analysis research, and education. GOSSYM was the first comprehensive crop management model developed in the U.S.

Chapter 4 presents applications, similar to those mentioned previously, of a soybean simulation model, GLYCM, for field management of soybean crop. Chapter 5 presents highly valuable experiences with different methods of agrotechnology transfer, including a decision support system, DSSAT, built around the CERES and CROPGRO family of crop models, in tropical and subtropical countries all over the world. Chapter 6 describes efforts of the International Fertilizer Development Center (IFDC) in using DSSAT and the associated global network of collaborators to develop and transfer fertilizer use and related technologies for sustainable agricultural production in developing countries. In Chapter 7, the authors present a comparison of the leading corn and soybean models for their performance and application under the most difficult water stress conditions in the U.S. Chapter 8 presents Australian experiences with using crop models to design better farming practices in the semiarid dry land farming systems, and the evolution of a new soil and crop-based Agricultural Production Systems Simulator, APSIM, and its application in farming system analysis and design. Chapter 9 presents an excellent review of the potential and current, mostly research, applications of models for a number of crops in the semiarid regions of the world.

Chapter 10 addresses the current need for having different models for different spatial scales, from individual plants or small plots to field, watershed, and basin scales, and example applications of four such models. Chapter 11 provides a good example of a distributed, multiple application of an agricultural system model to simulate spatial and temporal (year-to-year) variability of crop growth and nitrogen status in a field for site-specific fertilizer recommendations. Chapter 12 describes experiences with three approaches to using models for site-specific agriculture problems in spatially variable fields — making multiple model runs, using remote sensing of crop to adjust model inputs, state variables or parameters, and using optimization schemes to obtain variable model inputs. Chapter 13 reviews the literature on relationships of soil properties and crop yield to topographic attributes, and presents the hypothesis that topographic analysis and available soil map data can be combined with agricultural system models to improve spatial characterization of landscape processes within a field for precision management and for up-scaling results to watershed and larger scales.

Chapter 14 deals with the biggest and most difficult problem in modeling — how to determine model parameters for different components of the system and their change with environmental stresses and management practices. Chapter 15 describes a state-of-the-technology, object-oriented, modular modeling computer framework, the Object Modeling System, which is under development. This framework would enable future model developers to create and quickly update custom models specific to problems or scales of application from a library of modules in the computer. This framework would also help coordinate national and international efforts in modeling and serve as a reference library of quantified knowledge of system components to guide future research. Finally, Chapter 16 presents a thoughtful, competent list of agricultural concerns, that future research needs to address.

The editors are very grateful to the contributors for their best efforts in preparing and revising their chapters.

Lajpat R. Ahuja
USDA-ARS
Fort Collins, CO

Liwang Ma
USDA-ARS
Fort Collins, CO

Terry A. Howell
USDA-ARS
Bushland, TX

The Editors

Lajpat (Laj) R. Ahuja is a supervisory soil scientist and research leader of the USDA-ARS, Great Plains Systems Research Unit, Fort Collins, Colorado. He has made original and pioneering research contributions in several areas of agricultural systems: infiltration and water flow in soils, estimation of hydraulic properties, and scaling of their spatial variability; transport of agrochemicals to runoff and to groundwater through soil matrix and macropores; quantification of the effects of tillage and other management practices on above properties and processes; and modeling of the entire agricultural systems and application of system models in field research, technology transfer, and management decision support.

As development team leader, Ahuja guided the development, validation, and publication of the ARS Root Zone Water Quality Model (RZWQM), that is being widely used for evaluating effects of management on water quality and crop production. The Unit team has also developed a whole farm/ranch decision support system, GPFARM, for evaluating production, economics, and environmental impacts of alternative management systems.

Ahuja has authored or co-authored more than 200 publications; he is also senior editor of the book on RZWQM. He has served as associate editor (1987 to 1992) and technical editor (1994 to 1996) of the *Soil Science Society of America Journal*. He is an invited guest editor of an upcoming special issue of *Geoderma* on "Quantifying Agricultural Management Effects on Soil Properties and Processes," has served as advisor and consultant to several national and international organizations, and has organized several interagency/international workshops. Ahuja won the USDA-ARS, Southern Plains Area, Scientist of the Year Award in 1989. He was elected Fellow of the Soil Science Society of America in 1994 and the American Society of Agronomy in 1996.

Liwang Ma is a soil scientist with USDA-ARS, Great Plains Systems Research Unit, Fort Collins, Colorado. Dr. Ma received his B.S. and M.S. in biophysics from Beijing Agricultural University (now China Agricultural University) in 1984 and 1987, respectively, and his Ph.D. in soil physics from Louisiana State University in 1993. He is co-editor of three books and author or co-author of 50 publications. His research on agricultural system modeling extends from water and solute transport, to carbon and nitrogen dynamics, to plant growth. Dr. Ma is the recipient of the 1994 Prentiss E. Schilling Outstanding Dissertation Award in the College of Agriculture at Louisiana State University, Baton Rouge. He is now serving as associate editor for the *Soil Science Society of America Journal*.

Terry A. Howell is an agricultural engineer and research leader of the USDA-ARS, Water Management Research Unit at Bushland, Texas. He is noted for irrigation scheduling and evapotranspiration research, and for his original contributions on the design theory for drip irrigation laterals, design parameters for drainage systems based on crop performance, improved understanding of crop yield-water use relationships, use of remote weather stations to provide real-time irrigation scheduling information, improved weighing lysimeter designs for both reconstructed soil profiles and monolithic soil profiles, the importance of advective influences on evaporation in the Southern Great Plains, and improved understanding of the dynamics of irrigation application and efficiency of center-pivot and lateral-move sprinkler systems with various application methods. In particular, Howell led in developing and supplying the research data used in the North Plains ET (evapotranspiration) Network, which has been recognized by the State of Texas (1999 Environmental Excellence Award from the Texas Natural Resource Conservation Commission as well as the Governor of Texas and the Texas Legislature; the Texas A&M University Vice Chancellor's Award for Excellence in Research); a separate Federal Agency (the 1999 Federal Energy Management and Water Conservation Award from the U.S. Department of Energy); by ARS (1999 ARS Technology Transfer Award); and by the Federal Laboratory Consortium (the 2000 Mid-Continent Regional Laboratory Award for exceptional support and encouragement to the transfer of federal technologies to the private sector).

Howell is a Fellow of the American Society of Agricultural Engineers (1992) and the American Society of Agronomy (1999). He received the Person of the Year Award in 1995 from the Irrigation Association, the Royce J. Tipton Award in 1998 from the American Society of Civil Engineers, the Hancor Soil and Water Engineering Award in 2000 from the American Society of Agricultural Engineers, and the Senior Scientist Award for the Southern Plains Area from USDA-ARS in 2000.

Contributors

Lajpat R. Ahuja
Great Plains Systems Research Unit
USDA-ARS
Fort Collins, Colorado

Jeffry G. Arnold
Natural Resources Systems Research Unit
USDA-ARS
Temple, Texas

James C. Ascough, II
Great Plains Systems Research Unit
USDA-ARS
Fort Collins, Colorado

Walter E. Baethgen
International Fertilizer Development Center-
  Latin America
Montevideo, Uruguay

Donald N. Baker
Baker Consulting
Starkville, Mississippi

Edward M. Barnes
USDA-ARS
Phoenix, Arizona

William D. Batchelor
Agricultural and Biosystems Engineering
Iowa State University
Ames, Iowa

Tjark S. Bontkes
International Fertilizer Development Center-
  Africa
Lome, Togo

Walter T. Bowen
International Fertilizer Development
  Center/International Potato Center
Quito, Ecuador

Peter S. Carberry
CSIRO Sustainable EcoSystems
Toowoomba, Queensland, Australia

Olaf David
Great Plains Systems Research Unit
USDA-ARS
Fort Collins, Colorado

John R. Donnelly
CSIRO Plant Industry
Canberra, Australia

Hugh Dove
CSIRO Plant Industry
Canberra, Australia

Gale H. Dunn
Great Plains Systems Research Unit
USDA-ARS
Fort Collins, Colorado

Robert H. Erskine
Great Plains Systems Research Unit
USDA-ARS
Fort Collins, Colorado

Michael Freer
CSIRO Plant Industry
Canberra, Australia

Ariella F. Glinni
World Meteorological Organization
Geneva, Switzerland

Timothy R. Green
Great Plains Systems Research Unit
USDA-ARS
Fort Collins, Colorado

Jonathan D. Hanson
Northern Great Plains Research Laboratory
USDA-ARS
Mandan, North Dakota

Dean Hargreaves
CSIRO Sustainable EcoSystems
Toowoomba, Queensland, Australia

Jerry L. Hatfield
National Soil Tilth Laboratory
USDA-ARS
Ames, Iowa

Zvi Hochman
CSIRO Sustainable EcoSystems
Toowoomba, Queensland, Australia

Gerrit Hoogenboom
Department of Biological and Agricultural
  Engineering
University of Georgia
Griffin, Georgia

Terry A. Howell
Conservation and Production Research
  Laboratory
USDA-ARS
Bushland, Texas

Ayse Irmak
Agricultural and Biological Engineering
University of Florida
Gainesville, Florida

Attachai Jintrawet
Chiang Mai University
Chiang Mai, Thailand

James W. Jones
Agricultural and Biological Engineering
University of Florida
Gainesville, Florida

Vijaya Gopal Kakani
Department of Plant and Soil Services
Mississippi State University
Mississippi State, Mississippi

Brian A. Keating
Agricultural Production Systems Research Unit
CSIRO Sustainable Ecosystems
Indooroopilly, Brisbane, Queensland, Australia

Kurt Christian Kersebaum
Department for Landscape Systems Analysis
Center of Agricultural Landscape Research
Muencheberg, Germany

Bruce A. Kimball
Water Conservation Laboratory
USDA-ARS
Phoenix, Arizona

James R. Kiniry
Natural Resources Systems Research Unit
USDA-ARS
Temple, Texas

Karsten Lorenz
Department for Landscape Systems Analysis
Center of Agricultural Landscape Research
Muencheberg, Germany

Liwang Ma
Great Plains Systems Research Unit
USDA-ARS
Fort Collins, Colorado

Steven L. Markstrom
U.S. Geological Survey
Denver, Colorado

Ana Martinez
Department of Civil Engineering
Colorado State University
Fort Collins, Colorado

Robert L. McCown
CSIRO Sustainable EcoSystems
Toowoomba, Queensland, Australia

James M. McKinion
Genetics and Precision Agriculture Research
  Unit
USDA-ARS
Mississippi State, Mississippi

Andrew D. Moore
CSIRO Plant Industry
Canberra, Australia

David C. Nielsen
Central Great Plains Research Station
USDA-ARS
Akron, Colorado

Richard M. Ogoshi
Tropical Plant and Soil Science
University of Hawaii
Honolulu, Hawaii

Yakov Pachepsky
Hydrology Laboratory
USDA-ARS
Beltsville, Maryland

Joel Paz
Agricultural and Biosystems Engineering
Iowa State University
Ames, Iowa

Kambham Raja Reddy
Department of Plant and Soil Services
Mississippi State University
Mississippi State, Mississippi

Vangimalla R. Reddy
Alternate Crops Systems Laboratory
USDA-ARS
Beltsville, Maryland

Hannes I. Reuter
Department of Soil Landscape Research
Center of Agricultural Landscape Research
Muencheberg, Germany

Kenneth W. Rojas
Great Plains Systems Research Unit
USDA-ARS
Fort Collins, Colorado

E. John Sadler
Coastal Plains Soil, Water and Plants
  Research Center
USDA-ARS
Florence, South Carolina

Libby Salmon
CSIRO Plant Industry
Canberra, Australia

Ian W. Schneider
Great Plains Systems Research Unit
USDA-ARS
Fort Collins, Colorado

Marvin J. Shaffer
Great Plains Systems Research Unit
USDA-ARS
Fort Collins, Colorado

Richard J. Simpson
CSIRO Plant Industry
Canberra, Australia

Upendra Singh
International Fertilizer Development Center
Muscle Shoals, Alabama

Mannava V.K. Sivakumar
World Meteorological Organization
Geneva, Switzerland

Dennis J. Timlin
Alternate Crops and Systems Laboratory
USDA-ARS
Beltsville, Maryland

Andre du Toit
Agricultural Research Council
Potchefstroom, South Africa

Gordon Y. Tsuji
Tropical Plant and Soil Science
University of Hawaii
Honolulu, Hawaii

Goto Uehara
Tropical Plant and Soil Science
University of Hawaii
Honolulu, Hawaii

Ole Wendroth
Department of Soil Landscape Research
Center of Agricultural Landscape Research
Muencheberg, Germany

Frank D. Whisler
Department of Plant and Soil Services
Mississippi State University
Mississippi State, Mississippi

Paul W. Wilkens
International Fertilizer Development Center
Muscle Shoals, Alabama

Yun Xie
Beijing Normal University
Beijing, China

Contents

Whole System Integration and Modeling — Essential to Agricultural Science and Technology in the 21st Century

Lajpat R. Ahuja, Liwang Ma, and Terry A. Howell

CONTENTS

CURRENT STATUS

Agricultural system integration and modeling have gone through more than 40 years of development and evolution. Before the 1970s, a vast amount of modeling work was done for individual processes of agricultural systems and a foundation for system modeling was built. For example, in soil water movement, models and theories were developed in the areas of infiltration and water redistribution (Green and Ampt, 1911; Philips, 1957; Richards, 1931), soil hydraulic properties (Brooks and Corey, 1964), tile drainage (Bouwer and van Schilfgaarde, 1963), and solute transport (Nielsen and Biggar, 1962). In plant-soil interactions, models and theories were developed for evapotranspiration (Penman, 1948; Monteith, 1965), photosynthesis (Saeki, 1960), root growth (Foth, 1962; Brouwer, 1962), plant growth (Brouwer and de Wit, 1968), and soil nutrients (Olsen and Kemper, 1967; Shaffer et al., 1969).

Although in the early 1970s, a few models were developed to include multiple components of an agricultural system, such as the model developed by Dutt et al. (1972), agricultural system models were not fully developed and used until the 1980s. In the 1980s, several system models were developed, such as the PAPRAN model (Seligman and van Keulen, 1981), CREAMS (Knisel, 1980), GOSSYM (Baker et al., 1983), EPIC (Williams and Renard, 1985), GLYCIM (Acock et al., 1985), PRZM (Carsel et al., 1985), CERES (Ritchie et al., 1986), COMAX (Lemmon, 1986), NTRM (Shaffer and Larson, 1987), and GLEAMS (Leonard et al., 1987). In the 1990s, agricultural system models were more mechanistic and had more agricultural components, such as CROPGRO (Hoogenboom et al., 1992; Boote et al., 1997), Root Zone Water Quality Model (RZWQM) (RZWQM Team, 1992; Ahuja et al., 2000), APSIM (McCown et al., 1996), and GPFARM (Ascough et al., 1995; Shaffer et al., 2000). In addition, the new system models have taken advantage of current computer technology and come with a Windows™-based user interface to facilitate data management and model simulation. Some models are also linked to a decision support system (DSS), such as DSSAT which envelopes CERES and CROPGRO (Tsuji et al., 1994; Hoogenboom et al., 1999) and GPFARM (Shaffer et al. 2000). Agricultural system research and modeling are now being promoted by several international organizations, such as ICASA (International Consortium for Agricultural Systems Applications) and other professional societies.

The collective experiences from model developers and users show that, even though not perfect, the agricultural system models can be very useful in field research, technology transfer, and management decision making as demonstrated in this book. These experiences also show a number of problems or issues that should be addressed to improve the models and applications. The most important issues are:

1. System models need to be more thoroughly tested and validated for science defendability under a variety of soil, climate, and management conditions, with experimental data of high resolution in time and space.
2. Comprehensive shared experimental databases need to be built based on existing standard experimental protocols, and measured values related to modeling variables, so that conceptual model parameters can be experimentally verified.
3. Better methods are needed for determining parameters for different spatial and temporal scales, and for aggregating simulation results from plots to fields and larger scales.
4. The means to quickly update the science and databases is necessary as new knowledge and methods become available. A modular modeling approach will greatly help this process together with a public modular library.
5. Better communication and coordination is needed among model developers in the areas of model development, parameterization and evaluation.
6. Better collaboration between model developers and field scientists is needed for appropriate experimental data collection and for evaluation and application of models. Field scientists should be included within the model development team from the beginning, not just as a source of model validation data.
7. An urgent need exists for filling the most important knowledge gaps: agricultural management effects on soil–plant–atmosphere properties and processes; plant response to water, nutrient and temperature stresses; and effects of natural hazards such as hail, frost, insects, and diseases.

THE FUTURE VISION

Understanding real-world situations and solving significant agronomic, engineering, and environmental problems require integration and quantification of knowledge at the whole system level. In the 20th Century, we made tremendous advances in discovering fundamental principles in different scientific disciplines that created major breakthroughs in management and technology for agricultural systems, mostly by empirical means. However, as we enter the 21st century, agricultural research has more difficult and complex problems to solve.

The environmental consciousness of the general public is requiring us to modify farm management to protect water, air, and soil quality, while staying economically profitable. At the same time, market-based global competition in agricultural products is challenging economic viability of the traditional agricultural systems, and requires the development of new and dynamic production systems. Fortunately, the new electronic technologies can provide us a vast amount of real-time information about crop conditions and near-term weather via remote sensing by satellites or ground-based instruments and the Internet, that can be utilized to develop a whole new level of management. However, we need the means to capture and make sense of this vast amount of site-specific data.

Integration and quantification of knowledge at the whole-system level is essential to meeting all the above challenges and needs of the 21st century. Our customers, the agricultural producers, are asking for a quicker transfer of research results in an integrated usable form for site-specific management. Such a request can only be met with system models, because system models are indeed the integration and quantification of current knowledge based on fundamental principles and laws. Models enhance understanding of data taken under certain conditions and help extrapolate their applications to other conditions and locations. Models are the only way to find and understand the interrelationships among various components in a system and integrate numerous experimental results from different conditions.

System modeling has been a vital step in many scientific achievements. We would not have gone to the moon successfully without the combined use of good data and models. Models have been used extensively in designing and managing water resource reservoirs and distribution systems, and in analyzing waste disposal sites. Although a lot more work is needed to bring models of agricultural systems to the level of physics and hydraulic system models, agricultural system models have gone through a series of breakthroughs and can be used for practical applications, with some good data.

Integration of Modeling with Field Research

Integrating system modeling with field research is an essential first step to improve model usability and make a significant impact on the agriculture community. This integration will greatly benefit both field research and models in the following ways:

- Promote a systems approach to field research.
- Facilitate better understanding and quantification of research results.
- Promote quick and accurate transfer of results to different soil and weather conditions, and to different cropping and management systems outside the experimental plots.
- Help research to focus on the identified fundamental knowledge gaps and make field research more efficient, i.e., get more out of research per dollar spent.
- Provide the needed field test of the models, and improvements, if needed, before delivery to other potential users — agricultural consultants, farmers/ranchers, state extension agencies, and federal action agencies (NRCS, EPA, and others).

The most desirable vision for agricultural research and technology transfer is to have a continual two-way interaction among the cutting-edge field research, process-based models of agricultural systems, and decision support systems (Figure 1.1). The field research can certainly benefit from the process models as described above, but also a great deal from the feedback from the decision support systems (DSSs). On the other hand, field research forms the pivotal basis for models and DSSs. The DSSs generally have models as their cores (simple or complex).

Modeling of agricultural management effects on soil-plant-atmosphere properties and processes has to be a center piece of an agricultural system model, if it is to have useful applications in field research and decision support for improved management. An example is the ARS Root Zone Water Quality Model (RZWQM), which was built to simulate management effects on water quality and crop production (Figure 1.2, Ahuja et al., 2000).

After a system model has passed the field testing and validation and both modelers and field scientists are satisfied with the results, it should be advanced to the second step: application. Only through model application to specific cases can a model be further improved by exposure to differing circumstances. The field-tested model can be used as a decision aid for best management practices, including site-specific management or precision agriculture, and as a tool for in-depth analysis of

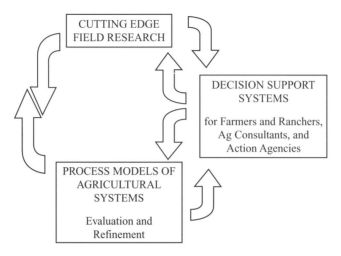

Figure 1.1   Interaction among field research, process-based system models, and decision support systems.

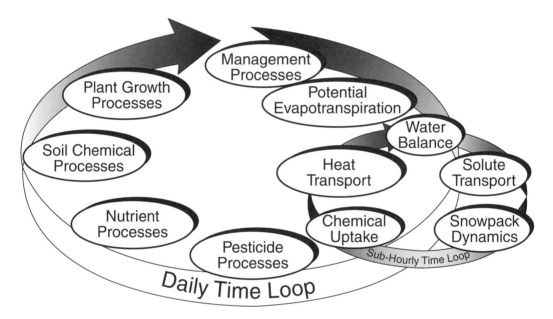

Figure 1.2   Management practices are the centerpiece of a process-based cropping system model RZWQM.

problems in management, environmental quality, global climate change, and other new emerging issues.

New Decision Support Systems

Decision support systems commonly have an agricultural system model at their core, but are supported by databases, an economic analysis package, an environmental impact analysis package, a user-friendly interface up front for users to check and provide their site-specific data, and a simple graphical display of results at the end. An example is the design of ARS GPFARM-DSS (Figure 1.3, Ascough et al., 1995; Shaffer et al., 2000). GPFARM (Great Plains Framework for Agricultural Resource Management) is a whole-farm decision support system for strategic planning — evaluation

GPFARM: A Farm Level DSS

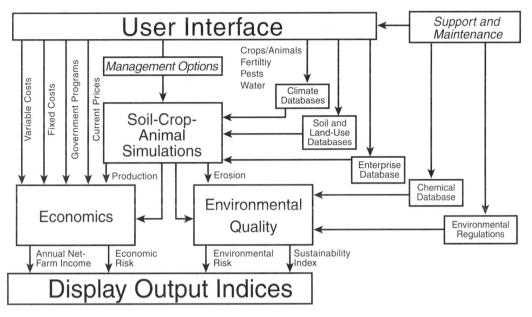

Figure 1.3 The design of the GPFARM decision support system.

of alternate cropping systems, range-livestock systems, and integrated crop livestock farming options, for production, economics, and environmental impacts.

Currently, process-level models may be difficult for agricultural consultants, NRCS field office personnel, and producers to use. A new approach toward a DSS is to create an integrated research information database as a core of the DSS in place of a model. A system model, validated against available experimental data, is used to generate production and environmental impacts of different management practices for all major soil types, weather conditions, and cropping systems outside the experimental limits. This model-generated information is then combined with available experimental data and the long-term experience of farmers and field professionals to create the database (Rojas et al., 2000). The database can be combined with an economic analysis package. It may also be connected to a so-called "Multi-Objective Decision Support System" for determining trade-offs between conflicting objectives, such as economic return and environmental quality. It is also very flexible in generating site-specific recommendations.

Collaborations for Further Developments

In the future, model developers need to work together to address the seven problem areas described in the previous section, and then train and work with field scientists to improve model usability and applicability in solving real world problems. Also, there is a need to document system models and simulated processes better, so that field scientists will be able to understand these processes without too much difficulty. We also need to document good case studies on model applications to serve as guides for field users. Any improvements to an existing model could be checked against these documented cases to see if these improvements are applicable to all situations. Since most field data are not collected for the purpose of evaluating with a system model, some good system-oriented experiments may be needed. International efforts are needed to coordinate system modeling and to encourage model developers and field scientists to work on identified knowledge gaps and research priorities.

An Advanced Modular Modeling Framework for Agricultural Systems

A modular modeling computer framework will consist of a library of alternate modules (or subroutines) for different sub-processes of science, associated databases, and the logic to facilitate the assembly of appropriate modules into a modeling package. The modeling package can be tailored or customized to a problem, data constraints, and scale of application. The framework will:

1. Enable the use of best science for all components of a model.
2. Allow quick updates or replacement of science or database modules as new knowledge becomes available.
3. Eliminate duplication of work by modelers.
4. Provide a common platform and standards for development and implementation.
5. Serve as a reference and coordination mechanism for future research and developments.
6. Make collaboration much easier among modelers by sharing science modules/components and experimental/simulated databases, so that specialties of each individual modeling group can be maximally utilized.

These actions will prepare the models for the important role in the 21st century, and take the agricultural research and technology to the next higher plateau.

REFERENCES

Acock, B., V.R. Reddy, F.D. Whisler, D.N. Baker, H.F. Hodges, and K.J. Boote. 1985. *The Soybean Crop Simulator GLYCIM, Model Documentation,* USDA, Washington, D.C.

Ahuja, L.R. , K.W. Rojas, J.D. Hanson, M.J. Shaffer, and L. Ma. , Eds. 2000. *Root Zone Water Quality Model,* Water Resources Publication, Englewood, CO. 1–360.

Ascough, J.C., II et al. 1995. The GPFARM decision support system for the whole farm/ranch management, *Proc. Workshop on Computer Applications in Water Management,* Great Plains Agricultural Council Publication No. 154 and Colorado Water Resources Research Institute Information series No. 79, Fort Collins, CO, 53–56.

Baker, D.N., J.R. Lambert, and J.M. McKinion. 1983. GOSSYM: A simulator of cotton crop growth and yield, *South Carolina Exp. Sta. Tech. Bull.,* 1089.

Boote, K.J., J.W. Jones, G. Hoogenboom, and N.B. Pickering. 1998. The CROPGRO model for grain legumes. In *Understanding Options for Agricultural Production,* G.Y. Tsuiji, G. Hoogenboom, and P.K. Thornton, Eds., Kluwer, Dordrect, The Netherlands.

Bouwer, H. and J. van Schilfgaarde. 1963. Simplified methods of predicting the fall of water table in drained land, *Trans. ASAE,* 6:288–291, 296.

Brooks, R.H. and A.T. Corey. 1964. *Hydraulic Properties of Porous Media. Hydrology Paper 3,* Colorado State University, Fort Collins, CO, 1–15.

Brouwer, R. 1962. Distribution of dry matter in the plant, *Neth. J. Agric. Sci.,* 10:361–376.

Brouwer, R. and C.T. De Wit. 1968. A simulation model of plant growth with special attention to root growth and its consequences. In *Proc. 15th Easter School Agric. Sci.,* University of Nottingham, Nottingham, England, 224–242.

Carsel, R.F., L.A. Mulkey, M.N. Lorber, and L.B. Baskin. 1985. The pesticide root zone model (PRZM): a procedure for evaluating pesticide leaching threats to groundwater, *Ecological Modeling,* 30:49–69.

Dutt, G.R., M.J. Shaffer, and W.J. Moore. 1972. *Computer Simulation Model of Dynamic Bio-Physicochemical Processes in Soils,* Tech. Bull. 196, Agricultural Experimental Station, the University of Arizona, Tucson, AZ.

Foth, H.D. 1962. Root and top growth of corn, *Agron. J.,* 54:49–52.

Green, W.H. and G.A. Ampt. 1911. Studies on soil physics: 1. Flow of air and water through soils, *J. Agr. Sci.*, 4:1–24.

Hoogenboom, G., J.W. Jones, and K.J. Boote. 1992. Modeling growth, development, and yield of grain legumes using SOYGRO, PNUTGRO, and BEANGRO: a review, *Trans. ASAE*, 35:2043–2056.

Hoogenboom, G., P.W. Wilkens, and G.Y. Tsuji, Eds. 1999. *Decision Support System for Agrotechnology Transfer Version 3*, Vol. 4, University of Hawaii, Honolulu.

Knisel, W.G., Ed. 1980. CREAMS: a field-scale model for chemicals, runoff, and erosion from agricultural management systems. *USDA-SEA Conservation Res. Rept. No. 26*. Washington, D.C.

Lemmon, H. 1986. Comax: an expert system for cotton crop management, *Science*, 233:29–33.

Leonard, R.A., W.G. Knisel, and D.S. Still. 1987. GLEAMS: groundwater loading effects of agricultural management systems, *Trans. ASAE*, 30:1403–1418.

McCown, R.L., G.L. Hammer, J.N.G. Hargreaves, D.L. Holzworth. D.M. Freebairn. 1996. APSIM: a novel software system for model development, model testing, and simulation in agricultural systems research, *Agric. Syst.*, 50: 255–271.

Monteith, J.L. 1965. Evaporation and the environment, *Symp. Soc. Exper. Biol.*, 19:205–234.

Nielsen, D.R. and J.W. Biggar. 1962. Miscible displacement: III. Theoretical consideration, *Soil Sci. Soc. Am. Proc.*, 26:216–221.

Olsen, S.R. and W.D. Kemper. 1967. Movement of nutrients to plant roots, *Adv. Agron.*, 20:91–151.

Penman, H.L. 1948. Natural evaporation from open water, bare soil, and grass, *Proc. Roy. Soc. Lond.*, A193:120–145.

Philips, J. 1957. The theory of infiltration 1: the infiltration equation and its solution, *Soil Sci.*, 83:345–357.

Richards, L. 1931. Capillary conduction of liquids through porous medium. *Physics*, 1:318–333.

Ritchie, J.T., D.C. Godwin, and S. Otter-Nacke. 1986. *CERES-Wheat: A Simulation Model of Wheat Growth and Development, CERES Model description,* Department of Crop and Soil Science, Michigan State University, East Lansing.

Rojas, K.W., P. Heilman, J. Huddleson, L. Ma, L. R. Ahuja, J. L. Hatfield, and S. Kasireddy. 2000. An integrated research information and decision support system for conservation planning and management, *Agron. Abstr.*, 419.

RZWQM Team. 1992. Root zone water quality model, Version 1.0: Technical Documentation, *GPSR Tech. Report No. 2,* USDA-ARS-GPSR, Fort Collins, CO.

Saeki, T. 1960. Interrelationship between leaf amount, light distribution, and total photosynthesis in a plant community. *Bot. Mag., Tokyo,* 73:55–63.

Seligman, N.G. and H. van Keulen. 1981. PAPRAN: A simulation model of annual pasture production limited by rainfall and nitrogen. In *Simulation of Nitrogen Behavior of Soil-Plant Systems, Proc. Workshop,* M.J. Frissel and J.A. van Veen, Eds., Centre for Agricultural Publishing and Documentation, Wageningen, Netherlands. 192–221.

Shaffer, M.J., G.R. Dutt, and W.J. Moore. 1969. Predicting changes in Nitrogenous compounds in soil-water systems, *Water Pollution Control Research Series 13030 ELY,* 15–28.

Shaffer, M.J. and W.E. Larson, Eds. 1987. NTRM, a soil-crop simulation model for nitrogen, tillage, and crop-residue management, *U.S. Department of Agriculture, Conservation Research Report No. 34-1.*

Shaffer, M.J., P.N.S. Bartling, and J.C. Ascough, II. 2000. Object-oriented simulation of integrated whole farms: GPFARM framework, *Comput and Electron. in Agriculture,* 28:29–49.

Tsuji, G.Y., G. Uehara, and S. Balas, Eds. *1994 DSSAT version 3,* University of Hawaii, Honolulu.

Williams, J.R. and K.G. Renard. 1985. Assessment of soil erosion and crop productivity with process models (EPIC). In *Soil Erosion and Crop Productivity,* R.F. Follett and B.A. Stewart, Eds., American Society of Agronomy, Madison, WI, 68–102.

CHAPTER 2

Forage-Livestock Models
for the Australian Livestock Industry

John R. Donnelly, Richard J. Simpson, Libby Salmon, Andrew D. Moore, Michael Freer, and Hugh Dove

CONTENTS

INTRODUCTION

Farm decision making and grazing management decisions in particular have always been risky in Australia because of extreme variability in seasonal weather, widely varying land capability, and uncertainty about future commodity prices. Even so, the best farmers are able to supply markets with quality products in most seasons although increasingly stringent requirements for timeliness and security of supply make this a difficult task. In some districts, farm profitability and sustainability of the environment are also threatened by issues such as soil acidification, rising water tables, and dry land salinity.

In 1967, the late Dr. Fred Morley initiated a research program at CSIRO, Australia's national research organization, to use computer models to reduce the guesswork in agricultural decision-making. Morley (1968) believed that agriculture had to find more effective ways to make good management decisions than merely relying on experience and common sense. He was concerned that, in a rapidly changing world, experience could quickly become irrelevant and common sense was too often based only on qualitative approximations. He knew computer models were used successfully in defense and business to explore the probable consequences of decisions, but saw no evidence or appreciation that the same technology could offer significant benefits to agriculture. He also recognized the great potential for better decision making in agriculture inherent in an early attempt by Arcus (1963) to simulate a grazing system.

By 1972, scientific interest and involvement in modeling grazing systems had increased dramatically (Morley, 1972). Thirty years later, Donnelly and Moore (1999) cited four decision support tools from several countries, including Australia, that were being used successfully to deliver the benefits of grazing systems research. A direct outcome of Morley's vision was the release by CSIRO of the GRAZPLAN family of decision support (DS) tools (Donnelly et al., 1997), which have changed the way farmers assess their pastures and manage their animals. Australia is at the leading edge of development and commercial implementation of this practical technology (Donnelly and Moore, 1999).

This chapter outlines the approach taken at CSIRO Plant Industry in Australia to develop models and DS tools for managing grazing enterprises. The different types of biological models used in DS tools are reviewed briefly, including the link between the purpose of the DS tool and the level of detail in the underlying models. Key features of several DS tools are described together with examples of their application in the grazing industries of temperate southern Australia. The impact this experience has had on future development goals at CSIRO is discussed as well as an exciting new development that will enable models produced by different research groups to be linked into any DS tool so that tailor-made applications are available for specific tasks. A preliminary assessment of success of the CSIRO program to achieve better technology transfer in agriculture is presented.

MODELS FOR GRAZING SYSTEMS RESEARCH

Building models of biological processes has always been an integral part of research methodology. Agricultural scientists routinely use mathematical equations and statistical models to summarize data from their experiments. This is used as the basis of many general recommendations or rules of thumb about management. Mostly this approach has served agriculture well and crop and animal yields have risen through use of improved genetics and better management practices. In recent times, cost increases and long-term decline in the real value of commodity prices mean that modern agriculture is becoming less profitable and more risky. Farm decision making needs to be much smarter to remain competitive. Targeted advice tailored to the operational circumstances of specific farms must replace the rules of thumb and generic advice of old. The emergence of user-friendly DS tools makes this change possible.

A Hierarchy of Models

Agricultural models can help evaluate tactical, short-term, day-to-day farm management as well as strategic, longer-term management options. The predictions of a tactical model can often be checked against an actual outcome after a relatively short passage of time. This contrasts markedly with predictions from a strategic model where a real-world outcome may take many years to eventuate. In this case the quantitative accuracy of the predictions can be difficult to establish, as many years of data collection may be needed to provide an adequate sample for testing the predictions. In general, however, predictions from tactical models are relatively easy to interpret and are more likely to gain user acceptance.

Thornley (2001) provides an excellent summary of the different types of models used in agriculture. He and other authors (e.g., Beever et al., 2000) also refer to an "organizational hierarchy" in biological systems where processes modeled empirically at one level are used to predict system responses at the next or higher level. Linking these empirical models at a lower level can give insight into system behavior at a higher level as the response integrates scientific knowledge about the lower-level processes. In general, but not exclusively:

- Empirical models are less complex biologically than mechanistic models.
- Static models do not have time as a variable and provide a "snapshot" of a system's response; they are less complex computationally than dynamic models where model state changes with the passage of time.
- Deterministic models can be static or dynamic but are less complex computationally than the equivalent stochastic models that require many repetitions of a simulation to account for variability in the estimates of parameters.

Most of the models described in this chapter are deterministic. The predictions from dynamic models are driven by actual historical daily weather data and can be presented as a frequency distribution. However, it is not practical at present to include a random element as part of each equation in these comprehensive models, due mainly to difficulties in interpretation and also the time required for computation.

Conceptual Models

A useful first step in modeling a farming or grazing system is to place the relevant component processes into context using a conceptual model, often a diagram, that shows the links between the processes as well as the flows of information or material between them. Figure 2.1 illustrates a representation of the fate of soluble phosphorus (P) applied as an annual fertilizer dressing of superphosphate to pastures grazed by sheep in southern Australia. P is the main fertilizer applied to pastures in Australia. Understanding the fate of P is a key to determining how much fertilizer is needed to maintain soil fertility and is vital information to keep investments in fertilizer on target. However, this conceptual model is not one that can be used to make specific decisions about annual fertilizer use on a particular farm because it hides the complex cycle of biological and physical interactions between fertilizer, soil, plant and animals. Instead, it shows in broad quantitative terms the current state of knowledge. It also serves as an approximate, quantitative template of the key processes that must be included in computer models of nutrient cycling in grazing systems.

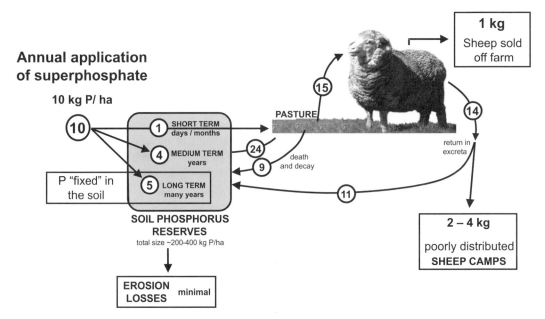

Figure 2.1   A conceptual model showing the approximate fate of P fertilizer applied annually to a pasture grazed by a flock of breeding ewes where 20% are replaced annually. The numbers within the circles show the partitioning of the annual flow of P (kg) to key pools and destinations in the cycle.

Empirical Models

At any level of system organization models can be based on empirical observation and statistical analysis. Good statistical models are parsimonious in terms of the number of predicting variables. They provide a precise description about the observations on which they are based and predictions have known errors. They do not imply causality or even knowledge of underlying processes although they may provide some insight into these. Predictions generally do not require validation, but extrapolation beyond that range of the data from which the model was derived requires caution.

Despite the limits on extrapolation, empirical models are valuable tools for guiding specific decisions rather than the operation of the whole farm or grazing enterprise. An example is a model developed by Donnelly (1984) to predict the probability of neonatal lamb deaths due to heat loss from exposure to chilling weather conditions, which is a major cause of lamb mortality in Australia. Field observations were used to predict the probability of neonatal mortality as a function of environmental chill, which was calculated from mean daily wind speed, mean daily temperature and rainfall. This simple model has been incorporated into a DS tool called LambAlive (see subsection on LambAlive) and released for commercial use (Donnelly et al., 1997). The predictions of lamb mortality under a wide range of weather conditions are sufficiently accurate to indicate to farmers whether it would be worthwhile adjusting the date for the start of the lambing period. In this case, no predictive advantage would be gained by attempting to model mortality at the more complex mechanistic level described below, although a mechanistic approach is necessary to predict animal production responses to cold weather.

Mechanistic Models

Livestock production is a complex of many dynamic processes that can be modified by management interventions. Mechanistic models can describe each process provided their mechanisms are understood and data are available for initialization. For example, the impact of cold weather on livestock production can be predicted from the extra heat an animal must generate to maintain body temperature. The amount of heat will be a function of the insulating properties of the body tissues, the area of skin surface from which heat can be lost, and the amount of external insulation provided by the animal's coat trapping a layer of air adjacent to the body (Freer et al., 1997). With this information it is possible to calculate the lower critical temperature of the animal; if the atmospheric temperature falls below this temperature, the additional metabolizable energy required to maintain body temperature can be calculated. A model or DS tool that includes these calculations gives a livestock manager an easy way to estimate the amount of extra feed animals will need to cope with the cold weather and still achieve their production targets. However, the manager will need to provide information about the current condition of the animals, details of their coat characteristics and the nutritional quality of the feed to be offered. This is significantly more information than that required by the more simple, empirical model predicting lamb mortality.

The purpose of a model, therefore, has a major bearing on the level of detail and number of processes that must be modeled for a particular application. Freer and Christian (1983) developed a model to estimate the intake of grazing animals and their need, if any, for feed supplements to meet specified production targets. The model has many predictive functions that are common to sheep and cattle, with specific parameters for a range of breed types and for all stages of growth and reproduction. Despite its generality, the model maintains a level of realism (Stuth et al., 1999) that has proved ideal for use in advisory situations. The model predicts the intake of metabolizable energy and protein by grazing sheep and cattle, and takes into account grazing selection and substitution of forage intake by supplements. The intake of the dietary protein and energy is partitioned for maintenance and production, but the model does not explicitly simulate processes involved in tissue metabolism. The model equations conform to the recommendations in the Australian feeding standards for ruminants (SCA, 1990), which are based on more than 50 years of research in Australia and elsewhere. Significant generality is achieved by scaling feed intake, body compositionand milk production to the mature size of the animal being simulated. As the animal develops, its productive attributes depend on its size and condition relative to its mature weight, rather than its current body weight. The model includes facilities for adjusting, in an empirical way, such responses as the effect of protein composition on the efficiency of wool growth, the effect of seasonal changes in the composition of digested herbage on the efficiency of weight gain, or the partition of absorbed nutrients between milk production and weight change.

More detailed models of animal nutrition and metabolism are required if the purpose is to advance research and understanding of ruminant nutrition. For example, it should be possible to predict the composition of weight gain in terms of subcutaneous, intramuscular and visceral fat and muscular and visceral protein from the amount and concentration of individual volatile fatty acids and amino acids produced from transactions in the rumen. Nagorcka et al. (2000) showed a significant improvement in modeling the rate of production of individual volatile fatty acids if substrate fermentation was linked to the three major groups of microbes found in the rumen. This opens a potential way for managers of beef feed lots to tailor feed composition so that desired carcass conformation is achieved to meet stringent market requirements for quality meat products. At present the approach is still at an early research stage but the mechanistic models are already providing greater understanding of links between underlying processes of production.

Although this highly mechanistic approach to modeling potentially offers significant advantages for intensive animal production, obstacles lie in the way of extending it to the management of grazing animals. Here, the main limitations, on a day-to-day basis, are the fluctuating quality of feed on offer and the difficulties in accurately characterizing the chemical composition of the diet selected by grazing animals. At present the more simple model based on the feeding standards has less scope for generating errors and is generally more reliable for making on-farm decisions about animal nutritional management.

Empirical or Mechanistic Models?

Both empirical and mechanistic models use mathematical equations to describe the quantitative responses of biological processes indicated in conceptual models. These mathematical models can range from mechanistic descriptions of cellular metabolism to empirical models of production responses of plants, herds, flocks, or even whole farms and regional catchments. Unlike mechanistic models, empirical models are not necessarily based on biological theory but are derived from statistical analyses of observed data. They do not need to draw on lower-level system attributes. Empirical models at one level can be coupled and this may provide increased insight into how lower-level processes interact and influence system response at the next higher level of organization. Perhaps the most important guidelines to building models and DS tools are clarity of purpose and minimum complexity in model content and structure consistent with achieving this purpose.

At present, empirical models of grazing system responses are more likely to give reliable predictions than models based on a series of mechanistic processes, where a greater detail of knowledge is required and where there is greater scope for generating errors. The cost of using an empirical model as a DS tool for guiding farm decisions is its lack of flexibility for applications that involve extrapolation beyond the source data. On the other hand, the detailed mechanistic model lacks well-defined statistical properties, and validation may be impractical or difficult and costly to undertake. The mechanistic model, however, has wider generality of application and can provide more insight into the sensitivity of different underlying processes to management intervention.

Data Requirements

Morley (1968) saw that agricultural modeling would require much data and the effort to collect it would be substantial. This would involve physicists, chemists, meteorologists, mathematicians and physiologists as well as agronomists. Dedicated experimental designs on centralized field stations would supply the data essential for model building, testing and revision. Analysis of data would be more penetrating and rigorous. Regional experiments would become less important. Unfortunately today the reality is different; there is more emphasis on regional and farmer-initiated experiments that often are more demonstration than true experiment. Rural industries seek quick returns from investment in research and dedicated experiments to meet the requirements of model building do not fit this mold. There is also a critical lack of data from long-term experiments that can be used to check the validity of model predictions.

The amount and accuracy of data required to initialize models is another critical limitation for realistic simulation particularly with detailed mechanistic models. Data inputs must be minimal and the data must be readily available if models or DS tools are to be used by farmers or their advisors. Where possible, default values must be provided. In general, extensive instrumentation to collect data at a site will not be feasible. For the model of Freer and Christian, the most critical input is the standard reference weight for the breed or strain of animal that is being simulated. This term is used to scale the animal's feed intake and production. If the standard reference weight is poorly estimated then the simulation of animal production will be inaccurate. For the pasture growth models discussed later in this chapter, critical data inputs are the physical properties of the soil such as bulk density, and the capacity of the soil to store plant-available water. Such data are not presently available for most farms or paddocks and the data are costly to collect. At present, we mostly rely on default descriptions of typical soil profiles for districts included as look-up tables in the program. Users of the models are also encouraged to invest in collecting input data.

MODEL AND DECISION SUPPORT TOOL DEVELOPMENT
FOR GRAZING SYSTEMS AT CSIRO AUSTRALIA

Scientists at CSIRO Plant Industry were early to recognize the advantage of a generalized structure for simulation models that allowed easy inclusion of new or extended modules of grazing system processes and flexible control of grazing management. This was seen as an essential requirement if a model were to contribute usefully to strategic management on real farms where there was wide diversity in enterprise structure and management (Christian et al., 1978). The original concept of a flexible scheme for management and optimization of the biological system underlying grazing enterprises has been further generalized and extended to include integration with cropping enterprises. This powerful new tool is called FarmWi$e (see subsection on FarmWi$e).

The original program was written in Fortran, and the approach was a marked departure from earlier models of grazing systems that mirrored the arbitrary and inflexible timetable of management events used in field experiments. A key issue in the design of the model was its operation at three levels of organization — the biological system, management of the biological system and optimization of management. A novel and highly systematic approach to coding was required to coordinate its execution. Although this early program lacked a user-friendly interface and was not used extensively by outside groups, it was the forerunner to the GRAZPLAN family of DS tools that has been released for commercial use in the grazing industries of temperate southern Australia.

The GRAZPLAN Family of Decision Support Tools

The GRAZPLAN family of DS tools developed at CSIRO is designed with user-friendly graphical interfaces that have evolved over time in response to user requests (Figure 2.2). The DS tools and their underlying models are described in detail by Donnelly et al. (1997), Freer et al. (1997) and Moore et al. (1997) and are outlined briefly in this chapter. The primary purpose of these tools is to enable better analysis of the consequences of management decisions for farm businesses, by using the power of the computer to integrate and evaluate all the effects that different management options have on a grazing enterprise. Analysis of business risk due to variable seasonal conditions is possible because potential farm performance can be evaluated across many years, using historical weather records to drive the simulations. Business risk due to market fluctuations in costs and prices can also be evaluated. The models reflect the current scientific understanding of the processes controlling on-farm production, including interactions between the processes.

The DS software is programmed with the Borland Delphi development tool and runs in a Microsoft Windows™ environment. The models are based on the best available experimental information but our understanding of some of the processes we are attempting to model is preliminary. Users are cautioned about the limitations that this may have on the accuracy of simulations. Successful application may also depend on users developing new practical skills in pasture recording and livestock husbandry so that essential data inputs are available.

CSIRO has a commercial contract with Horizon Pty. Ltd. (e-mail: horizontech@msn.com.au), which is the agent for the distribution of the GRAZPLAN software. This company provides after-sales service including a Help Desk and an organization of training courses. CSIRO provides technical training to users with financial support from Meat and Livestock Australia and Australian Wool Innovations Pty. Ltd.

A brief description of the GRAZPLAN DS tools follows.

LambAlive

This DS tool calculates the risk of death over the first 3 days of life of lambs as a result of exposure to chilling weather conditions. Likely reductions in mortality from shifting the date of the flock lambing period or of lambing in more sheltered paddocks can be evaluated.

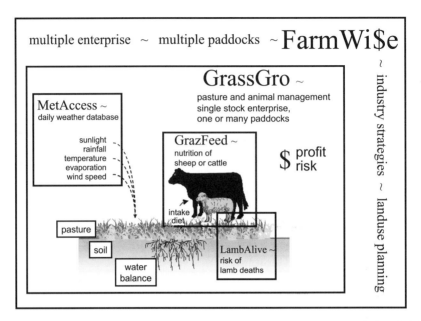

Figure 2.2   The GRAZPLAN family of DS tools is an integrated set that uses the identical underlying models where relevant. FarmWi$e is a whole-farm tool for evaluating mixed cropping and grazing enterprises. GrassGro can evaluate management decisions for a sheep or a cattle enterprise. GrazFeed is a nutritional management system for grazing sheep or cattle. LambAlive can be used to predict the mortality of newborn lambs from bad weather. MetAccess is a tool to analyze historical daily weather records. The other DS tools in the family access the MetAccess database of weather records.

MetAccess

MetAccess is a database tool that provides structured access to historical records of daily surface weather data collected by the Australian Bureau of Meteorology. The entire data set, for more than 6000 active locations where rainfall is recorded, is stored on a single CD-ROM. Data on up to 16 other daily surface weather observations are also included where available.

GrazFeed

GrazFeed is a software tool that assists farmers with feeding grazing livestock. Simple inputs supplied by the user describe the condition of a specified pasture and the animals grazing on it. Outputs include details of the nutritional requirements for energy and protein of the animals to meet specified levels of production. Because the program allows for substitution of pasture by supplement, the livestock manager can avoid expensive overfeeding of supplement. GrazFeed is suitable for use with any breed of sheep or cattle grazing on any type of pasture, but it is not designed for grazing systems based on semiarid rangelands where shrubs and forbs are components of the vegetation that are also consumed by animals.

The animal model in GrazFeed (Freer et al., 1997) has many predictive functions common to sheep and cattle, with specific parameters for a range of breed types and for all stages of growth and reproduction, mostly adapted from the feeding standards (SCA, 1990). Despite its generality, the model maintains a level of realism (see Stuth et al., 1999) that has blazed a new trail within Australia for the use of computers for advisory purposes in the grazing industries (Simpson et al., 2001).

GrassGro

The GrassGro DS tool is a related piece of software that links the generalized animal model described above to a generic model of pasture growth (Moore et al., 1997) covering a wide range of plant species and cultivars (Figure 2.3). GrassGro simulates production of temperate pastures and grazing animals through time as opposed to the single-day "snapshot" estimates of GrazFeed. Although the complexity of the underlying models is hidden from the user, the range of outputs generated allows the user to explore the biological relationships between the different processes operating in a grazing system. The user is free to concentrate on the problem to be solved. GrassGro also contains powerful facilities for analyzing business risk due to variability in weather and markets.

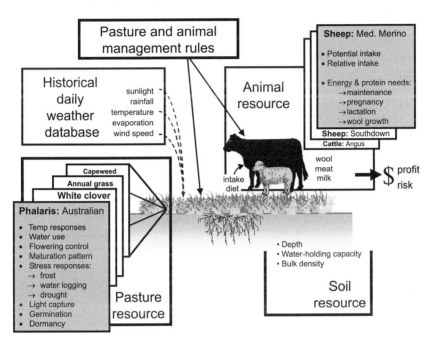

Figure 2.3 Grazing system resources as represented in the GrassGro decision support tool. Historical daily weather data drives pasture and animal production constrained by the soil, pasture species, animal enterprise, and management rules specified by the user. Production can be simulated on a daily basis over a number of years and environmental, production and business risks that are associated with climate variability can be evaluated.

Farm details are specified in GrassGro by selecting the relevant historical weather file and by describing soil characteristics specific to the farm. The flock or herd genotype is specified by selection of the appropriate breed, the mature weight and, for sheep, typical fleece characteristics. More than one species or pasture plant cultivar can be represented in a sward; selecting several pasture species from a menu specifies sward composition. When the user selects a pasture species, the pasture model opens a set of parameters that are used with generalized equations to uniquely characterize the genotype of the particular pasture species or cultivar (Moore et al., 1997). The equations respond to climate, soil and grazing management to produce the phenotype relevant to the location of the farm. Presently, pasture parameter sets are available to characterize about 12 species used in temperate pastures in Australia including perennial ryegrass, phalaris, annual grass, white clover, subterranean clover and lucerne. Additional pasture parameter sets are being developed and will include native pasture species and weeds. Colleagues at Saskatoon, Canada, have developed parameter sets for a broad range of species (c.15) found on the prairies (Cohen, 1995).

To develop new parameter sets, the same general form of equation is used where possible, with allowance for morphological and ecological differences between annuals and perennials, grasses, legumes and other forbs (Moore et al., 1997). Soil properties, weather events, grazing, and management interventions control expression of the phenotype. Developing parameter sets (about 100 parameters per plant cultivar) is a demanding task as data relevant to individual species may be difficult to find. Where data are missing, our approach is to use an existing parameter set as a template. For example, the set for *Phalaris aquatica*, where we have reasonable data to justify parameter values, would be used as the template for another species with similar growth characteristics but where there were few supporting data. Parameters are then modified based on experience and other relevant information; such changes being kept to a minimum. Though not ideal, the approach is working. It also allows us to model in an approximate way, weeds and other species that invade sown pastures and affect production, even though their biology may not have been studied extensively. To ensure the integrity of the pasture model, it is not possible for users to change these "genotypic" parameters.

GrassGro can be used for two types of simulation: historical and tactical. Historical simulations predict output over a series of years to assess the effect of climatic variability on production and form the basis of risk analysis. They are most useful for longer-term strategic planning of a grazing enterprise. Tactical simulations are used to project forward from a known starting point to assess the probability of production outcomes that result from changes to the management plan over the short to medium term.

Management rules covering pasture composition, grazing practices, herd or flock reproduction, lambing, calving and weaning policies, policies for sale and purchase of additional or replacement livestock following culling, and supplementary feeding are set for each run of the program. These rules are not sufficiently flexible for the management of dairy herds but extensions to do this are under consideration. An example of a user interface to GrassGro is shown in Figure 2.4. The dialogue shows how reproduction in a breeding cow herd is specified.

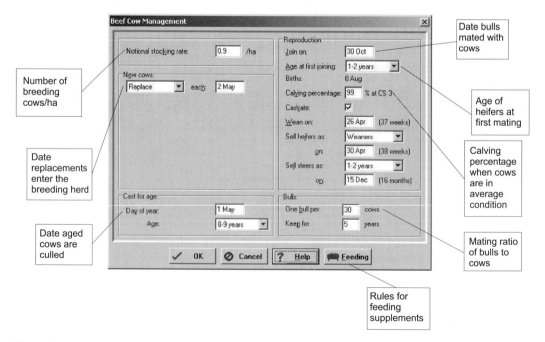

Figure 2.4   A typical dialogue box from the user interface of GrassGro. This dialogue allows the user a great deal of flexibility to specify the precise management for reproduction in a breeding cow herd.

The yield of a pasture can be simulated reasonably well by models that treat the sward as a single entity and do not model the mix of plant species present. Such models are empirical and have been used to establish seasonal patterns of herbage yield. The pasture model used in GrassGro goes further and attempts to model the individual species in the mixture in a more mechanistic way. However, we are not yet satisfied that the representation of the processes involved in competition between the plant species or cultivars accurately models change in composition over time. This applies to plant species that were either sown or have subsequently volunteered in the pasture.

The usefulness of GrassGro as a decision support tool depends greatly on the way model outputs are displayed. This helps users to easily visualize how accurately the biological processes of the grazing system are modeled. The series of outputs shown in Figure 2.5, for example, summarize the major responses of a simulated breeding ewe enterprise. This output can be scanned quickly to check whether the simulation appears reasonable.

FarmWi$e

FarmWi$e is a flexible DS tool for grazing systems management. It uses the same daily time-step simulation models that are the basis of GrazFeed and GrassGro. It extends the domain of GrassGro to include, for example, mixed sheep and cattle enterprises, soil fertility management, soil acidification, sequences of different crops, and pastures grown in rotation and irrigation.

FarmWi$e is configurable so that the user can tailor a simulation to represent precisely the grazing or mixed grazing system of interest. It is also extendable by the addition of submodels from other sources that follow the CSIRO programming protocol (Moore et al., 2001). The specification of management is flexible using a special purpose scripting to specify the rules that schedule events. The rules can use model variables so that management can respond to the state of the system at relevant times in the simulation.

Component submodels in FarmWi$e are stored as dynamic link libraries (DLL) (i.e., executable code). Written in any language that can be compiled and stored as a DLL, component models can be deleted or replaced with an alternative without the need for the lengthy process of recompilation. The protocol also allows coupling of multiple replicates of any component model so that it is simple, for example, to set up both a multi-paddock and multiple enterprise farm.

A graphical interface allows component models to be placed within a structure that visually represents a farm. This visual representation can show any level of detail that is necessary to describe the farm's operation, and the structure can be modified by simple rearrangement of the components on the graphical interface.

Management of the farm is customized through a management script written by the user. This allows, for example, the allocation of different livestock classes to specific paddocks, the scheduling of irrigation, the sale or purchase of livestock, or sowing of a crop. The script is basically a set of commands that is read by FarmWi$e to initialize and control the operations of the farm.

The output of this flexible modeling tool can range from detailed daily values to summaries of user-selected variables that describe the system. These can be displayed as charts or tables. Variables from different farming systems or simulation runs can be displayed in a common chart or table to compare outputs.

APPLICATION OF MODELS — RESEARCH

As illustrated by the following examples, there is considerable information supporting a modeling approach to guide the direction of research before committing resources to undertake expensive experiments.

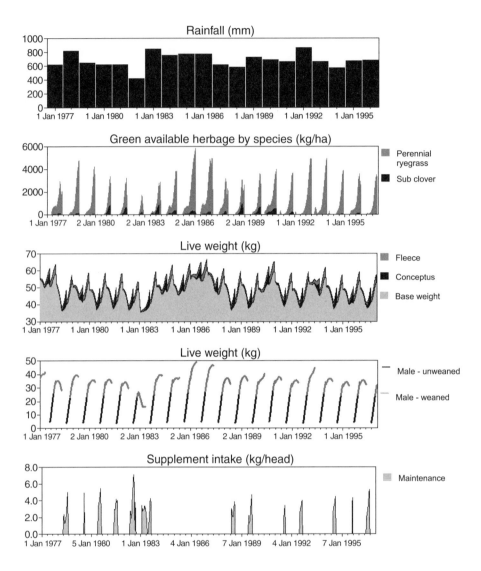

Figure 2.5 Typical output from a GrassGro simulation over 20 years of a ewe-breeding enterprise. The time course data are generated daily and can be quickly interpreted to see if the model output is within realistic bounds. For example, the bottom trace of supplement intake indicates substantial hand-feeding during the early 1980s. This coincides with a severe drought at that time. Otherwise, feeding at this level and frequency would alert the user to a possible anomaly or an error in the set up for the simulation.

First, at CSIRO Plant Industry, researchers are particularly interested in being able to predict the effects of organic acids and enzymes excreted into the rhizosphere by plants, to assess their capacity to mobilize P and to alleviate aluminum toxicity. Although the capacity to model rhizosphere dynamics is limited, a simple modeling approach was used successfully by Hayes (1999) to simulate the potential capabilities of a novel clover plant (*Trifolium subterraneum*) with a genetically improved capacity to access insoluble organic P found in soils. The model was used to determine the enzymatic characteristics that transgenic clover plants expressing a phytase gene in their roots would need to enhance plant P nutrition to achieve optimal growth. These analyses supported existing empirical knowledge, even though the representation of the rhizosphere environment was greatly simplified. In addition to the immediate application of this model to assist in the development of a genetically modified plant, the model provides insight about the way to describe these processes in farming systems models. The ability to predict the amount of P that can be mobilized from various soil P fractions by different plant species, apart from genetically modified ones, would be extremely beneficial.

Second, decision support tools such as GrassGro are useful for evaluating the likely impact on enterprise gross margins of achieving various plant breeding objectives. Donnelly et al. (1994) investigated, in simulated pastures based on *Phalaris aquatica*, the impact of breeding cultivars with increased winter growth. Although their analyses demonstrate a likely financial advantage in breeding for this trait, they also demonstrate a marked interaction with grazing management. For example, in a wool-producing enterprise running at 10 Merino wethers/ha, the use of a 'winter-active cultivar' increased predicted gross margin/ha by only 12% (due entirely to reduced supplement costs), whereas at 15 wethers/ha, the increase was more substantial (28%) and resulted from increased income from animals as well as reduced supplement costs. Similar results were obtained from analyses based on a hypothetical cultivar which maintained a higher digestibility of dead material during the summer months.

Dove (1998) extended these assessments by looking at the additivity of these two separate plant breeding objectives, and also the impact of the presence of legume in the sward on the financial advantage offered by the improved grass cultivar. As shown in Figure 2.6, the presence in the simulated grass sward of a cultivar with higher digestibility over the summer months resulted in increases in gross margin/ha in a ewe breeding enterprise, which were highly dependent on stocking rate, as in the earlier simulations of Donnelly et al. (1994). At 12.5 ewes/ha, the increase in gross margin was $53/ha, only $19 of which was attributable to decreased summer feeding costs.

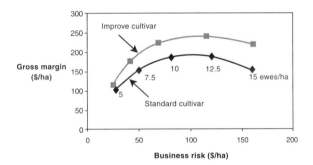

Figure 2.6   The effect of stocking rate on the simulated economic value of an improved cultivar of perennial ryegrass compared with a standard cultivar. The improved cultivar, which had a higher digestibility over summer, increased gross margins ($/ha) and marginally reduced business risk (standard deviation of gross margins).

Further simulations demonstrated two important points:

- The inclusion of both plant breeding traits in the simulated improved cultivar resulted in increases in gross margin/ha which were, on average, 80 to 90% additive, that is, the advantage of having both traits was 80 to 90% of the sum of the advantages of the separate traits.
- The inclusion of legume in the simulated sward meant that the advantage of the improved grass cultivar was reduced to some extent, partly because there was less of it in the sward and partly because the legume "filled in gaps" in the supply of feed from the grass component. Nevertheless, the gross margins in simulations involving the "improved" grass cultivar were still higher than those for the unimproved.

APPLICATIONS OF MODELS — ON-FARM CASE STUDIES

A major objective of the initiatives to develop models and DS tools at CSIRO Plant Industry is to provide producers with practical tools to help them strategically position their farm enterprises for future profit and sustainability. The tools can also be used to provide tactical guidance for day-to-day management within a broader strategic plan. The following case studies demonstrate the application of GRAZPLAN DS tools to practical on-farm problems.

One way to manage a farm successfully is to estimate the likely profit margins of different strategies and judge these in the light of likely business and environmental risks. The process is demanding, as the farm manager must identify the key management variables that control production and product quality and estimate any impact on the environment. The variables listed in Table2.1 are the key profit drivers of grazing enterprises of temperate Australia. Although the list is relatively short, setting an appropriate level and predicting the outcome of interactions between drivers is a skill that can be difficult to master. Computer models can help guide this process, because they provide a framework and a systematic way to investigate and integrate the impact of these key profit drivers on the farm business. However, of 90 Australian models designed for investigation of farm production and the impact of farming practice on water catchments (Hook, 1997), most were built for research use, few focus on the profit drivers, and few are used outside the research groups that built them. Moreover, many models require large inputs of data and need a high level of expertise to operate. Seligman (1993) and others have commented that early predictions about the use of models to guide decisions in agriculture were too optimistic. The situation is now changing and DS tools are starting to have significant impact on the management of "real-world" grazing enterprises (Donnelly and Moore, 1999; Stuth et al., 1999; Bell and Allan, 2000).

Table 2.1   The Key Drivers of Profit in Temperate Grazing Systems in Australia

Stocking rate
Time of calving/lambing
Fertilizer use
Animal genetics
Animal health
Supplementary daily feeding policy
Herd/flock structure

Tactical Farm Planning

Tactical farm planning is about short-term decisions that must be made as a consequence of prevailing weather conditions or to take advantage of market opportunities for farm products. The following case studies show how the GRAZPLAN tools can help farm managers improve profits under adverse as well as favorable seasonal conditions.

Alleviating the Impact of Drought with Fodder Crops

Case Study: A livestock producer wanted to assess the chances of growing a fodder crop over summer after failure of winter and spring rains.

Rainfall over substantial parts of southeast Australia is nonseasonal, but pastures for sheep and cattle enterprises are typically based on winter-growing annuals and perennials. Failure of winter or spring rains can cause crippling losses over the following summer and autumn due to the need to feed large amounts of supplements. This situation was experienced in 1994 on a wool growing property near Braidwood in southern New South Wales. By mid-spring pastures had ceased to grow, and property owners believed the only remaining option to provide desperately needed feed for livestock over the coming summer and autumn was to sow a forage crop of turnips. At the same time, the prevailing negative value of the Southern Oscillation Index (SOI), promoted in Australia as a drought indicator, suggested that dry conditions would continue until autumn. Fearing an increased risk of erosion, the property owners, therefore, were uncertain about the wisdom of sowing summer forages and asked for an analysis using MetAccess. It was pointed out that success or failure for the specific fodder crop could not be predicted; however, an indication of the odds that sowing the fodder crop would help overcome the critical feed shortage could be based on historical daily rainfall records at Braidwood. Based on those records, the prospect of adequate rain for a safe sowing of grazing turnips in late spring was quite promising, and the chance of good follow-up rains after sowing was excellent. Moreover, the correlation between the SOI and summer rainfall is not particularly strong at that time of the year, andscientists at CSIRO considered that it could be ignored for the growth of the turnip crop. As it turned out, this analysis proved correct, the sowing and subsequent growth of the forage crop was successful, and owners avoided a prohibitively expensive supplementary feed bill. If conditions had remained dry and the crop failed, the decision would have still been the correct one based on the information available at the time, although the outcome would have been adverse.

This case study shows how a relatively simple DS tool such as MetAccess can give a livestock producer sound information on which to base a decision. The owners were further advised to cope with climate variability in the Braidwood area by placing greater reliance on perennial pasture species wherever they can be sown, rather than using pastures based on winter-growing annuals.

Timely Drought Decisions for Breeding Cow Herds

Case Study: As the green pick from summer rain in February disappeared cattle producers on the southern tablelands of New South Wales met with their livestock advisor to discuss options for managing breeding cow herds.

Cows with calves at foot were running out of feed in February 1998 following one of the driest spring and summer seasons on record. There was uncertainty about when the autumn rains were likely to arrive and the amount of feed that would be needed to grow the calves to weights suitable for the on-property weaner sales in May.

The first task of the advisor was to calculate when autumn rains were likely to occur based on historical rainfall records. From past experience, the producers agreed with the advisor that a total of at least 30 mm of rain over a week would be needed before the beginning of April for reasonable pasture growth to commence. If rain were delayed after this period, there would be insufficient time for calves to fatten to the desired weight for the sales. Analysis of the rainfall records for the district over the past 56 years with MetAccess showed that there was only a "1–year–in–2" chance of drought breaking rains occurring before the beginning of April (Figure 2.7). On the basis of this analysis, the producers considered the chance of getting adequate pasture to fatten calves from an early break was very slim and the only alternative was to embark on an expensive supplementary feeding program, requiring careful calculation to ensure cost containment. It would be important to keep the total number of cattle handfed on property to a minimum.

The advisor helped the producers use GrazFeed to test a range of supplements that would allow the calves to reach a target sale weight of 260 kg from their current live weight of 180 kg (Figure 2.8 and Table 2.2).

Feed budgets for full supplementary feeding until the end of April then gave the graziers a clear indication of additional income needed from the sales to recover the cost of feeding.

It was uneconomical to hand-feed the lighter calves on the cows so a decision was made to wean these calves earlier than usual and send them to fatten on agistment in another region of New South Wales. A reasonable price was secured well before the rest of the district started seeking agistment, and the agisted calves did well and were sold direct to northern markets.

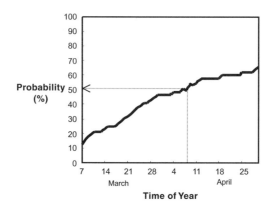

Figure 2.7 Probability of 30 mm or more of rainfall in any 7-day period after March 1 [weather data from Australian Bureau of Meteorology Station, Tharwa Store (1939–96)].

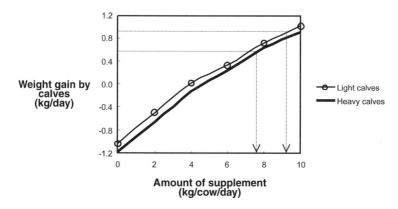

Figure 2.8 Weight gain by calves at foot after feeding barley and lupins.

Table 2.2 Comparison of the Costs of Agistment and a Supplementary Feed Mix (80% barley and 20% lupins) with ME:DM = 13.6 MJ/kg DM

Weight of Calf	Starting Live Weight	Required Weight Gain (kg/day)	Amount Fed (kg/cow/day)	Daily Feed Cost ($)	Total Cost Hand-Feeding ($/calf)	Total Cost Agistment Including Transport ($/calf)	Difference ($/calf)
Light	180	0.89	9.16	1.81	162.90	61.00	101.90
Heavy	210	0.56	7.63	1.51	135.90	61.00	74.90

The small group of heavier calves (over 210 kg) was kept on the cows and handfed, despite the high cost, to maintain the viability of the local weaner sales. These heavier calves made the grade for the weaner sales by feeding the cows at the correct level for production.

By adopting this strategy, cash flows were maintained and costs contained (Table 2.2). The advisor commented, "This is about separating fact from hope. MetAccess and GrazFeed gave these producers a set of realistic costed options based on hard facts. The producers combined their own knowledge and expectations with the additional information to make a timely decision."

Cattle Fattening Options in Southern New South Wales

Case Study: In favorable seasons, when there is a surplus of pasture, livestock producers can purchase additional animals for fattening and sale; however, this opportunity also involves significant business risks if normal seasonal rains fail and the additional animals do not reach their target sale weight.

Simpson et al. (2001) developed this case study of beef producers in southern New South Wales who occasionally have sufficient feed on hand in late summer to consider purchasing weaner steers to fatten for the domestic retail trade, or for sale to a feedlot at the end of the following spring. If seasonal conditions deteriorate, however, the additional animals may not reach sale weight and the main farm enterprise could be placed under increased pressure, threatening normal farm income. The risk of these outcomes can be assessed quickly with a DS tool, such as GrassGro, before a decision is made to invest in the opportunity. GrassGro can be set up to represent the prevailing pasture conditions and the genotype and condition of the weaner steers available for immediate purchase. The possible pasture and animal production outcomes that may occur are simulated from the day of purchase to the day of sale using the historical weather record over a long run of years (1958 to 1997).

Figure 2.9 shows the probability of achieving differing live weight outcomes given three alternative stocking rates. The simulations suggest that there is a 95% chance of reaching the minimum live weight for the domestic retail trade (330 kg) if the producer grazes the animals at two steers/ha. This reduces to 85% and 73%, respectively, at three and four steers/ha. To sell into an export feedlot the steers must reach 400 kg. This can be achieved 1 year in 2, given similar seasonal starting conditions if grazing at two steers/ha, but there is only a one-in-five chance at four steers/ha.

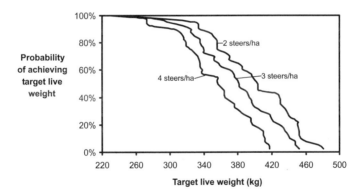

Figure 2.9   Probability of achieving target steer live weights by November 30 at Holbrook, New South Wales at three stocking rates. Minimum target live weight for domestic trade is 330 kg, and for export feedlot entry 400 kg.

This form of business risk assessment gives a measure of the likely variability in feed supply and would normally be combined with further economic analysis when making a business decision. Tactical simulations are not just about capturing new markets and production opportunities. They can also be very helpful when preparing a business for adverse conditions. For instance, Alcock et al. (1998) report tactical preparations for an anticipated feed shortage due to drought.

Strategic Farm Planning

A clear understanding of how the key biological, managerial and economic drivers interact among all enterprises on a farm is a fundamental requirement for profitable and sustainable land use. Research shows that the primary profit drivers for the grazing enterprises are usually stocking rate, time of lambing or calving, fertilizer and supplement use, choice of suitable plant species and animal genotypes, and attention to animal health. Level of subdivision and rotational management to rest pastures from grazing are generally of secondary importance. Rural consultants (e.g., Lean et al., 1997) find that attention to these key profit drivers makes it possible to shift typical grazing enterprises in temperate Australia from severe financial loss to a modest level of profit even when commodity prices are low. Given the dramatic reversal of profitability that consultants like Lean and his colleagues can achieve for their clients, why then are many grazing enterprises in Australia currently unprofitable? There is no single answer, but it is evident that clear messages must be provided to graziers about the role of each profit driver. Advice also needs to be specific to the soil, weather, pasture species and grazing management of individual farms, particularly when profit margins are low.

Estimating Production Risk for a Grazing Lease

Case Study: Leasing additional land is a strategy that enables an efficient producer to increase the scale of a grazing enterprise and reduce the cost of production to lift profits, although the opportunity to expand a grazing business also carries risks.

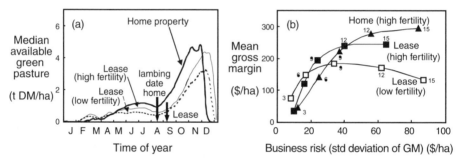

Figure 2.10 **(a)** Simulated median pasture availability over 15 years on the high fertility home property stocked at 12 ewes/ha and the lease property stocked at two possible rates: 9 ewes/ha (if low fertility) and 12 ewes/ha (if high fertility). The arrows indicate the optimum lambing date on the home and lease properties respectively. **(b)** Response of average gross margins (GM) and business risk to stocking rate and soil fertility at each property. The numbers associated with each symbol refer to stocking rates in ewes/ha.

A wool-growing partnership on the southern tablelands of New South Wales was interested in leasing a property in an unfamiliar district. The area where the lease was located had a 750 mm annual rainfall and acidic granite and basalt soils; it had not been fertilized for many years. The original owners had stocked the lease, which was sown to phalaris and subterranean clover pastures, at 11 dry sheep/ha. To make a decision to undertake the lease the partners wanted a more quantitative assessment of the restrictions on carrying capacity of the feed supply and the likely responses by the pastures to inputs of fertilizer. They also wanted to know the best time to lamb down their flock for optimum utilization of the pasture supply and labor. They thought that carrying capacity might be increased to 15 to 20 dry sheep/ha with moderate to heavy applications of phosphorus and lime. GrassGro simulations were seen as an objective way to assess the production potential of the lease property.

GrassGro simulations using local weather records over the period 1984 to 1998 and soil profile data at the lease showed that at the relatively low stocking rate of 9 ewes/ha the yield of pasture dry matter was below 650 kg/ha from start of growth in autumn until mid-September when spring growth commenced (Figure 2.10). This concurred with the lease owner's comment that livestock relied on the feed produced in "big springs with not much in between." Other test simulation runs confirmed that the likely cause of the restricted pasture growth was low temperatures.

Time of lambing between early August and late October was also tested with GrassGro. The optimum date to start lambing on the home property was mid-August and about four weeks later on the lease property. This indicated a potential advantage to share labor resources between the two properties.

Simulations also indicated that with improved soil fertility, 12 ewes/ha could be run profitably on the lease property, approximately the same carrying capacity as on the home property. At 15 ewes/ha (approx. 20 dry sheep/ha), supplementary feed costs increased to unacceptable levels on both properties.

These analyses with GrassGro added to the information that the woolgrower had about potential production of the lease provided by local consultants, regional trial data and his own experience. GrassGro had provided an objective framework to assess the resources of the lease that were critical to profitability.

Fine Tuning the Time of Lambing in Spring: Is It Profitable?

Case Study: A producer of fine wool wanted information about the likely impact on profits of "fine-tuning" his time of lambing in the spring. GrassGro was used to examine lamb mortalities and lamb weaning weights for different lambing dates in spring but the analyses showed that the path to bigger profits lay in modifying other aspects of the management plan.

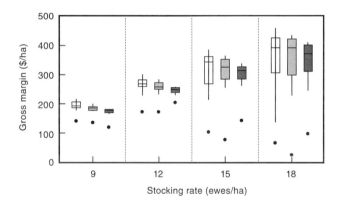

Figure 2.11   Box plots of gross margins for lambing dates in mid August (☐), mid September (▨) and mid October (▧) at four stocking rates. Vertical bars represent the range of all values except 1982 (a drought year), which is indicated by the dots. The upper line of the box is the 75th percentile, the middle line is the median and the lower line is the 25th percentile.

Selection of an appropriate time of lambing is a major profit driver in most pasture-based sheep breeding enterprises in most of southern Australia. Lambing in late winter or early spring generally ensures the best match between ewe nutrient demand and pasture supply (McLaughlin, 1968; Reeve and Sharkey, 1980; Lloyd Davies and Devaud, 1988). But the question remains: What are the gains, if any, to be made from lambing at different times in this period?

The property in question was situated on the central tablelands of New South Wales at an elevation of 1000 m with a mean annual rainfall of 850 mm. Low temperatures limited pasture growth until late September and the sparse feed supply threatened both lamb and ewe survival.

GrassGro was used to estimate annual gross margins over 20 years for three lambing dates for a self-replacing, fine-wool Merino flock. In the simulations, the flock was stocked at four rates (nine, 12, 15 and 18 ewes/ha) on highly fertile pastures of cocksfoot, annual grass and subterranean clover.

These analyses showed that time of lambing in spring had minor effects on lamb mortality, lamb weaning weight and the need for ewe supplements. The differences in predicted gross margins were not significant (Figure 2.11). Increasing the stocking rate from nine to 18 ewes/ha had a large impact on profitability, showing that in this environment the primary profit driver was stocking rate rather than lambing date; however, at the highest stocking rate there is greater downside risk associated with August lambing compared to September or October lambing.

As a result of these analyses, the producer recognized the overriding importance of stocking rate as a driver of profit. He is now increasing his stocking rate and he will monitor performance of the enterprise closely (Behrendt et al., 2000). This case study shows how GrassGro provided a framework for the producer to explore the outcomes of several management options over a range of seasonal conditions that would otherwise be difficult to do.

PROGRESS IN ACHIEVING INDUSTRY ACCEPTANCE OF MODELS

Experience indicates that the tactical management tool, GrazFeed, has already had a far-reaching impact on the profitability and environmental sustainability of grazing enterprises (Bell and Allan, 2000). For both cropping and grazing enterprises the benefits to farmers from the use of the strategic management tools are less certain. Nevertheless, CSIRO is committed to ongoing revision and further development of the underlying biological models and the user interfaces to achieve better industry acceptance and to ensure the successful DS tools maintain their relevance for industry use.

Adoption of GrazFeed

GrazFeed is now a well-established management support tool used widely by growers and extension workers in much of southern Australia. GrazFeed was first released for commercial use in 1990, and since then more than 1200 producers and advisors have obtained licenses. It was used to develop guidelines relating pasture characteristics with livestock production for the successful PROGRAZE extension package that has been delivered to more than 4000 producers (Mason and Kay, 2000). GrazFeed has undergone numerous upgrades in response to feedback from users and the user interface has been updated as computer technology advances. Currently, the authors are evaluating a version that includes breed types that may make it suitable for use in animal production enterprises in North America.

Reasons for GrazFeed's Success in Commercial Use

Reasons for the success of DS tools include simple user interfaces that hide the complexity of the underlying models, minimal requirements for input data, the provision of default values wherever possible, and flexibility to describe and test real life management options. Although direct use of DS tools by Australian farmers is still relatively rare, indications are clear that some of the early promise for this technology is at last being realized and a revolution is occurring in the way farmers manage their pastures and animals. There is also a slowly growing appreciation in agribusiness that computer models and DS tools may provide the only objective and feasible way to study how the whole farm environment responds to management interventions. Financial support to continue development of DS tools, however, remains difficult to secure.

Commercial release of GrazFeed was preceded by the development of a well-researched and published feeding standard for ruminants (SCA, 1990). This standard gave GrazFeed a huge advantage in gaining credibility within the research and extension communities as a DS tool for the nutritional management of grazing flocks and herds. The standard provides an agreed framework and a set of equations with which a model can be built. Unfortunately, a basis for developing an analogous standard for the growth of pasture and crop plants has not emerged, so an agreed framework for developing models of crops and pastures is not available. Two decades ago, there were already sound reasons for building a DS tool such as GrazFeed, and several of these were later outlined in Stuth et al. (1999). First, Freer and Christian (1983) recognized the advantage of developing a tactical model for testing the current feed requirements of a flock or herd. Second, Freer was a member of the sub-committee, appointed by the Australian Standing Committee on Agriculture, charged with developing feeding standards for ruminants. GrazFeed, seen as an easy way for livestock producers to implement the feeding standards, makes the calculations of the amount of feed needed for drought feeding or maintaining animals simple to undertake. Moreover, the calculations are accurate for all classes of stock and all types of feed used in a grazing enterprise.

When CSIRO Plant Industry commenced development of GrazFeed, the grazing industry did not see a need for it and was reluctant to fund it. CSIRO identified a product champion in the government extension service of New South Wales Agriculture (NSW Agriculture). This was a vital step for its widespread adoption. NSW Agriculture commenced training its extension officers in the necessary pasture assessment skills and in using these assessments as inputs to GrazFeed. Subsequently, similar workshops were established for livestock producers through the PROGRAZE project.

The adoption of GrazFeed has resulted in massive savings for livestock producers, particularly during drought, by rationalizing the need for feeding supplements to livestock. Advisory groups in all the southern Australian states now use GrazFeed routinely in preparing advice for producers. The leading advisors, however, want to extend the utility of GrazFeed to tackle questions that are more concerned with strategic rather than tactical management, for example, questions about stocking rate. CSIRO has released GrassGro for commercial use as a more appropriate way to do that.

Progress with the Commercial Adoption of GrassGro

GrassGro was released for commercial use in 1997. It is a much more comprehensive tool than GrazFeed, and it is only sold as a package that includes an intensive 3-day training course. At present about 100 advisors and research users have been licensed to use GrassGro in Australia. It has been adapted for teaching undergraduate courses in the rural and natural sciences at the University of New England in northern New South Wales. Twelve teaching staff attended intensive training courses and are using GrassGro in 24 teaching units. By late 2000, more than 700 student contact hours were spent working with GrassGro. GrassGro is also used for teaching at the University of Adelaide and at the University of Melbourne, where it is also used for delivering extension courses to farmers. The authors expect that the use of both GrazFeed and GrassGro for teaching will lead to a growing body of users in the next generation of farmers and advisors.

GrassGro was designed so that research findings could be tailored more closely to the land capability and grazing enterprises of individual farms. The authors thought that significant benefits would accrue to individual growers if advice could be tailored to their precise needs. This was a marked departure from the conventional approach in extension where advice is usually of a more general nature; however, feedback from a committee appointed to evaluate how GrassGro is used in industry suggested that our design goal did not match their expectations. Several committee members, who were also professional rural advisors, used GrassGro to evaluate grazing enterprises in districts where the considerable investment in time to do this could be spread over as many clients as possible. To achieve this expectation and assist advisors using GrassGro, CSIRO is developing templates based on thoroughly worked analyses where the focus is on the key profit drivers in grazing systems.

CONCLUSION

This chapter has described the development of an integrated set of DS tools that are now used, by farm advisors in particular, to remove uncertainty from the outcome of management decisions on the profitability and environmental sustainability of grazing enterprises in temperate southern Australia. The models operate with a daily time-step at the level of the whole animal and whole plant. The generality of the equations in the models is based on the use of parameters that quantify the genotypes of animals and plants commonly used in Australian grazing systems. Production on an individual farm is simulated by nominating the appropriate animal and plant genotypes, operating within specified managerial constraints and driving the model with available local environmental data. This approach enables users to simulate with the one model, most breeds of sheep, cattle and a wide range of pasture species used in Australia's temperate grazing systems. Local historical daily weather records can be used to predict the outcome of both tactical and strategic management decisions. Hence, a powerful facility has been developed to estimate how the business risk for an enterprise changes with alternative management options.

Detailed examples outline the use of these tools in case studies undertaken on the properties of collaborating farmers. The challenge of modeling pasture and crop growth in GrassGro and FarmWi$e for planning strategic management is greater than guiding tactical management decisions with GrazFeed. One advantage is confidence in the animal model so thoroughly tested in GrazFeed, although the main challenges lie in issues such as modeling competition between companion species in a pasture sward and responses of pastures to applied fertilizer. The authors' approach, therefore, is to make incremental improvements to the underlying biological process models, to increase the scope of management issues that can be investigated and to improve the ease of use of the DS tools. The authors' viewpoint is that these models provide a framework for effective decision support; they are not intended to determine management decisions. It is likely that many individuals who say that models do not work or do not deliver probably do not understand their purpose.

Quite apart from the huge cost of developing and testing DS tools that model pasture and crop systems, a major investment of time is needed to acquire the skill to use them. The benefit from use must be able to repay this investment. For strategic applications the skilled advisor who uses the tool or model on a regular basis is the primary target. Casual users who attend a course and then use the DS tool once or twice a year will be prone to making serious errors that may prove costly. Predictions in the 1970s that all farmers would have computers on their desks and use models to make decisions have proved wide of the mark. The reality of today is that DS tools have started to deliver their promise although rural advisors rather than farmers are presently the prime users.

ACKNOWLEDGMENTS

This chapter is dedicated to the late Dr. F.H.W. Morley. His pioneering investigations into the biology of grazing systems in Australia and the use of computer simulation to evaluate grazing management issues have had a lasting impact. We are also indebted to numerous colleagues at CSIRO Plant Industry, past and present, who have contributed their time and skills as scientists, technicians and programmers. The authors thank the GrassGro Advisory Group for its guidance, and Meat and Livestock Australia, Australian Wool Innovations Pty. Ltd., and Australian and Pacific Science Foundation for financial support.

Horizon Agriculture Pty. Ltd. (e-mail: horizontech@msn.com.au) is the agent for the distribution of the GRAZPLAN software.

REFERENCES

Alcock, D.J. et al. 1998. Using GrassGro to support tactical decisions on grazing farms: a case study at "Yaloak," Ballan, Victoria. In *Proc. 9th Aust. Agron. Conf.,* Wagga Wagga, Australia, 298.

Arcus, P.L. 1963. An introduction to the use of simulation in the study of grazing management problems, *Proc. N.Z. Soc. Anim. Prod.,* 23:159–168.

Beever, D.E., J. France, and G. Alderman. 2000. Prediction of response to nutrients by ruminants through mathematical modelling and improved feed characterization. In Feeding Systems and Feed Evaluation Models, M.K. Theodorou and J. France, Eds., CAB International, Wallingford, U.K.

Behrendt, K., A. Stefanski, and E.M. Salmon. 2000. Fine tuning the time of lambing in spring: is it profitable? In *Proc. 15th Ann. Conf. Grassl. Soc.,* Armidale, NSW, Australia, 127.

Bell, A.K. and C.J. Allan. 2000. PROGRAZE — an extension package in grazing and pasture management, *Aust. J. Exp. Agric.,* 40:325–330.

Christian, K.R. et al. 1978. *Simulation of Grazing Systems,* Centre for Agricultural Publishing and Documentation, Wageningen, The Netherlands.

Cohen, R.D.H. et al. 1995. GrassGro — a computer decision support system for pasture and livestock management, *Proc. Am. Soc. Anim. Sci., West. Sec.,*46:376.–379.

Donnelly, J.R. 1984. The productivity of breeding ewes grazing in lucerne or grass and clover pastures on the tablelands of southern Australia. III. Lamb mortality and weaning percentage, *Aust. J. Agric. Res.,* 35:709–721.

Donnelly, J.R., M. Freer, and A.D. Moore. 1994. Evaluating pasture breeding objectives using computer models, *N.Z. J. Agric. Res.,* 37:269–275.

Donnelly, J.R. and A.D. Moore. 1999. Decision support: delivering the benefits of grazing systems research (CD-ROM computer file). In *Proc. 18th Int. Grassl. Congr.,* Winnipeg and Saskatoon, Canada.

Donnelly, J.R., A.D. Moore, and M. Freer. 1997. GRAZPLAN: decision support systems for Australian grazing enterprises. I. Overview of the GRAZPLAN project, and a description of the MetAccess and LambAlive DSS, *Agric. Syst.,* 54:57–76.

Dove, H. 1998. Pastures and animal performance: principles and predictions. In *Pasture Technology to Improve Livestock Profit,* Wrighton Seeds Pty. Ltd., Melbourne, Australia.

Freer, M. and K.R. Christian. 1983. Application of feeding standards system to grazing ruminants. In *Feed Information and Animal Production,* G.E. Robards and R.G. Packham, Eds., CAB, Farnham Royal, U.K., 333–355.

Freer, M., A.D. Moore, and J.R. Donnelly. 1997. GRAZPLAN: decision support systems for Australian grazing enterprises. II. The animal biology model for feed intake, production and reproduction and the GrazFeed DSS, *Agric. Syst.,* 54:77–126.

Hayes, J.E. 1999. Phytate as a source of phosphorus for the nutrition of pasture plants, Ph.D. thesis, Australian National University, Canberra, ACT, Australia.

Hook, R.A. 1997. *A Directory of Australian Modelling Groups and Models,* CSIRO Publishing, Collingwood, Victoria, Australia.

Lean, G.R., A.L. Vizard, and J.K. Webb Ware. 1997. Changes in productivity and profitability of wool-growing farms that follow recommendations from agricultural and veterinary studies, *Aust. Vet. J.,* 75:726–731.

Lloyd Davies, H. and E. Devaud. 1988. Merino ewe and lamb performance in the Central Tablelands of New South Wales following joining in March–April or June–July. *Aust. J. Exp. Agric.,* 28:561–565.

Mason, W.K. and G. Kay. 2000. Temperate pasture sustainability key program: an overview, *Aust. J. Exp. Agric.,* 40:121–123.

McLaughlin, J.W. 1968. Autumn and spring lambing of merino ewes in south-western Victoria. In *Proc. 7th Aust. Soc. Anim. Prod.,* 223–229.

Moore, A.D., J.R. Donnelly, and M. Freer. 1997. GRAZPLAN: decision support systems for Australian grazing enterprises: III. Pasture growth and soil moisture submodels, and the GrassGro DSS, *Agric. Syst.,* 55:535–582.

Moore, A.D. et al. 2001. Specification of the CSIRO Common Modelling Protocol. *LWRRDC Project CPI 9,* Final report to Land and Water Australia, CSIRO, Canberra, Australia.

Morley, F.H.W. 1968. Computers and designs, calories and decisions, *Aust. J. Sci.,* 30(10):405–409.

Morley, F.H.W. 1972. A systems approach to animal production; what is it about? *Proc. Aust. Soc. Anim. Prod.,* 9:1–9.

Nagorcka, B.N., G.L.R. Gordon, and R.A. Dynes. 2000. Towards a more accurate representation of fermentation in mathematical models of the rumen. In *Modelling Nutrient Utilisation in Farm Animals,* J.P. McNamara, J. France and D.E. Beever, Eds., CAB International, Wallingford, U.K.

Reeve, J.L. and M.J. Sharkey. 1980. The effect of stocking rate, time of lambing and inclusion of lucerne on prime lamb production in north-east Victoria, *Aust. J. Exp. Agric. Anim. Husb.,* 20:637–653.

SCA. 1990. *Feeding Standards for Australian Livestock,* Standing Committee on Agriculture and CSIRO, Melbourne, Australia.

Seligman, N.G. 1993. Modelling as a tool for grassland science progress. In *Proc. 17th Int. Grassl. Cong.,* Hamilton, New Zealand, 743–748.

Simpson, R.J. et al. 2001. Towards a common advisory toolkit for managing temperate grazing systems. In *Proc. 10th Aust. Agron. Conf.,* Hobart, Australia. Available at http://www.regional.org.au/au/asa/2001/plenery/3/simpson.htm (verified August 30, 2001).

Stuth, J.W., M. Freer, H. Dove, and R.K. Lyons. 1999. Nutritional management for free-ranging livestock. In *Nutritional Ecology of Herbivores, American Society of Animal Science,* H.J.G Jung and G.C. Fahey, Eds., Savoy, Illinois, 696–751.

Thornley, J.H.M. 2001. Modelling grassland ecosystem. In *Proc. 19th Int. Grassl. Congr.,* São Pedro, Brazil, 1029–1035.

CHAPTER **3**

Applications of a Cotton Simulation Model, GOSSYM, for Crop Management, Economic, and Policy Decisions

Kambham R. Reddy, Vijaya Gopal Kakani, James M. McKinion, and Donald N. Baker

CONTENTS

INTRODUCTION

Computer simulation of cotton growth and yield began at a meeting held at the University of Arizona sponsored by the Departments of Agronomy and Agricultural Engineering and organized by H.N. Stapleton in 1968. From this meeting, several modeling projects emerged. Stapleton's group developed a computer program called COTTON (Stapleton et al., 1973). Dr. W.G. Duncan, who held a joint appointment between the University of Kentucky and the University of Florida, collaborated with researchers at Mississippi State University to develop a physiological cotton simulation model called SIMCOT (Duncan, 1971). It used average plant data file from which the cotton plant was simulated. The model was then modified to incorporate a nutritional theory of plant growth. Carbon and nitrogen supply demand ratios were used as stress factors to calculate organ growth and developmental responses to those nutrients, and the upgraded model named SIMCOT II (McKinion et al., 1975).

Over the next two years, several improvements were made in the area of cotton physiology using the data of Hesketh and Baker (1967), Hesketh et al. (1971) and Hesketh (1972). During the middle and late 1970s, the SIMCOT II model was integrated with a two-dimensional gridded soil model called RHIZOS (Lambert et al., 1977), and the new model was called GOSSYM, an acronym coming from the word *Gossypium*, the genus of cotton (Baker et al., 1977). With progress in developing systems for understanding plant responses to the environment, it was realized that the commonly collected field data had limited value in model development because:

1. Field data were too confounded with covariates to allow one to separate cause and effect as most field experiments were designed to test differences between means.
2. In most field experiments, at least one critical factor needed in the modeling process was not measured, i.e., solar radiation.

Phene et al. (1978) recognized the importance of unambiguously determining the role of specific environmental factors on plant processes, and they were the first to design naturally lit plant growth chambers with realistic soil volumes. These became known as Soil–Plant–Atmosphere–Research (SPAR) units. They were used for developing physiological process rate equations for cotton, soybean and wheat simulation models. Since that time, extensive data sets have been obtained that are unique and instrumental in developing improved cotton model and other crop simulation models (Phene et al., 1978; Marani et al., 1985; K.R. Reddy et al., 1997a, 2000, 2001).

In early 1984, the GOSSYM research team was approached by Dr. Andy Jordan of the National Cotton Council about using the GOSSYM model on commercial cotton farms as a decision aid. As a result of that effort, it was realized that the program and user interface were harder to understand and use. Therefore, an expert system was specifically designed for the GOSSYM model called, COMAX (CrOp MAnagement eXpert, Lemmon, 1986; McKinion and Lemmon, 1985). With the help of State Cotton Specialist for Mississippi Cooperative Extension Service and the National Cotton Council, the model was delivered to 70 cotton farms in several states in the Midsouth by late 1987.

By 1990, the model had grown from a pilot program on two farms in 1984 to an ongoing program used by over 100 farmers in 12 states. The extension specialists and consultants served an additional 200 to 300 farmers (Ladewig and Thomas, 1992). This success was primarily due to continuous on-farm testing and developing new and improved algorithms as the model was being tested within the U.S. Cotton Belt, and abroad (Marani and Baker, 1978; McKinion et al., 1989; Ladewig and Thomas, 1992; Pan et al., 1994; K.R. Reddy et al., 1995; Jallas et al., 2000). Research was conducted in different parts of U.S. and other parts of the world to improve the model for effective prediction of crop growth under field conditions (Gertsis and Symeonakis, 1998; Gertsis and Whisler, 1998; Marani and Baker, 1978; K.R. Reddy et al., 1997a, 2000; Jallas et al., 2000). Basic processes were simulated with data collected from carefully controlled experiments, while field data were used to help refine the responses under multi-variant field conditions. After a brief description of the GOSSYM/COMAX, we concentrate in this chapter on the applications of the model for crop management and its usefulness in providing both the farmers, crop production managers, and policy makers with economic and policy decisions.

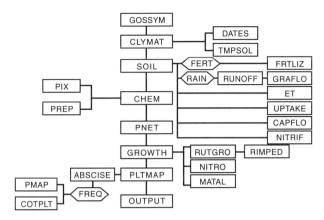

Figure 3.1 Flow diagram of the subroutines and structure of GOSSYM, a cotton crop simulator with organization and program flow of the model. See text for details.

THE COTTON DECISION SUPPORT SYSTEM, GOSSYM-COMAX

The development, characteristics, and some applications of GOSSYM have been previously described (Baker et al., 1983; McKinion et al., 1989; Boone et al., 1995; K.R. Reddy et al., 1997a; Hodges et al., 1998). GOSSYM is a mass balance dynamic simulation model that accounts for carbon, nitrogen, and water in the plant and soil root-zone. It simulates crop responses to the environmental variables such as solar radiation, temperature, rain/irrigation, and wind, as well as to variation in soil properties and cultural practices. The model estimates growth and development rates by calculating potential rates for the observed daily temperatures assuming other conditions are not limiting, then it corrects the potential rates by intensity of environmental stresses (Baker et al., 1983; K.R. Reddy et al., 1997a; Hodges et al., 1998). Each day, the model provides the user with the plant size and growth stage as well as growth rate and the intensity of the stress factors. A grower can assume certain future weather conditions (days and weeks) to determine yield estimates and impact of alternative cultural practices on the maturity of the crop.

A flow chart of GOSSYM shows the general organization of the model and program flow (Figure 3.1). GOSSYM is the main program from which all of the subroutines vertically below it in the diagram are called. CLYMAT reads the daily weather information and calls DATES, which keeps track of both day of the year and the calendar date being simulated; and calls TMPSOL, which calculates the soil temperatures by soil layer. SOIL is a mini-main program, which calls the soil sub-programs (Boone et al., 1995). The soil routines provide the plant model with estimates of soil water potential in each grid cell, as described in the following paragraphs, in both the rooted and non-rooted portion of the soil profile, an estimate of the nitrogen entrained in the transpiration stream available for growth, and an estimate of metabolic sink strength in the root system.

The belowground processes are treated in a two dimensional grid. The mass balances of roots in three age categories, water, nitrate and ammonia, and organic matter are maintained and updated several times per day. FERTLIZ distributes ammonium, nitrate, and urea fertilizers into the soil matrix when applications are needed. GRAFLO simulates the movement of both rain and irrigation water into the soil profile. Evaporation (E) estimates from the soil surface and transpiration (T) from the plant are summed as evapotranspiration (ET). UPTAKE calculates the amount of soil water taken from the soil region where roots are present. CAPFLO estimates the rewetting of dry soil from wetter soil by capillary flow. NITRIF calculates the conversion of ammonium to nitrates by bacterial action in the soil medium. CHEM is also a mini-main program, which calls subprograms that calculate the effect of chemicals on plant physiological processes (PIX® and PREP®). PIX deals with the effects of the plant growth regulator, mepiquat chloride, that will be discussed further and PREP deals with the effect of the boll opener, ethephon.

In PNET, leaf water potential, canopy light interception, photosynthesis, and respiration are calculated. Then, in the GROWTH subroutine, potential dry matter accretion of each organ is calculated from temperature. These potential organ growth rates are adjusted for turgor and nitrogen availability. Then, photosynthates and any reserve carbohydrates are partitioned to the various organs in proportion to the total growth requirements. The partition-control factor is the carbohydrate supply:demand ratio. RUTGRO calculates the simulated potential and actual growth rates of roots. RIMPED calculates the effect of root penetration resistance on the capability of roots to elongate. NITRO calculates the partitioning of nitrogen in the plant. MATBAL keeps track of the nitrogen and carbon material balance in all parts of the plant and soil complex. In PLTMAP, stress induced fruit loss and developmental delays are calculated using both carbohydrate and nitrogen supply:demand ratios. These developmental delays are used to delay the simulation of plastochrons or other developmental intervals that are calculated as functions of temperature (Hodges et al., 1998). ABSCISE estimates the rate of abscission of fruit, squares, and leaves due to stress and age. PMAPS, COTPLT, and OUTPUT print various user-selected reports from the simulation model. The program cycles through these subroutines daily from emergence to either the current in-season date or to the end of the season. In-season simulations may be augmented with forecasted weather data for the remainder of the season depending on the user's choice.

The COMAX system is an expert system that was explicitly developed for working with GOSSYM model (Lemmon, 1986; McKinion et al., 1989). COMAX is a forward-chaining, rule-based system that contains an inference engine, a file maintenance system for the simulation model requirements, a database system for the knowledge base, and "user friendly" menu-driven system for user interactions. The inference engine applies rules to:

1. Organize weather and cultural practices input data files, including plant growth regulator applications used by the GOSSYM program.
2. Execute the GOSSYM program.
3. Interpret the model results making recommendations on timing and amounts of irrigation, fertilizers, plant growth regulators, and harvest-aid chemicals.

For more detailed information on COMAX, see Lemmon (1986) and Hodges et al. (1998).

The model has been continuously updated as new information became available (K.R. Reddy et al., 1995, 1997a, 1997b, 2000; K.R. Reddy and Boone, 2001). Recent improvements include the use of phytomer concept for plant height simulation and leaf area development (K.R. Reddy et al., 1997b). Simulation of plant height involves the temperature-controlled rates of internode initiations, duration of extension, and elongation and knowledge of internode lengths at node initiation under optimum growing conditions. Similarly, potential leaf area development was simulated by the time required to initiate a new leaf on the mainstem and branches, and growth rates and duration of expansion and leaf sizes at leaf unfolding as functions of observed temperature under optimum water and nutrient conditions (K.R. Reddy et al., 1997b). The effect of nitrogen and water deficiencies, and the influence of plant growth regulators on leaf area development were also incorporated using appropriate stress-specific reduction factors (K.R. Reddy et al., 1995, 1997a, 1997b). In addition, enhancements were made to simulate boll abscissions due to high air temperatures (K.R. Reddy et al., 1997c). These modifications have increased the model's sensitivity to a wide range of environmental conditions including future climatic conditions. Simulations for future climatic conditions have assumed historic daily weather patterns plus or minus certain perturbation amounts for the various weather variables.

MODEL VALIDATION

The model has been validated extensively across a wide range of environmental conditions and cultural practices (Fye et al., 1984; V.R. Reddy et al., 1985; V.R. Reddy and Baker, 1988, 1990; Boone et al., 1993; K.R. Reddy et al., 1995; K.R. Reddy and Boone, 2001). Models are not complete enough during their initial inception for making crop management decisions. Validation helps in the continuous evolution of the model by providing information feedback from researchers testing it under new environments, and also from farmers and farm managers using it in variable climate and soil conditions. Validation can be defined as "comparison of the predictions of a verified model with experimental observations other than those used to build and calibrate it, and identification and correction of errors in the model until it is suitable for its intended purpose" (Whisler et al., 1986). Validation is usually done in areas where the model has not been tried before. This validation has included data from areas of the USA cotton belt, and also from other cotton growing countries like China (Pan et al., 1994), Greece (Gertsis and Symeonakis, 1998; Gertsis and Whisler, 1998) and Israel (Marani and Baker, 1978).

The initial focus for GOSSYM validation was its response to water stress. As the model was developed using crop data under Mississippi conditions, validation under Arizona conditions (Fye et al., 1984) suggested that it needed alterations in the maximum reduction in photosynthesis due to water stress to simulate an apparent hardening process in the cotton plants. This study also helped to modify the growth rates of roots and plant height and leaves as affected by water stress. In 1985, V.R. Reddy and co-workers further validated data collected from a cotton crop grown in two locations of Mississippi. They found that under stress conditions 70% of the carbohydrates were partitioned to squares and bolls. This feature was incorporated into the model making subsequent model predictions closer to the observed values. A field study with cotton conducted at Lubbock, Texas, in 1994 was used to validate the model evapo-transpiration (ET) subroutines (Staggenborg et al., 1996). The simulations showed that the model underestimated cumulative evaporation by 18%, while cumulative transpiration was 8% lower than observed values. These predictions were attributed to the overestimation of Leaf Area Index (LAI) by the model, thus reducing simulated incident solar radiation at the soil surface. They suggested that the measured environmental humidity should be taken into account for calculating the potential ET, in order to improve its predictive capabilities.

Increase in adoption of the model by farmers necessitated a more precise prediction of growth and development. Atwell (1995) collected more detailed crop growth and development information for model validation with several modern Upland and Pima cotton cultivars. An example of the model performance and accuracy is shown in Figure 3.2 for plant height and mainstem nodes. This study led to the development of cultivar-specific genetic coefficients for modern cultivars and extended GOSSYM use across a wider geographic area and genetic base.

A model is successful if it can effectively predict the crop growth at places other than at its origin. Collaborative studies were conducted by the Cotton Research Institute, the Chinese Academy of Agricultural Sciences, Henan, People's Republic of China, and the Crop Simulation Research Unit, USDA-ARS at Mississippi State University to adapt the GOSSYM model to cotton production systems of China. Field experiments were conducted in the single cropping district of the Hanghuai cotton belt of China between 1991 and 1993 (Pan et al., 1994). The model accurately predicted the key developmental stages within acceptable limits (±4 d). Plant height, leaf area, squares, and fruiting sites were accurately simulated by the model, but the model could not account for the damage caused by cotton bollworm infestation during certain periods of this study. In the cotton production system of China, vegetative branches are removed and main stem tips are pruned manually. Thus, modification of appropriate functions in GROWTH and PLTMAP subroutines to account for these local cultural practices was necessary.

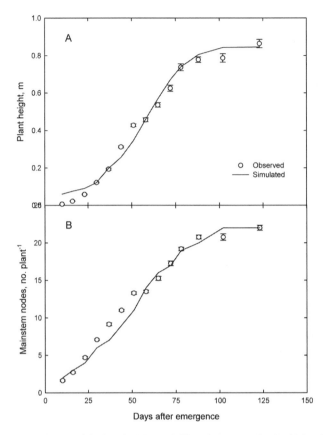

Figure 3.2 Observed vs. simulated (A) plant height and (B) mainstem nodes for Upland cotton, cv. DPL 5415 during the 1992 growing season at Mississippi State University. Vertical bars are standard errors. ($r^2 = 0.98$). (From Atwell, K.D., Calibration and validation of GOSSYM (thesis), Mississippi State University, Mississippi State, MS, 1995. With permission.)

MODEL APPLICATIONS

Farm Management

Preseason Decisions

The uses of the model for preseason decisions have not been well documented. Farmers and farm managers have reported making numerous model runs with cotton varieties of different maturities using many years of local weather and the pertinent soil types. The results of such an exercise allow the producer to see the interaction of weather and crop maturity on yield in a given production environment. It also allows the producer to estimate more accurately the value of adding practices not used in the past such as irrigation or fertilizer applications. Farm mangers also used the model to help determine whether to lease a particular farm and more importantly to determine its yield potential using historical weather data and cultural practices. Experienced users have learned much about the way cotton grows, develops, and responds to different environmental conditions and cultural practices.

In-Season Decisions

Timely decisions are the key to successful harvest. An analysis of GOSSYM usage by farmers in 12 American states was conducted by a group independent of model developers (Ladewig and

Powell, 1989; Ladeweig and Thomas, 1992). The survey revealed that the majority of the farmers used the GOSSYM model as a decision support system for determining crop termination, nitrogen utilization, and irrigation practices. On an average, users of GOSSYM earned U.S. $80 more per hectare when compared with users who did not use any simulation models. McKinion et al. (1989) reported that benefits of using GOSSYM-COMAX as a management tool were $100 to 350 ha^{-1}.

The use of GOSSYM/COMAX in farm decision making can be well illustrated by the following example. A pilot test was conducted in 1985 on the Mitchner farm, Sumner (Mississippi), to get a realistic experience of GOSSYM/COMAX operation. GOSSYM/COMAX suggested that the farmer apply an additional 56 kg (N) ha^{-1} and predicted an increase in cotton lint yield of 224 kg ha^{-1}. The farmer, who had not planned to apply any fertilizer, applied 22 kg (N) ha^{-1}. Cotton was picked both by hand and machine in this study. The hand–picked area showed a net increase in yield of 202 kg ha^{-1} of cotton lint, while the machine picked recorded a 129 kg ha^{-1} increase in lint yield. The difference between the hand–picked and machine–harvested yield is attributable to losses in mechanical harvest. The additional economic value of the machine–picked cotton was about U.S. $161/ha^{-1}, where the cost of fertilizer was $10 and the application cost of fertilizer was $15. This led to a net increase of U.S. $135/ha^{-1} on this 2700 ha farm.

Irrigation and Nitrogen Management

Timely irrigation and maintaining soil fertility are important in sustaining cotton productivity and profitability. Cotton plants are sensitive to both water stress and reduction in nitrogen supply caused by water stress (Radin and Mauney, 1986; K.R. Reddy et al., 1997a; Gerick et al., 1998).

GOSSYM simulates soil water and nitrogen present in the two-dimensional array of cells. Water and nitrogen uptake are calculated in cells containing roots. GOSSYM calculates daily E, T, and ET using modified routines of Ritchie (1972). These values (E, T, and ET) and plant water demand are calculated from potential ET rates, canopy light interception, and soil water content. Staggenborg et al. (1996) evaluated GOSSYM at Lubbock, TX, and determined that it underestimated E by 18% during the 12-day period of measurement. This underestimation was due to overestimation of LAI, thus reducing incident radiation at the soil surface; however, the simulated ET over the entire crop duration of 102 days was within 10% at the end of the measuring period, despite overestimation of LAI. It was concluded from this study that GOSSYM could be used to assess water use by cotton, and as a tool for scheduling irrigation in a semiarid region, provided the current algorithms used to calculate potential ET are modified to include air humidity.

Crop simulation models that include soil processes are the only tools that simultaneously integrate the interacting soil, water, plant, and weather factors, which determine soil-N availability and current and future N needs. Wanjura and McMichael (1989) used simulation analysis instead of costly field experimentation to study the impact of N fertilization on cotton productivity. Simulated preplanting application of N resulted in 4% higher yields compared to side dressing at first square and first bloom under rainfed conditions, but when supplemental irrigations totaling 204 mm was provided along with the N source, the yield of preplant fertilized cotton was 4% less than that when N was applied at first square and first bloom (Figure 3.3). Thus, crop models can be used to study crop performance with simultaneous imposition of various factors.

Adoption of the model for on-farm use required the GOSSYM model to simulate the optimum nitrogen supply under specific sets of farm conditions (soil, weather and cultivar). A survey of GOSSYM users (Albers et al., 1992) found that 76% of the farmers who used the model changed their N-management practices. Stevens et al. (1996) validated the nitrogen dynamics in cotton crops. The GOSSYM simulated lint yields on the Loring soil (fine-silty, mixed, thermic Typic Fragiudalfs) were greatest with 90 kg ha^{-1}. The study revealed that GOSSYM simulated responses to N fertilizer were similar to actual data but were lower over the whole range of applied N. GOSSYM over-estimated soil N availability by 10 to 30 kg N ha^{-1}, overestimated fertilizer N recovery, and underestimated cotton yield. This was attributed to the inability of the model to simulate mineralization

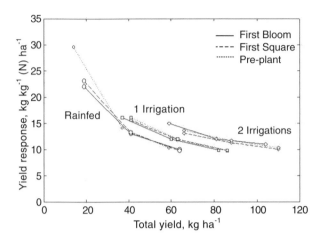

Figure 3.3   Simulated lint yield response per unit of total applied for three nitrogen application strategies (PREPLANT — all nitrogen applied as basal dose; FIRST SQUARE — 50% of total N applied as basal dose and remaining at first square; FIRST BLOOM - 50% of total N applied as basal dose and remaining at first bloom) using three soil moisture regimes; rainfed — no supplemental irrigation; 1 IRRIGATION — rainfed plus one 102 mm summer irrigation; 2 IRRIGATIONS — rainfed plus two 102 mm summer irrigations) 1965–1986. (From Wanjura, D.F. and McMichael, B.L., *Trans. ASAE,* 1989. With permission.)

and immobilization processes or ammonia-volatilization losses from the soil or the plants (Boone et al., 1995), which could explain the overprediction of fertilizer N recovery by plants in GOSSYM. Thus, improving our understanding of the processes controlling the N dynamics of the plant and soil system is essential to improving the model simulations and its wider utility for that purpose.

Excessive N application in farmlands is a major cause for the eutrophication of groundwater and also an unnecessary cost for the farmer. Hunt et al. (1998) reported that the GOSSYM model could be used to avoid excessive N fertilizer application on cotton farms. They conducted a study to determine if seed yields or excess N application were affected by timing of N application via buried microirrigation tubing, tubing spacing or peanut rotation. Rotation did not have any affect on the measured parameters. GOSSYM/COMAX management did not improve seed yield, but it did reduce the excess N (fertilizer N–seed N) to <20 kg ha^{-1}/yr^{-1}. Hence, the GOSSYM/COMAX system may be used to tailor the fertilizer needs of individual fields.

Herbicide, Growth Regulator, and Crop Termination Applications

Weed control in cotton is largely achieved through the use of herbicides. Applying the plant growth regulator, mepiquat chloride (1,1-dimethypiperidinium chloride), checks vegetative growth in cotton. Crop termination chemicals are used to defoliate the plants and open bolls at the end of the season. Successful yield and biomass predictions are possible if GOSSYM accounts for the effects of these chemicals on cotton growth and development and lint yield in a wide array of environmental and cultural conditions.

Yields across the Cotton Belt declined between 1965 and 1980, despite improvements in technology and introductions of improved higher yielding cultivars (Meredith, 1987). One suggested cause for the lower yields was the increased use of herbicides and their toxic effects on root growth in the herbicide zone in the soil surface layer. Root reduction is often observed when herbicides are applied on cotton (Anderson et al., 1967; Oliver and Frans, 1968; Bayer et al., 1967; Pavlista, 1980), and this is known to cause a reduction in cotton plant height and yield (Hayes et al., 1981; Kappelman and Buchanan, 1968). Using the cotton model GOSSYM, V.R. Reddy et al. (1987) demonstrated how the response of cotton to herbicide damage could be simulated. A series of simulations were carried out to study the effects of root growth inhibition and reduction in the permeability of roots to water and nutrients on cotton growth, development and lint yield. This analysis demonstrated that herbicides, if improperly applied (deeper than 50 mm) or in excessive amounts, would result in reduced root growth, water and nutritional shortages, and lower yields. Using weather, soil and cultural practices input data at five locations in the U.S., for more than 20 years, V.R. Reddy et al. (1990) simulated cotton production and analyzed the effect of root damage caused by herbicides on cotton yield trends. The reduction in root growth also decreased the uptake of water and nitrogen by the cotton crop. The simulated adverse effects of herbicides on lint yield varied from location to location and the lint yield decrease ranged from 14 kg ha^{-1} to 137 kg ha^{-1}.

Mepiquat chloride (MC) is a plant growth regulator shown to reduce vegetative growth in cotton (York, 1983a, 1983b; V.R. Reddy et al., 1990). MC suppresses excessive plant growth by decreasing plant height (main-stem internode length); number of nodes, fruiting, and vegetative branch internode lengths; and leaf area (York, 1983a, 1983b; Stuart et al., 1984; Zummo et al., 1984; V.R. Reddy et al., 1990). The on-farm success of the GOSSYM is being attributed to continuous development and upgrading the model as new information becomes available. An example in that direction is the development of a subroutine dealing with a plant growth regulator, mepiquat chloride. A detailed description of the MC model development, its incorporation into GOSSYM and validation were given by V.R. Reddy (1993) and K.R. Reddy et al. (1995). The initial MC subroutine was developed with a single rate of MC applied to flowering cotton plants. This subroutine has not worked satisfactorily in all growing conditions and over a range of MC application rates. The new MC subroutine was developed from leaf expansion, stem elongation, and photosynthetic rates data of plants containing different MC concentrations in the tissues (K.R. Reddy et al., 1995). The model with the new subroutine predicted with greater accuracy where, stem elongation rate was reduced by 38% due to 30 µg MC a.i. g^{-1} of MC, while leaf area and photosynthesis was reduced by 30%. The new model also performed well with data sets from a wide range of environmental conditions, a variety of cultural practices, and diverse genetic resources (Figure 3.4). The data sets comprised both single and multiple rates of MC applications applied on different dates during the growing season.

GOSSYM is also used to determine the optimum MC application strategy. Watkins et al. (1998) evaluated 12 different MC application strategies under two different soil types (Bosket sandy loam and Dundee silty clay loam) and three different weather scenarios (normal, cold-wet, and hot-dry) in the Mississippi Delta using GOSSYM/COMAX management system. The simulations revealed quantitatively what most growers knew intuitively, but could not predict *a priori*. The soil type and weather conditions determine the type of MC application strategy that should be used as opposed to using a blanket MC application strategy for all weather conditions.

Precision Agriculture Management

Precision farming and agriculture aims to improve crop production efficiency and reduce environmental pollution by adjusting production inputs (e.g., seed, fertilizer, and pesticide) to the specific conditions within each area of the field. This depends on the successful development, integration, as well as utilization of hardware, software, and people for data collection, planning, and execution. The multidisciplinary field of precision farming requires expertise in remote sensing, geographic information systems, global positioning system, and crop modeling.

Currently, efforts to integrate these systems into a single unit for planning and improving the efficiency of cotton production systems are being investigated. McCauley (1999) used GOSSYM/COMAX integrated with GRASS, a geographic information system, to produce spatially variable outputs. Inputs to the model were collected from a 3.9 ha cotton field. Soil nitrate, a primary driver for fertilizer recommendations was sampled on a 15.2 m regular grid for depths to 150 mm and on a 30.5 m rectangular grid at six 150 mm depth intervals (down to 900 mm). COMAX was used to determine spatially variable fertilizer recommendations. GOSSYM was used

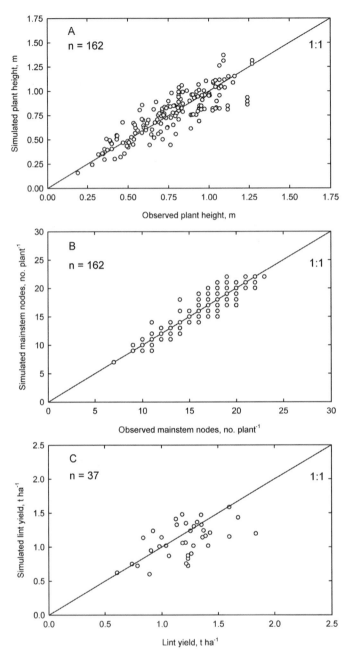

Figure 3.4 Comparison of observed and simulated (A) plant height, (B) mainstem nodes, (C) lint yield using the new MC routine in GOSSYM. The data collected from plants grown on various soil types, weather conditions, and management practices across the U.S. Cotton Belt. (From K.R. Reddy et al., *Agron. J.,* 1995. With permission.)

to simulate the application of these recommendations and predicted spatially variable yield and residual nitrogen. This study concluded that crop simulations and geographic information systems are a valuable combination for modeling the effects of precision farming and planning variable rate applications. Simulations from this study indicated that excessive fertilization, although potentially damaging to the environment, might also have negative impacts on yield. This conclusion illustrates one of the advantages of precision farming, especially in fields with high variability of soil properties.

In a recent study, McKinion et al. (2001) combined the GOSSYM/COMAX system with the Arc View GIS software. It was used to evaluate nitrogen and water stress experiments on cotton conducted during 1997 on the Kenneth Hood Farm in Bolivar County, Mississippi. The total actual N applied was 160 kg (N) ha^{-1}. The amount of irrigation water applied was 83 mm in three applications. During the crop growth period of May to October of 1997, a total of 754 mm of rainfall was also received. The entire selected area for the study was divided into grids containing 1-ha areas and 88 simulations were carried out which were based on the variation in soil types.

A whole field simulation based on the summation of the above simulations predicted an average yield of 1133 kg ha^{-1}, higher by 4.3%, compared with growers actual yield of 1084 kg ha^{-1}. The precision agricultural system (GOSSYM/COMAX + Arc View GIS) recommended nitrogen application rates from 7.8 to 199 kg (N) ha^{-1}. Further evaluation is needed for the higher rate of 199 kg ha^{-1}, while the remaining rates of 16.8 to 108 kg (N) ha^{-1} were reasonable. The irrigation totals range from 0 to 176 mm of water. The number of irrigations called for varied from one to seven. Obviously, growers will not be able to apply water on a per hectare basis as addressed in this analysis, but the numbers are included here to show the range in variability with just having soil type information as a variable.

The yield predictions were shown as differences between the precision agriculture optimized yield and the yield predicted using the grower's actual cultural practices. The yield differences ranged from –112 to 561 kg ha^{-1} (–0.2 to 1 bale per acre) across the field. The negative values show that the expert system is not infallible. When an event such as this occurs, the user should conduct manual simulations to determine if improvements can be made or an appropriate interpretation of the problem can be obtained, although the amount of computer work and data analysis likely would be prohibitive in terms of the user's time for making decisions on a per hectare basis. The predicted yield improvement using the precision agriculture tool showed that the grower could expect an increase of 286 kg ha^{-1} (0.51 bales per acre) for this field even with the few negative results.

The study dramatically shows that there is potential for both increasing yields and decreasing the use of agricultural chemicals by the adoption of simulation model driven precision agriculture technology. A tool for generating the irrigation and nitrogen rates by soil type and/or soil site sample as demonstrated in this study can be used to automate the calculation of optimum water and N rates.

Another valuable tool in precision agriculture is remote sensing which may also be combined with crop modeling. The agricultural research community is currently engaged in identifying cotton plant spectral reflectance signals associated with growth and development stages, nutrient and water status of cotton plants. Studies conducted so far using spectral reflectance have attempted to identify plant responses to different stresses by using plant chlorophyll content as the primary indicator. Bowman (1989) reported that in cotton, significant correlations were obtained between spectral reflectance at 810 nm, 1665 nm, and 2210 nm and leaf relative water content, total water potential and turgor pressure. Efforts at Mississippi State University through the Remote Sensing Technologies Center are to identify cotton plant signals for deficiency of various plant nutrients (N, P, and K) and plant water status (total water potential). Recently, Tarpley et al. (2000), compared predicted and actual leaf N concentrations by regression for a validation set of field-grown samples from diverse genotypes. Ratios obtained from combining the red-edge measure (700 or 716 nm) with a waveband of high reflectance in the very near infrared region (755 to 920 and 1000 nm) provided good precision (correlation) and accuracy (one-to-one relationship between predicted to actual values). The other indices based on chlorophyll reflectance feature also had good precision but were less accurate, presumably due to the influence of other factors associated with chlorophyll concentration (Carter, 1994; Masoni et al., 1996; Sunderman, 1997).

Once the relevance of signals as described above is accurately validated, sensors can be developed to predict cotton crop nutrient and water status under field conditions. Such sensors may be used via ground-based aircraft or satellites. Remote sensing may then be combined with crop model predictions of yield response for real-time in-season crop management.

Research and Policy Management

Yield Decline Assessment Analysis

Meredith (1987) reported an analysis of data that cotton yields had declined slightly between 1960 and 1980 despite the introduction of better and improved varieties, effective pesticides applications and continued increases in atmospheric carbon dioxide concentrations. Cotton scientists discussed these issues at several cotton research conferences to identify causes and to recommend corrective measures. Weather (Davis and Gallup, 1977; Orr et al., 1982), increasing soil compaction (Brooks, 1977), increased nematode populations (Orr et al., 1982), soil herbicide accumulations (Hurst, 1977), untimely use of pesticides (Leigh, 1977), excess and/or limited nitrogen supply (Maples, 1977; Gerik et al., 1998), and rising atmospheric ozone levels (V.R. Reddy et al., 1989) were all implicated as causative agents for the yield decline, but there was no comprehensive effort to determine the exact causes for the temporal yield declines because of the complexities of site-specific soil–plant–environment, and management practices.

The National Cotton Council in 1984 contracted the GOSSYM-COMAX group to investigate the causes for yield decline across the U.S. Cotton Belt. The cotton model was used to evaluate, retrospectively, the influence of environmental conditions and cultural practices on cotton yields. The weather issue was analyzed first by taking weather, soil, and cultural input data at five locations across the U.S. Cotton Belt for more than 20 years (V.R. Reddy and Baker, 1990). The results of the simulation analysis showed that weather varied greatly from year to year, causing large fluctuations in yields. The model tracked yield variations fairly well, but overpredicted yield by 20%, indicating that other yield-reducing factors were in play. Weather was ruled out as a yield-decline factor.

Addressing the issue of herbicides, described previously in the herbicide application section, V.R. Reddy et al. (1987, 1990) identified and hypothesized two herbicide effects on cotton growth and development: root pruning at various depths and reduced root water and nutrient permeability. The simulation analysis indicated that both factors reduced simulated cotton yields. They concluded that improper applications of herbicides was one of the causative factors for the cotton yield decline from 1960 to 1980 at all locations across the U.S. Cotton Belt.

The effects of compaction on soils were analyzed next (Whisler et al., 1993). Brooks (1977) suggested that greater soil compaction was caused by the increasing size and weight of farm equipment in the 1970s and 1980s compared with equipment used in earlier years. As in the previous yield-decline studies, weather, soil, and cultural input data from six locations across the Cotton Belt were used in this analysis, but there were no consistent trends traceable to soil compaction at all locations. However, compaction effects were masked and complicated by weather and the effects varied from location to location. For example, prior to 1974, compaction was found to have a negative effect in Florence, South Carolina, but after annual in-row subsoiling became a common cultural practice, it alleviated this effect. Soil compaction effects in Stoneville, Mississippi, were generally detrimental, but they were often masked by weather, according to the simulation analysis. In years with abundant water, wheel traffic compaction had little negative effects on yields, since shallow root systems could extract sufficient moisture for crop growth and yield.

Finally, the effects of changes in atmospheric ozone and carbon dioxide were evaluated (V.R. Reddy et al., 1989). The photosynthetic module was modified to accommodate the influence of rising ozone and atmospheric carbon dioxide levels (Miller et al., 1998) from the 1960s to the 1980s. The simulated effects of the two environmental factors on cotton yields varied among locations because of the interactions of soil, crop, and atmospheric variables and with nutrient levels. Under well-fertilized conditions, it was found that the increase in atmospheric carbon dioxide from 1960 to 1985 would have increased lint yields by an accumulated average of 10%. The inclusion of 23 years of summertime surface mean ozone concentrations along with the increased carbon dioxide concentrations showed a 17% decrease in the corresponding simulated mean yield of cotton lint in California, but not in other locations where ozone concentrations were lower. They demonstrated that a physiologically, physically, and mechanistically–based model such as GOSSYM is the only available tool that can be used to study the effects of such environmental factors on crop growth and yield, but that the affected physiological processes in the model must be appropriately calibrated for each variable tested.

Tillage and Erosion Studies

The model has been used to evaluate the effects of soil erosion and erosion-related activities on cotton lint yields (Whisler et al., 1986). One soil profile, 1 m deep, was assumed to have a traffic pan 170 to 240 mm below the soil surface and that the surface soil was eroded by 50 or 100 mm. Weather for a relatively dry year, 1980, and wet year, 1982, were used and compared. For the dry year, 50 mm of erosion reduced the simulated yield by 9%, and 100 mm of erosion reduced simulated yield by 19%. For the wet year, the maximum simulated yield reduction was only 2%. For a 0.3 m deep profile of the same soil, but where the traffic pan was reformed each year at 170 to 240 mm below the soil surface, the reductions in yield were greater. On the shallower soil, the predicted yield was reduced 32% in a dry year and increased erosion further reduced the yield another 10 to 20%. In a wet year, simulated yields on the shallower soil were only reduced by 14%, but more erosion further reduced the yields 20 to 40%.

The model has been used to investigate the effects of simulated tillage and wheel traffic on cotton crop growth and yield (Whisler et al., 1986, 1993). The soil compaction due to wheel traffic and subsequent loosening of the soil surface due to cultivation can change the root distribution patterns and water and nutrient extraction patterns, especially in lighter, sandier textured soils. In looking at overall effects of wheel traffic compaction, there were no consistent trends. The interacting effects of weather that varied from location to location masked compaction trends. For the Norwood silty clay soil at College Station, Texas, wheel traffic may have enhanced yields by changing the root/shoot partitioning in response to water stresses. In other areas and soils, such as the Dundee silt loam of the Mississippi Delta, it appears that compaction generally reduced yields. The erosion and tillage studies could only be done in a meaningful and quantitative way by using a process-level model such as GOSSYM. Thus, Whisler et al. (1993) concluded that the model, GOSSYM, could be used to show the interaction of soils and weather on crop yields.

In addition to tillage and erosion, soil temperature is another important factor affecting cotton growth, development and lint yield. The empirical soil temperature subroutine (TMPSOL) of the GOSSYM did not perform well under bare and cotton-cropped surface conditions in the field. A more mechanistic soil temperature subroutine of a soybean growth simulator (GLYCIM) was incorporated into GOSSYM (Khorsandi and Whisler, 1996). The resulting new soil temperature subroutine was called HEAT. Under bare surface conditions, HEAT underpredicted the average daily soil temperatures at all locations and depths (Khorsandi et al., 1997); however, under cotton cropped surface conditions, HEAT calculated the soil temperature adequately, especially after canopy closure. Thus, incorporation of HEAT into GOSSYM improved the accuracy of cotton lint yield predictions.

Insect Damage Assessment

Validation of GOSSYM against a commercial crop grown in 1976 (Baker et al., 1993) revealed that a 32% yield loss occurred due to lygus (*Lygus lineolaris*) damage to squares and to fruiting branch development at the eighth and ninth nodes on the plant. Along with lygus, more than 20 species of arthropods attack cotton beltwide. In the Mississippi Delta, the estimated average cost of insecticides was \$131 U.S. ha^{-1} (Mississippi Cooperative Extension Service, 1990). The GOSSYM can determine the need for insect control or recommend the best control strategy by inputting plant maps that reflect the location and degree of fruit loss caused by insects. Advances in understanding the pest attack and population build-up processes in cotton led to the compilation of models that predict pest population and damage. The COTFLEX (Stone and Toman, 1989; Stone et al., 1987), CALEX/Cotton (Plant, 1989, Plant et al., 1987), CIC-EM (Bowden et al., 1990), WHIMS (Olson and Wagner, 1992), TEXCIM (Willers et al., 1990), HELDMG (Thomas, 1989), MOTHZV (Hartstack and Witz, 1983), TEXCIM (Witz et al., 1990), and CIM (Brown et al., 1983) are some of the cotton pest models that were linked to GOSSYM to predict reduction in yield due to insect pests. These models were also helpful in designing control measures.

The insect models use information about the pest damage rate, the spatial distribution of damage, and the plant's response to damage. Among these models, the HELDMG (Thomas, 1989) differentiates damage caused by different larval stages of *Heliothis* as well as the damage based on the fruit position on the plant. Therefore, estimating the value of each fruit on the plant and the probability of valuable fruit being damaged makes it possible to dynamically quantify the economic loss caused by the pest, thus leading to improved pest management decisions. Another well developed and tested pest model is rbWHIMS (Olson and Wagner, 1992). It contains about 700 rules and formulates recommendations for 11 species including boll weevil, thrips, whiteflies, aphids, cutworms, and armyworms. The rules in the model also account for the seasonality of occurrence of these various pests, as different pests occur at different developmental stages of the crop. Insect input for the model is from weekly scouting information obtained in the field. The model gives three recommendations:

1. Spray with appropriate insecticide (larvicide or ovicide or both).
2. Scout in three days (indicates that insect population is at potential economic damage levels).
3. Scout in seven days (no immediate pest damage).

Effective integration of such an insect model with a physiology–based cotton model such as GOSSYM greatly improves the management decisions made by cotton farmers and farm managers.

Cultivar/Genetics Improvement Research

Several opportunities are available for improving crop performance and productivity through optimization of cultural practices, plant breeding, and new technological developments including biotechnology. A number of plant breeders have envisioned models as tools for predicting the effects and economic benefits of various genetic combinations. Breeders are also interested in predicting the responses of particular genotypes to specified environments and how the crop can be best managed to maximize yields (Landivar et al., 1983a, 1983b)

GOSSYM was used to investigate the lingering question of why the okra type (deeply lobate leaf-type) of cotton underperforms compared to other types of cotton even though it produces more bolls per plant. Landivar et al. (1983a), using GOSSYM, tried to analyze the inconsistencies in lint yield of okra type of cotton by varying nitrogen application rates and plant carbohydrate supply. They concluded that higher irrigation and nitrogen supply are required for the plant to supply sufficient carbohydrates to maintain the fruit load. Thus, in the Mississippi cotton-growing areas, normal leaf-type cotton varieties perform better as the crop is grown under rainfed conditions. Increased carbohydrate supply through increased photosynthetic efficiency also retained a higher percentage of fruits at all N application rates, which suggests that photosynthesis is the critical known limiting factor in okra leaf type cotton.

Specific leaf weight (SLW) has been used as a criterion to select for improved photosynthetic performance in crops (Barnes et al., 1969; Dornhoff and Shibles, 1970; Kerby et al., 1980). The GOSSYM model was used in the selection of physiological characters for yield improvement

(Landivar et al., 1983b) and to determine whether SLW can be used to predict higher photosynthetic efficiency. In this study, an increase in lint yield of 54% was obtained by increasing the photosynthesis by 30%, if adequate N and water were available. Increase in SLW did not improve crop yield as most of assimilates produced were simply utilized for increasing the leaf thickness. Thus, increased photosynthesis is a superior yield selection criterion compared to specific leaf weight.

Future Climate Scenarios

Increased precision in predicting future climates is used by crop models to predict plant growth, development, and yield more accurately under anticipated conditions. Working Group I of Intergovernmental Panel on Climate Change (IPCC) in its recent report (IPCC, 2001), *Climate Change 2001: The Scientific Basis*, concluded that surface temperatures increased globally by an average of 0.6 ± 0.2°C in the 20th century. The IPCC suggests that by 2050, surface temperatures will likely increase by 0.8 to 2.6°C, CO_2 concentration in the atmosphere is expected to rise to 463 to 623 μL L^{-1}, sea level will likely rise because of melting ice and water expansion by 50 to 320 mm (IPCC, 2001). Crop models have been used to predict the impact of these climate changes on agriculture. Even though increased CO_2 in the atmosphere is known to increase crop yields, the interaction of crop, genotype, soil, nutrients and atmosphere would modify the CO_2 effects.

Simulation studies with GOSSYM conducted with increased atmospheric CO_2 and O_3 (V.R. Reddy et al., 1989; V.R. Reddy and Baker, 1990) have shown that cotton yields varied with location and there was considerable interaction with soil, plant, and nutrient variables. Based on the simulation results, the authors concluded that at sufficient soil N levels, an increase in 10% of cotton lint yield would occur under predicted atmospheric CO_2 concentrations. At Stoneville, Mississippi, with current practices, the crop could not utilize the increased CO_2 concentration due to N stress. They also concluded that yield decreases in Phoenix, Arizona, and Fresno, California, were due to the increased of O_3 concentrations. Results from a 30-year simulation study (Prashant, 2000) show that increase carbon dioxide had a positive effect on cotton production. There was a 54% increase in cotton lint yield with increase in CO_2 concentration from 200 μL L^{-1} to 900 μL L^{-1} (Figure 3.5) due to increased carbon availability.

In the cotton growing delta region of Mississippi, an increase in CO_2 concentration along with increased air temperatures averaging 4°C is predicted in future climates (National Center for

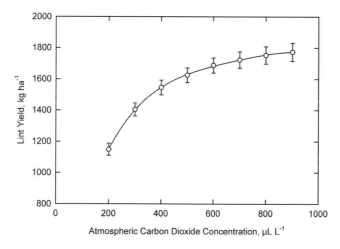

Figure 3.5 Cotton response climate change: simulated lint yield response to carbon dioxide enrichment. The data are means of 30-year simulations (1964–1993). (From Prashant, R.D., Simulating the impacts of climate change on cotton production in Mississippi Delta (thesis), Mississippi State University, Mississippi State, MS, 2000. With permission.)

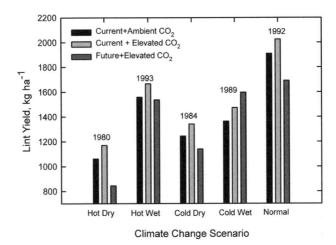

Figure 3.6   Cotton response to climate change: simulated yields for different years with varying weather patterns including current weather with ambient CO_2, current weather with elevated CO_2, and future weather with elevated CO_2. (From Prashant, R.D., Simulating the impacts of climate change on cotton production in Mississippi Delta (thesis), Mississippi State University, Mississippi State, MS, 2000. With permission.)

Atmospheric Research, Boulder, CO). In this study, the future climate scenarios were generated by adding the predicted changes in mean monthly temperature (minimum and maximum) and rainfall to the daily values of the past 30 years, based on the assumption that changes in daily weather parameters will be constant for each month (Prashant, 2000). The overall effect of such a climate change for the Mississippi delta was a 6% decrease in cotton lint yield. Simulated lint yield in different climate scenarios is presented in Figure 3.6. Increased temperature reduced the crop-growing period by 11 days. Also, a decrease in number of days with temperatures less than 15°C increased the number of growing days available per season in future weather. Altering the planting date and providing irrigation would mitigate the negative temperature and water stress effects on crop growth and yield (Prashant et al., 2000).

Educational Applications

Models have a useful role to play as tools in education, both as aids to learning principles of crop and soil management, and in helping students and commercial users develop a "systems" way of thinking that enables them to appreciate their specialty as part of a larger system (Matthews et al., 2000). They listed the following as strengths of a model:

1. Provides a framework for understanding a system and for investigating how manipulating it affects its various components.
2. Evaluates long-term impact of particular interventions.
3. Provides an analysis of the risks involved in adopting a particular strategy.
4. Provides answers more quickly and cheaply than is possible with traditional experimentation.

Crop models are the outcome of sustained integration of research by agricultural scientists. These models are used in university education systems in different parts of the world to provide students a holistic understanding of plant growth and development. They are also used to explain and solve simple problems like selecting a seeding rate for a given field or soil type and to understand complex phenomenon such as flower production and responses of growth and development to various environmental factors. Such an education system will breed a research community that can integrate various fields of agriculture and use the advances in computer hardware and software for the benefit of the farming community.

The main sources for improvement of crop models have been the postgraduate research students of the university system. Supervised by the research staff, studies were conducted to fill knowledge gaps in crop models and improve model performance. Since 1979, Mississippi State University has accepted a total of 22 theses (15 Ph.D. and seven M.S.), which have contributed to the continuous improvement of GOSSYM model, for research degrees in agriculture and other related disciplines. The research topics encompassed fields of agricultural engineering, agronomy, climate change, computer science, economics, entomology, and extension education, meteorology, and soil science. The GOSSYM model is also being used to create better understanding of environmental plant and crop physiology concepts. Improvements in crop model performance along with the ability to graphically present the changes in plant growth and development makes the model an inseparable part of classroom teaching.

The GOSSYM model is already being used to educate farmers to improve farm productivity and crop consultants to provide valuable information to farmers for reaping richer harvests.

CONCLUSIONS

The availability of computers has made it feasible to make many calculations rapidly. Other technologies such as rapid availability and automation of weather information facilitates the information gathering process essential to simulating real time crop growth. Several USDA scientists and colleagues from state universities developed a cotton crop model, which they deemed mechanistically realistic. It was validated with the assistance of agricultural scientists from around the world encompassing different cotton production practices. The model was released to U.S. farmers on a limited basis in 1984 and gradually extended to wider applications. The producer found it to be helpful for a number of preseason applications as well as in-season applications. The in-season model applications involved the timing and amount of N fertilizer application, the timing of irrigation, the timing and amount of plant growth regulator and mature crop termination chemical applications. Perhaps equally important to producers, the model helped growers to examine their crop in a more timely manner and in ways they had not previously considered.

Scientists and production-support infrastructure personnel also found the model educational, providing insight into crop responses to environmental conditions in a more fundamental way than was previously feasible. Agricultural scientists studied the impact of various environmental conditions in ways considered impossible with any other techniques. As more is learned, the possibilities and opportunities for model applications for providing production solutions as well as providing insight into the physical and physiological processes associated with crop responses to the environment appear limitless.

ACKNOWLEDGMENTS

The authors wish to thank Drs. Harry Hodges, Vara Prasad Pagadala, and Dennis Timlin for their valuable suggestions. Part of this research was funded by the U.S. DOE National Institute for Global Environment Change through the South Central Regional Center at Tulane University (DOE cooperative agreement no. DE-FCO3-90ER 61010) and the National Aeronautical and Space Administration-Funded Remote Sensing Technology Center at Mississippi State University (NASA grant no. NCC13-99001).

REFERENCES

Albers, D.W., J. Mobley, and P.W. Tracy. 1992. Evaluation of GOSSYM-COMAX in Missouri 1987–1991. In *Proc. Beltwide Cotton Conf.,* National Cotton Council, Memphis, TN, 1350–1351.

Anderson, W.P., A.B. Richards, and J.W. Whiteworth. 1967. Trifluralin effects on cotton seedlings, *Weeds,* 15:224–227.

Atwell, K.D. 1995. Calibration and validation of GOSSYM (Cotton, *Gossypium hirsutum, Gossypium barbadense*). Thesis. Submitted to Mississippi State University, Mississippi State, MS.

Baker, D.N., J.R. Lambert, C.J. Phene, and J.M. McKinion. 1977. GOSSYM: a simulator of cotton crop dynamics. In *Proc. Conf. Appl. Comput. Mgmt. of Large Scale Agric. Complexes,* Scientific Research Institute of Planning, Latvian GOSPLAN, Riga, Latvia, 39.

Baker, D.N., V.R. Reddy, J.M. McKinion, and F.D. Whisler. 1993. An analysis of the impact of lygus on cotton, *Comput. Elec. Agric.,* 8:147–161.

Baker, D.N., J.R. Lambert, and J.M. McKinion. 1983. GOSSYM: a simulator of cotton crop growth and yield. *S.C. Agric. Expt. Stn. Tech. Bull.,* No. 1089, Clemson, SC.

Barnes, D.K. , R.B. Pearce, G.E. Carlson, R.H. Hart, and C.H. Hanson. 1969. Specific leaf weight differences in alfalfa associated with variety and plant age, *Crop Sci.,* 9:421–423.

Bar Yosef, B., J.R. Lambert, and D.N. Baker. 1982. RHIZOS: a simulation of root growth and soil processes. Sensitivity analysis and validation for cotton, *Trans. ASAE,* 25:1268–1273.

Bayer, D.E. , C.L. Foy, T.E. Mallory, and E.G. Cutter. 1967. Morphological and histological effects of trifluralin on root development, *Am. J. Bot.,* 54:945–952.

Boone, M.Y.L., D.O. Porter, and J.M. McKinion. 1993. Calibration of GOSSYM: theory and practice, *Comput. Electron. Agric.,* 9:193–203.

Boone, M.Y.L., D.O. Porter, and J.M. McKinion. 1995. RHIZOS 1991: a simulator of row crop rhizospheres. *USDA, ARS Bulletin 113,* Government Printing Office, Washington, D.C.

Bowden, R.O., R.G. Luttrell, and L.G. Brown. 1990. Cotton Insect Consultant for Expert Management (CICEM): An expert system for managing cotton insects. Working paper No. 90-09. Department of Industrial Engineering, Mississippi State University, MS.

Bowman, W.D. 1989. The relationship between leaf water status, gas exchange, and spectral reflectance in cotton leaves, *Remote Sens. Envi.,* 30:249–255.

Brooks, O.L. 1977. Effects of soil environment on cotton yields. In *Proc. Beltwide Prod. Mech. Conf.,* Atlanta, GA, January 10–13, 1977, National Cotton Council, Memphis, TN, 68–69.

Brown, L.G., R.W. McClendon, and J.W. Jones. 1983. Cotton and insect management simulation model. In *Cotton Insect Management with Special Reference to the Boll Weevil,* R.L. Ridgway, E.P. Lloyd, and W.H. Cross, Eds., USDA-ARS Agriculture Handbook No. 589, 437–479.

Carter, G.A. 1994. Ratios of leaf reflectance in narrow wavebands as indicators of plant stress, *Intl. J. Remote Sens.,* 15:697–703.

Davis, D.R. and J. Gallup. 1977. Have weather patterns changed and affected cotton yields? In *Proc. Beltwide Cotton Prodn. Mech. Conf.,* January 10–13, 1977, Atlanta, GA, National Cotton Council, Memphis, TN, 70–71.

Dornhoff, G.M. and R.M. Shibles. 1970. Varietal differences in net photosynthesis of soybean leaves, *Crop Sci.,* 10:42–45.

Duncan, W.G. 1971. SIMCOT: a simulator of cotton growth and yield. In *Proc. Workshop on Tree Growth Dynamics and Modeling,* October 11–12, 1971, Duke University, Durham, NC, 115–118.

Fye, R.E., V.R. Reddy, and D.N. Baker. 1984. The validation of GOSSYM: Part 1 — Arizona conditions, *Agric. Sys.,* 14:85–105.

Gerik, T.J., D.M. Oosterhuis, and H.A. Torbet. 1998. Managing cotton nitrogen supply, *Adv. Agron.,* 64:115–147.

Gertsis, A.C. and A.G. Symeonakis. 1998. Efficient cotton production in Greece with the use of GOSSYM — a cotton crop simulation model, *Acta Hort.,* 476:307–312.

Gertsis, A.C. and F.D. Whisler. 1998. GOSSYM: a cotton crop simulation model as a tool for the farmer, *Acta Hort.,* 476:213–217.

Hartstack, A.W. and J.A. Witz. 1983. Models for cotton insect pest management. In *Cotton Insect Management with Special Reference to the Boll Weevil,* R.L. Ridgway, E.P. Lloyd, and W.H. Cross, Eds., USDA-ARS Agriculture Handbook No. 589, 359–381.

Hayes, R.M. , J.R. Hoskinson, J.R. Overton, and L.S. Jeffery. 1981. Effect of consecutive annual applications of fluometuron on cotton. Weed Sci. 29:120–123.

Hesketh, J.D. 1972. Simulation of growth and yield in cotton: II. Environmental control of morphogenesis, *Crop Sci.,* 12:436–439.

Hesketh, J.D. and D.N. Baker. 1967. Light and carbon assimilation by plant communities, *Crop Sci.,* 7:285–293.

Hesketh, J.D., D.N. Baker, and W.G. Duncan. 1971. Simulation of growth and yield in cotton: respiration and the carbon balance, *Crop Sci.,* 11:394–398.

Hodges, H.F. , F.D. Whisler, S.M.Bridges, K.R. Reddy, and J.M. McKinion. 1998. Simulation in crop management: GOSSYM/COMAX. In *Agricultural Systems Modeling and Simulation*, R.M. Peart and R.B. Curry, Eds., Marcel Dekker Inc., New York, 235–281.

Hunt, P.G., P.J. Bauer, C.R. Camp, and T. Matheny. 1998. A Nitrogen accumulation in cotton grown continuously or in rotation with peanut using subsurface microirrigation and GOSSYM/COMAX management, *Crop Sci.,* 38:410–415.

Hurst, H.R. 1977. Are herbicides cutting yields? In *Proc. Beltwide Cotton Prodn. Res. Conf.,* Atlanta, GA, Jan. 10–13, 1977. National Cotton Council, Memphis, TN, 61–63.

Intergovernmental Panel on Climate Change. 2001. Climate Change 2001: The Scientific basis. http://www.ipcc.ch/ (verified August 28, 2001).

Jallas, E. , M. Cretenet, S. Turner, R. Sesquira, and P. Martin. 2000. *COTONS release 2.0.3,* CIRAD, BP 5035, Montpellier, France.

Kappelman, A.J., Jr. and G.A. Buchanan. 1968. Influence of fungicides, herbicides and combinations on seedling growth of cotton, *Agron. J.,* 60:660–663.

Kerby, T.A., D.R. Buxton, and K. Matsuda. 1980. Carbon source-sink relationships with narrow-row cotton canopies, *Crop Sci.,* 20:208–213.

Khorsandi, F. , M.Y.L. Boone, G. Weerakkody, and F.D. Whisler. 1997. Validation of the soil temperature subroutine HEAT in the cotton simulation model GOSSYM, *Agron. J.,* 89:415–420.

Khorsandi, F. and F.D. Whisler. 1996. Validation of the soil-temperature model of GOSSYM-COMAX, *Agric. Sys.,* 51:131–146.

Ladewig, H. and E.T. Powell. 1989. *An Assessment of GOSSYM/COMAX as a Decision System in the U.S. Cotton Industry,* Texas Agric. Extn. Ser., College Station, TX, 51.

Ladewig, A.M. and J.K. Thomas. 1992. *A Follow-Up Evaluation of the GOSSYM/COMAX Cotton Program,* Texas Agric. Extn. Ser., College Station, TX, 47.

Lambert, J.R., D.N. Baker, and C.J. Phene. 1977. Dynamic simulation of processes in the soil under growing row crops: RHIZOS, In *Proc. Conf. Application of Comput. to the Management of Large Scale Agricultural Complexes,* Scientific Research Institute of Planning, Latvian GOSPLAN, Riga, Latvia.

Landivar, J.A., D.N. Baker, and J.N. Jenkins. 1983a. Application of GOSSYM to genetic feasibility studies: I. Analyses of fruit abscission and yield in okra-leaf cottons, *Crop Sci.,* 23:497–504.

Landivar, J.A., D.N. Baker, and J.N. Jenkins. 1983b. Application of GOSSYM to genetic feasibility studies: II. Analyses of increasing photosynthesis, specific leaf weight and longevity of leaves in cotton, *Crop Sci.,* 23:504–510.

Leigh, T.F. 1977. Insecticides and cotton yield. In *Proc. Beltwide Cotton Prodn. Mech. Conf.,* Atlanta, GA, January 10–13, 1977. National Cotton Council, Memphis, TN, 63–65.

Lemmon, H. E. 1986. COMAX: An expert system for cotton crop management, *Science,* 223:9–33.

Maples, R. 1977. Are we using too much fertilizer? In *Proc. Beltwide Cotton Prodn Mech. Conf.,* Atlanta, GA, January 10–13, 1977. National Cotton Council, Memphis, TN, 65–68.

Marani, A. and D.N. Baker. 1978. Development of predictive dynamic simulation model of growth and yield of Acala cotton. Science Report to U.S.–Israel Binational Science Foundation.

Marani, A. 1985. Effect of water stress on canopy senescence and carbon exchange rates in cotton, *Crop Sci.,* 25:798–802.

Masoni, A., L. Ercoli, and M. Marioti. 1996. Spectral properties of leaves deficient in iron, sulphur, magnesium and manganese, *Agron. J.,* 88:937–943.

Matthews, R. , W. Stephens, T. Hess, T. Mason, and A. Graves. 2000. Applications of crop/soil simulation models in developing countries. *Report No. PD 82.* Institute of Water and Environment, Cranfield University, Silsoe, Bedfordshire, U.K., 181.

McCauley, J.D. 1999. Simulation of cotton production for precision farming, *Preci. Agric.* 1:81–94.

McKinion, J.M. , D.N. Baker, J.D. Hesketh, and J.W. Jones. 1975. SIMCOT II: a simulation of cotton growth and yield. In *Computer Simulation of a Cotton Production System – Users Manual,* Washington, D.C., ARS-S-52, 27–82.

McKinion, J.M. , D.N. Baker, F.D. Whisler, and J.R. Lambert. 1989. Application of the GOSSYM/COMAX system to cotton crop management, *Agric. Sys.,* 31:55–65.

McKinion, J.M. and H.E. Lemmon. 1985. Expert systems for agriculture, *Comput. Elec. Agric.,* 1:31–40.

McKinion, J. M., J.N. Jenkins, D. Akins, S.B. Turner, J.L. Willers, E. Jallas and F.D. Whisler. 2001. Analysis of a precision agriculture approach to cotton production, *Comput. Elec. Agric.,* 32:213–228.

Meredith, W.R., Jr. 1987. Final report of the cotton foundation yield and productivity study. In *Proc. Beltwide Cotton Prodn. Conf.,* Dallas, TX, January 4–8, 1987. National Cotton Council, Memphis, TN, 33–37.

Miller, J. E., R.P. Patterson, A.S. Heagle, W.A. Pursely, and W.W. Heck. 1998. Growth of cotton under chronic ozone stress at two levels of soil moisture, *J. Environ. Qual.,* 17:635–643.

Mississippi Cooperative Extension Service (MCES). 1990. *Farm Management Handbook,* MCES, Mississippi State, MS.

Oliver, L.R. and R.E. Frans. 1968. Inhibition of cotton and soybean roots from incorporated trifluralin and persistence in soil, *Weed Sci.,* 16:199–203.

Olson, R.L. and T.L. Wagner. 1992. WHIMS: a knowledge-based system for cotton pest management, *Arti. Intelligence Appl.,* 6:41–58.

Orr, C.C., A.F. Robinson, C.M. Heald, J.A. Veech, and W.W. Carter. 1982. Estimating cotton losses to nematodes. In *Proc. Beltwide Cotton Prodn. Res. Conf.,* Las Vegas, NV, 22.

Pan, X., Deng, S., and J.M. McKinion. 1994. Modification of GOSSYM for predicting the behavior of cotton production systems in China. Report submitted to Cotton Research Institute, Chinese Academy of Agricultural Sciences, Anyang, Henan, Peoples Republic of China, 7.

Pavlista, A.D. 1980. A comparative study of pendemethalin and trifluralin, In *Proc. South. Weed Sci. Soc.,* Hot Springs, AR, Jan. 15–17, *1980, 33:*257–267.

Phene, C.J., D.N. Baker, J.R. Lambert, J.E. Parsons, and J.M. McKinion. 1978. SPAR-a soil–plant–atmospheric–research system, *Trans. ASAE,* 21:924–930.

Plant, R.E. 1989. An integrated decision support system for agricultural management, *Agric. Sys.,* 29:49–66.

Plant, R.E., L.T. Wilson, L. Zelinski, P.B. Goddell, and T.A. Kerby. 1987. CALEX/Cotton: an expert system-based management aid for California cotton growers. In *Proc. Beltwide Cotton Prodn. Res. Conf.,* January 4–9, 1987, Dallas, TX, 203–206.

Prashant, R.D. 2000. Simulating the impacts of climate change on cotton production in Mississippi Delta. Thesis. Submitted to Mississippi State University, Mississippi State, MS.

Prashant R.D., K.R. Reddy, and M.Y.L. Boone. 2000. Simulating the effects of climate change in cotton production in the U.S. midsouth. In *Proc. Beltwide Cotton Conf. National Cotton Council,* Memphis, TN, 591.

Radin, J.W. and Mauney, J.R. 1986. The nitrogen stress syndrome. In *Cotton Physiology,* J.R. Mauney and J.McD. Stewart, Eds., The Cotton Foundation, Memphis, TN, 91–105.

Reddy, K. R., and M.Y.L. Boone. 2001. Modeling and validating cotton leaf area development and stem elongation, *Acta Hort.,* in press.

Reddy, K.R., M.Y.L. Boone, A.R. Reddy, H.F. Hodges, S. Turner, and J.M. McKinion. 1995. Developing and validating a model for a plant growth regulator, *Agron. J.,* 87:1100–1105.

Reddy, K.R., H.F. Hodges, and J.M. McKinion. 1997a. Crop modeling and applications: a cotton example, *Adv. Agron.,* 59:225–290.

Reddy, K.R., H.F. Hodges, and J.M. McKinion. 1997b. Modeling temperature effects on cotton internode and leaf growth, *Crop Sci.,* 37:503–509.

Reddy, K.R., H.F. Hodges, and J.M. McKinion. 1997c. A comparison of scenarios for the effect of global climate change on cotton growth and yield, *Aust. J. Plant Physiol.,* 24:707–713.

Reddy, K.R., H.F. Hodges, and B.A. Kimball. 2000. Crop ecosystem responses to global climate change: cotton. In *Climate Change and Global Crop Productivity,* K.R. Reddy and H. F. Hodges, Eds., CAB International, Wallingford, U.K., 161–187.

Reddy, K.R., H.F. Hodges, J.J. Read, J.M. McKinion, J.T. Baker, L. Tarpley, and V.R. Reddy. 2001. Soil–plant–atmosphere–research (SPAR) facility: a tool for plant research and modeling, *Biotronics,* 30:1–24.

Reddy, K.R., H.F. Hodges, and J.M. McKinion. 1993. A temperature model for cotton phenology, *Biotronics*, 2:47–59.

Reddy, V.R., D.N. Baker, and J.N. Jenkins. 1985. Validation of GOSSYM: Part II — Mississippi conditions, *Agric. Sys.*, 17:133–154.

Reddy, V.R. and D.N. Baker. 1988. Estimation of parameters for the cotton simulation model GOSSYM: cultivar differences, *Agric. Sys.*, 26:111–122.

Reddy, V.R. and D.N. Baker. 1990. Application of GOSSYM to analysis of the effects of weather on cotton yields, *Agric. Sys.*, 32:83–95.

Reddy, V.R., D.N. Baker, F.D. Whisler, and R.E. Fye. 1987. Application of GOSSYM to yield decline in cotton: I. Systems analysis of effects of herbicides on growth, development, and yield, *Agron. J.*, 79:42–47.

Reddy, V.R., D.N. Baker, and J.M. McKinion. 1989. Analysis of effects of atmospheric carbon dioxide and ozone on cotton yield trends, *J. Envi. Qual.*, 18:427–432.

Reddy, V.R., D.N. Baker, F.D. Whisler, and J.M. McKinion. 1990. Analysis of the effects of herbicides on cotton yield trends, *Agric. Sys.*, 33:347–359.

Reddy, V.R. 1993. Modeling mepiquat chloride — temperature interactions in cotton: the model, *Comput. Elec. Agric.*, 8:227–236.

Ritchie, J.T. 1972. Model for predicting evaporation from a row crop with incomplete cover, *Water Resources Res.*, 8:1204–1213.

Staggenborg, S.A., R.J. Lascano, and D.R. Krieg. 1996. Determining cotton water use in a semiarid climate with the GOSSYM cotton simulation model, *Agron. J.*, 88:740–745.

Stapleton, H.N., D.R. Buxton, F.L. Watson, D.L. Nolting, and D.N. Baker. 1973. COTTON: a computer simulation of cotton growth, *Arizona Agriculture Experiment Station Technical Bulletin,* No. 206, Arizona, 124.

Stevens, W.E., J.J. Varco, and J.R. Johnson. 1996. Evaluating cotton nitrogen dynamics in the GOSSYM simulation model, *Agron. J.*, 88:127–132.

Stone, N.D., R.E. Frisbie, J.W. Richardson, and C. Sansone. 1987. COTFLEX, a modular expert system that synthesizes biological and economic analysis: the pest management advisor as an example. In *Proc. Beltwide Cotton Prodn. Res. Conf.*, Dallas, TX, Jan. 4–9, 1987, National Cotton Council, Memphis, TN, 194–197.

Stone, N.D. and T.W. Toman. 1989. A dynamically linked expert-database system for decision support in Texas cotton, *Comput. Elec. Agric.*, 4:139–148.

Stuart, B.L., V.R. Isbell, C.W. Wendt, and J.R. Abernathy. 1984. Modification of cotton water relations and growth with mepiquat chloride, *Agron. J.*, 76:651–655.

Sunderman, H.D. 1997. Variability in leaf chlorophyll concentration among fully fertilized corn hybrids, *Comm. Soil Sci. Plant Anal.*, 28:1793–1803.

Tarpley, L., K.R. Reddy, and G.F. Saseenrath-Cole. 2000. Reflectance indices with precision and accuracy in predicting cotton leaf nitrogen concentration, *Crop Sci.*, 40:1814–1819.

Thomas, W.M. 1989. Modeling within-plant distribution of *Heliothis* spp. (*Lepidoptera noctuidae*) damage in cotton, *Agric. Sys.*, 30:71–80.

Wanjura, D.F. and B.L. McMichael. 1989. Simulation analysis of nitrogen application strategies for cotton, *Trans. ASAE,* 32:627–632.

Watkins, K.B., Y.C. Lu, and V.R. Reddy. 1998. An economic evaluation of alternative pix application strategies for cotton production using GOSSYM/COMAX, *Comput. Elec. Agric.*, 20:251–262.

Whisler, F.D., B. Acock, D.N. Baker, R.E. Fye, H.F. Hodges, J.R. Lambert, H.E. Lemmon, J.M. McKinion, and V.R. Reddy. 1986. Crop Simulation models in agronomic systems, *Adv. Agron.*, 40:142–208.

Whisler, F.D. et al. 1993. Analysis of the effects of soil compaction on cotton yield trends, *Agric. Sys.*, 4:199–207.

Willers, J.M., R.A. Sequeira, and S. Turner. 1994. The integration of insect models and plant models: Some considerations. In *Proc. Beltwide Cotton Engr. Sys. Conf.*, San Diego, CA, January 5–8, 1994, National Cotton Council, Memphis, TN, 592–593.

Witz, J.A. et al. 1990. Programming the combination of cotton insect model, TEXCIM, with the cotton physiology mode, GOSSYM, using the *Heliothis* damage model, HELDMG: a status report, In *Proc. Beltwide Cotton Prodn. Res. Conf.*, Las Vegas, NV, Jan. 9–14, 1990, Memphis, TN, 343.

York, A.C. 1983a. Cotton cultivar response to mepiquat chloride, *Agron. J.*, 75:663–667.

York, A.C. 1983b. Response of cotton to mepiquat chloride with varying N rates and plant populations, *Agron. J.,* 75:667–672.

Zummo, G.R., J.H. Benedict, and J.C. Segers. 1984. Effect of plant growth regulator mepiquat chloride on host plant resistance in cotton bollworm (*Lepidoptera noctuidae*), *J. Econ. Entomol.,* 77:922–924.

CHAPTER 4

Experience with On-Farm Applications of GLYCIM/GUICS

Dennis J. Timlin, Yakov Pachepsky, Frank D. Whisler, and Vangimalla R. Reddy

CONTENTS

INTRODUCTION

Agricultural producers need to make informed decisions in order to manage their enterprises efficiently. These decisions are based on experience and available information as well as on input from agricultural consultants. As their enterprises grow in size and complexity, it becomes more difficult for growers to manage large amounts of information. Furthermore, uncertain weather conditions increase risk. As personal computers have become more widely available, there has been an effort among agricultural researchers to provide computer-based tools to help manage and synthesize information. The main reason for using tools such as computer models on farms is to increase profit and manage resources, although learning more about how crops respond to environmental factors, and help in complying with governmental regulations are also important. Furthermore, the ability to compare the probable outcomes of different decisions can help a producer make a more informed choice and reduce risk in the face of future uncertainties. These computer-based tools have become known as decision support systems (DSS).

Early DSS tools, known as expert systems (McKinion and Lemmon, 1985), resulted from an effort to encapsulate information and experience so that the computer program could answer questions by synthesizing heuristic information that had been input and choosing the correct answer from a knowledge base. Crop simulation models were incorporated in DSS to account for dynamic seasonal and interannual interdependencies of weather, plant characteristics, and soil. Computer simulation models mimic crop response to environmental variables because they calculate photosynthesis, carbon partitioning, water and nutrient uptake, and yield using equations developed from experimental data. The level of detail varies from empirical/mechanistic (Hammer et al., 1995; Jones and Kiniry, 1986) to highly mechanistic models [GOSSYM (Baker et al., 1983)]. Crop simulation models can estimate the growth of a crop from emergence to maturity, account for major physiological and morphogenic processes, and describe primary relationships in the soil–plant–atmosphere system.

GOSSYM was one of the first simulation model-based DSS for crops. Since the early 1980s it has been widely used by cotton producers for water and nitrogen management as well as for timing harvest operations (Baker et al., 1983; Landivar et al., 1989). GOSSYM was also combined with an expert system and called GOSSYM-COMAX (Lemmon, 1986), where rule-based reasoning was used to interpret simulation results.

The soybean model, GLYCIM, was developed after GOSSYM and shared some design components and modules. GLYCIM has highly mechanistic, dynamic representations of plant growth, development and yield, and soil and weather processes. The mechanisms involved in the physical and physiological processes in soybean and its environment are mathematically described in GLYCIM (Acock et al., 1985). These processes include light interception, carbon and nitrogen fixation, organ initiation, growth and abscission, and flows of water, nutrients, heat and oxygen in the soil. GLYCIM is organized into modules in accordance with a generic modular structure and runs in hourly time steps. Documentation, including the FORTRAN listing, definition of variables, description of theory, and details of input and output files has been published elsewhere (Acock et al., 1985; Acock and Trent, 1991).

The model GLYCIM has been designed to simulate the growth of any cultivar on any soil and at any location and time of year. Simulations are initiated at the cotyledonary stage with appropriate data on the number, size, and weight of organs on the plant. Plant growth in size and phenological stage are predicted by the model. During the simulation, GLYCIM provides predicted values for most of the physiological variables. It also simulates nitrogen contents of various organs on the plant and water and nitrogen status of the soil. The model provides the dry weights of all plant parts and final seed yield.

The environmental inputs necessary to run GLYCIM are solar radiation, maximum and minimum air temperature, rainfall, and wind speed. The model also uses wet and dry bulb temperature if available and has the capability to use either hourly or daily data. GLYCIM also needs information on the physical and hydraulic properties of the soil, maturity group of the cultivar, latitude of the field, date of emergence, row spacing, plant population within a row, row orientation, irrigation amount, method and date, and CO_2 concentration in the atmosphere.

Currently, 26 parameters define growth and development rates, and yield components (Table 4.1). Three parameters define the rate of vegetative development, 12 parameters define the rates of reproductive stage progress from R0 (floral initiation) to R8 (podset), six parameters define the rate of stem extension and number of branches, one parameter defines root growth, and four parameters define dry matter partitioning.

Since the 1991 growing season, GLYCIM has been used by farmers for crop management and input optimization in the Mississippi Valley. The model is being used for selecting cultivar, row spacing, plant population and planting date prior to planting, and for post-planting decisions such as irrigation scheduling, insect control, harvest timing, and forecasting of final yield (Reddy et al.,

Table 4.1 Vegetative and Reproductive Development Rate and Yield Component Parameters Used in GLYCIM

Vegetative

Slope of the vegetative (V) stage dependence on growing degree days (GGD day^{-1})
Maximum vegetative (V) stage
Correction factor for the early vegetative stage progress rate to account for clay content

Reproductive

Progress rate toward floral initiation (R0) at solstice (day^{-1})
Daily rate of the progress to R0 before solstice (day^{-1})
Daily rate of the progress to R0 after solstice (day^{-1})
Progress rate from R0 toward R2 (day^{-1})
Slope of the dependence between the end of R2 and emergence date
Intercept of the dependence between the end of R2 and emergence date
Progress rate from R2 toward R6 (day^{-1})
Length of the plateau R5 (day)
Length of the plateau R6 with no water stress (day)
Rate of decay of the R6 plateau as the water stress increases (day^{-1})
Rate of the progress toward R7 (day^{-1})
Reproductive (R) stage to stop vegetative growth

Stem Elongation and Height

Potential stem elongation per dry weight increase of petioles (cm g^{-1})
a in the dependence ($h = a(v)^b$) between height and vegetative stages
b in the dependence ($h = a(v)^b$) between height and vegetative stages
Stem weight per unit stem elongation (g)
Increment in leaf area per increment in vegetative stage
Number of branches per unit plant density

Roots

Potential rate root weight increase (g day^{-1})

Yield Components

Increase in pod weight as a function of progress in R stages
Increase in seed weight per seed fill rate
Number of seeds per bushel
Seed fill rate g day^{-1}

1995). The model helps farmers to optimize inputs and maximize profits. Since 1991, USDA scientists have collected 156 data sets on soils, crop growth and development, weather, and management conditions (planting and harvest date, and irrigation). These data come from the fields of cooperating growers and include numerous cultivars grown under various soils, and weather and management conditions. As GLYCIM was used on-farm, an interface was developed and evolved over time. This paper describes the experiences of the cooperating growers and researchers involved in the GLYCIM on-farm project, and the current interface (GUICS — Graphical User Interface for Crop Simulators) developed to facilitate use of GLYCIM by the growers.

ON-FARM TESTING OF GLYCIM

GLYCIM was originally evaluated using data collected on the soybean cultivar "forrest" at the Plant Science Farm at Mississippi State University (Acock, et al., 1985; Gertis, 1985). All model equations and parameters came from experiments in growth chambers and greenhouses; soybean plants were mainly grown in pots and in small plots. On-farm testing began in 1991 when a soybean grower, Kenneth Hood of Pershire Farms, Gunnison, Mississippi, agreed to allow researchers to plant strips of different soybean cultivars and take measurements on growth, development and yield. The purpose of the field trials was to collect data on phenology, dry matter production, and yield to validate the model.

On-farm research with GLYCIM was expanded to the Mississippi Delta farms of Edward Hester and Jay Mullens in 1992 and to the Fletcher Clark farm in 1993. Because two of these growers had been involved in the project to evaluate the use of GOSSYM/COMAX, they had weather stations close by and some familiarity with simulation models for irrigation scheduling. An ongoing program to characterize soil hydraulic properties for the cotton model, GOSSYM was continued for the GLYCIM testing (Whisler, 1982). Soil samples have been taken across the Cotton Belt and analyzed for their physical and hydrological properties. After 1993, additional growers from Mississippi, Alabama, Missouri, Louisiana, Tennessee, and Arkansas joined the project. Some of these growers had experience with GOSSYM. Others came by word of mouth based on contact with growers already using GLYCIM. By 1997, we had 12 growers in the program, and all requested the model again the next year. This allowed us to test GLYCIM at a wide variety of locations.

Participants represented a variety of farm operations for the Mississippi Valley (Reddy et al., 1997). Farm size ranged from medium (300 ha) to large (6000 ha), and ages from 23 to 75. All operations were family farms where cotton and rice were the primary crops. Soybeans were usually a secondary crop.

Field Sampling Protocol

The sampling program was carried out to collect data on vegetative and reproductive stages, dry matter and yield. Most of the plots were irrigated because that was the growers' practice, but, as researchers, we tried to include non-irrigated plots, when available, in order to compare irrigated and non-irrigated yields. Data were also collected on as many soybean varieties as possible to provide a wide variety of input data for growers.

The field plots were laid out in the spring of each year after the soybean plants emerged. The plots were usually 12.1 m wide and 176.8 m to 192 m long (0.12 to 0.16 ha) in the form of a transect. Three replicates (transects) were laid out in each field. Crop management was the same at all plots but different varieties were planted. Planting dates were also varied. This allowed data collection for a number of different varieties in the same location. Sampling was bi-weekly to weekly depending on growth stage. Height and developmental stages were measured on the same plants to minimize variability due to soil and plant. Destructive samples for dry matter determination were taken from outside the plots. A 7.6 m (25 ft) strip was harvested from each transect using a plot combine trailered from Mississippi State University.

Modifications to GLYCIM Based on On-Farm Testing

The on-farm experience with GLYCIM in farmers' fields helped identify several weaknesses in the model from 1991 to 1993. These weaknesses were in the prediction of soybean phenology, especially floral initiation and anthesis and soybean response to short-term cold injury. A series of experiments were conducted in controlled-environment plant growth chambers and in the field to supplement data collected from growers' fields. As a result, we incorporated new algorithms in the model that improved GLYCIM's predictive capability for phenology (Reddy et al., 1995; Acock et al., 1997; Reddy et al., 1998) and yield under a range of conditions. More than 80 data sets were assembled during the period of 1993 to 1996.

Initially, GLYCIM had only two user-selectable parameters for crop growth and development. These were parameters for maturity group and determinacy. Parameters for growth, development, and yield processes such as progression of vegetative stages and seed fill rates were hard coded in the program. There were a total of 24 of these. In 1993, many of the parameters were made into variables and grouped by plant cultivar (Reddy et al., 1995). The number of user-selectable parameters was reduced to 18 (PARM1 to PARM18 in Reddy et al., 1995) of which only 15 were varied among cultivars (Table 4.1). The remaining three parameters included an evapotranspiration pan factor, and parameters that defined the number of branches per unit plant density, and root growth rate that are not varied unless data are available. The use of cultivar related parameter files allowed us to simulate cultivar specific rate processes better and easily modify parameters for new cultivars. Later, after 1996, based on results in growers' fields, four new parameters were added. One was used to adjust the vegetative stage for soil clay content and another three were used to calculate floral initiation (R0) based on summer solstice (Table 4.1) (Acock et al., 1997).

Other modifications were made to GLCYIM to adapt the model better to user requirements as the project progressed. Irrigation was originally simulated by augmenting the rainfall data. Several growers used flood or furrow irrigation that was not compatible with the capacitance-based infiltration component of GLYCIM. When infiltration is modeled this way, water input as rainfall or irrigation fills the surface soil layer instantaneously to some upper limit, usually termed field capacity. Additional water moves to the next layer finally becoming drainage if the soil's water-holding capacity is exceeded. To handle furrow or flood irrigation, the water infiltration code was modified to include gradient-driven infiltration using the Green-Ampt equation (Pachepsky and Timlin, 1996).

Several growers preferred to plant narrow row soybeans with a grain drill but the model could not simulate the yields and phenology well. Two years of experiments used different population densities (10 to 60 plants per m^2) to develop equations to describe the effects of plant population density on branching (Reddy et al., 1999). These were added to GLYCIM and several growers now experiment with row spacing in the simulations to evaluate planting strategies for different soils.

The GLYCIM validation study was an on-farm trial, not a complex research-designed experiment. On-farm experiments can be difficult to manage with potential problems in logistical support, analytical needs and farmer participation (Lightfoot and Barker, 1988). The farmers were left to experiment with GLYCIM, and the research goal was to collect validation data sets and improve the model. Researchers monitored the irrigation schedules closely and duplicated model runs of GLYCIM made by the farmers and conveyed the results back to them. Meetings were held with the participants to insure that improvements in GLYCIM performance would be relevant to their needs.

Graphical User Interfaces

GLYCIM was used on farms in 1991, before it had a user interface. The model ran from a command line, and input and output data were supplied in files edited by a simple text editor. The primary use of GLYCIM was to provide a tool for growers to estimate yield using current weather data and a standard weather file for forecasting, and to gain further data for GLYCIM with different soils and varieties. Initially the intended use of GLYCIM was to predict crop yield and harvest date.

Developed as a research model, the initial version required a user to edit the input files manually and did not provide a visual representation of the output or easily interpreted tables. There was an obvious need to develop a user interface. Without the interface, the potential of a model may be lost and assembling input data can be formidable. Most users would probably have neither the time nor inclination to learn and/or perform tedious procedures for entering data and displaying results.

In 1993, a Microsoft Windows™ 3.1-based user-friendly interface, called WINGLY, with a "point and click" design was released to the growers. There was little user input. Most of the input in the WINGLY design was through manually editing files. The interface managed the arrangements of the input files to allow a user to run simulations with different soils and varieties. It was also programmed to call the weather station, and download and manage the weather data, and it had a simple graphical output interface to assemble tables and summary data that allowed users to view the output information in a readily understandable format.

WINGLY was also designed as a simple DSS. Each day during the growing season a grower would begin by downloading daily weather data over a telephone line. Then WINGLY would add weather data from a standard weather file that contained averages of 20 years of daily weather data for a nearby location. These average weather data were used to fill in rainfall, radiation and air temperature for the remainder of the season. The grower would enter the file name corresponding to the proper cultivar, initialization, soils, and irrigation data. These files were prepared by the authors, the model developers. GLYCIM would predict yield given the weather data to date, no rainfall in the next five days, and a selected year's rainfall until the end of the season. This would allow the grower to project yield given no irrigation and to decide when to irrigate.

An alert to the user would be triggered if a rule-based analysis of the simulation results detected moisture stress. This expert rule was developed from field trials of GLYCIM on farms in the Mississippi Valley and grower experience from 1991 to 1993. While running the command-line version of GLYCIM, grower Kenneth Hood noticed that one of the output files contained information on water stress and tried to irrigate according to the indication of stress. He found this was successful. Based on this experience, the model developers defined a trigger point based on a period when the plant could not take up sufficient water to meet transpiration needs for three or more consecutive hours and continuing for three or more consecutive days. If irrigation were necessary within the next five days, the model would alert the user: *"Irrigate before mm/dd if there is no rainfall in the next three days."* The grower would add the irrigation amount to an irrigation file and rerun the model with varying irrigation amounts until the water stress was alleviated. Additional information provided by WINGLY included warnings of the time to harvest and a summary of the seasonal simulation that a grower could use to evaluate an irrigation or other management strategy such as cultivar selection or row spacing.

In 1995, the authors began development of a new interface, GUICS (Graphical User Interface for Crop Simulators) (Acock et al., 1998). Previously, WINGLY relied on entering text and not on drop-down menus, and could not manage multiple scenarios nor easily allow a user to make comparisons of scenarios. It was a product of Microsoft Windows 3.1, 16-bit technology and was also limited to one model — GLYCIM. Surveys of on-farm use of computerized DSSs, both simulation- and nonsimulation-based, have shown that the complexity of DSS use is one of the most limiting factors (Greer, 1994). Furthermore, the Microsoft Windows 95 operating system provided a new environment for data management and visualization as well as 32-bit processing.

GUICS was designed as a usable generic user interface that could manage several models. The ability to manage several different models provides a tool to study the effects of crop rotations, or to compare models. The usability paradigm includes features such as:

1. Effectiveness of task performance or user productivity
2. Learnability, including the primary learning time and relearning time in intermittent use
3. Flexibility, adaptable to changing tasks or a changing environment
4. Attitude, defined in terms of the users liking or disliking the interface (Acock et al., 1999)

Icons help users recognize by pattern rather than by recalling information. Wizards are used to guide the user through the task of assembling data into a scenario and making a simulation run. One data category is shown at a time. The interface also features forgiveness to facilitate exploration, allowing a user to back out of a selection without losing information. Further, the interface has automated weather data downloading and a minimum number of active buttons to reduce memorization and display a minimum amount of text to reduce clutter.

The data presentation in GUICS is greatly simplified over that presented in WINGLY. Output is presented in tables and summary files rather than a detailed listing used for research and debugging. GUICS also stores the output from the scenarios separate from the input data. In order to run the model, the user selects the button for "View Results." There is no "Run" button; the path to view the output or run the model is the same button. This was an attempt to isolate the user from the mechanics of running the model and decouple the concepts of a simulation run and simulation result. The results would be available only after the simulation ended.

GUICS has tools to help a user to obtain weather data through phone lines, assemble a simulation scenario and view results. Developing GUICS involved research in the hierarchies (i.e., projects contain scenarios and scenarios contain model runs) of information use in simulators, and in system requirements for different groups of users. GUICS has a fully object-oriented design and implementation (Acock et al., 1999). It is open to enhancements, e.g., using maps, using databases to store datasets, and working with suites of models. Users of crop simulation models often need to work with several models to study the effects of crop rotations, to compare models, or to obtain information for decision making within a farm operation.

A survey was carried out during the early design stage of GUICS (Reddy et al., 1997) to:

1. Assess user satisfaction with WINGLY.
2. Predict user acceptance of the new interface.
3. Research future user needs.

Hand-drawn panels of the interface's initial screen design were prepared to help evaluate the appearance of the interface. Later, a computer mockup was designed and presented to users while the program itself was plastic enough to allow major changes without requiring extensive modifications to code. The GUICS prototype was evaluated by giving hands-on experience to a group of end users, including seven farmers from five southern states who already had experience with WINGLY.

Evaluation of the GUICS Interface

Employing the general guidelines of usability testing (Rubin, 1994), the authors evaluated GUICS. In observational interviews (Martin and Eastman, 1996), they

1. Outlined the new features of the interface.
2. Demonstrated how to get warnings of stress effects on yields.
3. Asked the users to go through the whole process of crop simulation on their own and recorded all difficulties experienced.
4. Asked users to point out any inconveniences and discussed with them the usability of GUICS.
5. Asked whether the users would prefer to keep WINGLY or to use GUICS.

Acceptance was of concern, because users often prefer a familiar interface to the one that is supposedly improved (Rudisill et al., 1996). The interviewers also asked about the need for mapping tools for input and output, and about the need for resources to be accessed through the Internet.

GUICS was amended as a result of these interviews, and the amended version was used on 15 farms in 1997 and 1998. The only serious problems encountered were errors in downloads of weather data leading to faulty weather files being used in simulations.

GROWERS' EXPERIENCES

Experience with the Interface, GUICS

Two of the growers were able to put together a scenario and obtain simulation results immediately after the demonstration. Wizards appeared to be a big help. Lack of consistency in implementing Windows shortcut conventions and not including consistent error messages in the wizards were reported as problems by the growers. Two users found the icons confusing. Guidelines on naming datasets and scenarios and on writing memos were requested. None of the users saw an advantage in combining various scenarios into projects for on-farm use. One of the users indicated that the accumulation and collection of garbage data might become an issue as data manipulation becomes easier. The ability to have several crop models running under the same interface was welcomed (although this has not been implemented yet). None of the users objected to replacing WINGLY with GUICS, provided they were given a converter to transform WINGLY data files to GUICS data files.

Many of the requested enhancements centered around the need to manage weather files. An automated update of ASCII weather files was requested. Tools to generate several predicted weather files were desired. Most users felt an advisory system on weed control would be useful. Some of the growers had yield monitors and all the growers agreed there was a need for mapping tools in the DSS. A mapping unit from NRCS soil survey reports could be used as a kernel to link soil, weather, management, and cultivar data.

Discussion of the need for mapping tools revealed a variety of interests, mostly related to the familiarity of the users with precision farming technology. All agreed it would be convenient to use a mapping unit as the kernel of a project relating soil, weather, management, and cultivar data. Two users thought soil mapping units could be kernel units, whereas one thought a field might be the more appropriate unit. Three users had yield monitors and thought that crop simulation should be related to yield map analysis, and that the appropriate tools should be integrated into GUICS. One user pointed out that the accumulation of data eventually might make desirable a database to support complex queries. None of the users were aware of Internet resources that could help them use GUICS as a decision support tool, although all them expressed interest in information on such resources.

Despite the development of GUICS, many users still find the output too time-consuming to interpret. There is still, probably, too much reliance on graphs displaying a time course of the simulation, a holdover from the paradigm of research and a scientist's eye toward information. The problem is time, and most growers are limited in the amount of time they can devote to under-standing all the information. Growers are also looking for diagnostics and other information that will help them meet specific goals. Based on results of a survey carried out with the participating growers (Reddy et al., 1997), the output variables given of most interest were irrigation timing, yield, and maturity date. The preferred output was a single number. A short summary of the main parameters of crop development was the second choice, and the full model output was least desired. The users mentioned that the percent canopy and early warning of impending moisture stress would be useful but are not now available in the graphical and summary output of the DSS. GLYCIM gives output data on more than 20 characteristics of the developing crop, but only one grower was interested in this information. Two users expressed interest in seed protein content. Economic information was mentioned but the users were not enthusiastic about bookkeeping with a DSS. Heavy users of GLYCIM/GUICS felt that graphs were useful as they allowed a user to compare scenarios. The users were primarily concerned that the initial compilation of the information needed for a simulation would be beyond the resources of most farmers.

On-Farm Applications of GLYCIM/GUICS

The growers use GLYCIM for preplant (strategic) planning decisions such as the selection of cultivar/soil type combination, planting date, row spacing, and postplant (tactical) decisions such as irrigation scheduling, harvest timing, and yield prediction. Researchers did not envision the use of GLYCIM for cultivar selection in the early project development stages. One grower found that the interface allowed him to make yield and growing season comparisons among varieties and soils and began using the model to make preplant decisions (Remy, 1994). Another grower reported he does not plant a field that he has not tested with model runs beforehand (Remy, 1994). He runs scenarios and compares estimated yields and harvest dates to test for soil, cultivar, and row spacing interactions. The grower uses different weather records and irrigation schedules from his farm to optimize simulated production.

According to the cooperating growers, the use of the GLYCIM/GUICS for crop management decision making, and input optimization has increased profits and resulted in more efficient water use by the growers (Remy, 1994). In a survey by Mississippi State University, the soybean growers using GLYCIM/GUICS attributed an increase in soybean yields of up to 29% and irrigation use efficiency of up to 400% to the use of GLYCIM (Remy, 1994). Many of the soils in the Mississippi Valley are shrinking and swelling clays (i.e., Sharkey series, very fine smectitic thermic Chromic Epiaquerts). Large cracks form as these soils dry. Traditionally, growers would not irrigate until they began to observe cracks, although the soybean plants were already beginning to be stressed. The model alerts the farmers to irrigate earlier than their traditional practice. Growers reported that before using GLYCIM to schedule irrigation they started watering too late and quit too early (Manning, 1996). By irrigating earlier, water stress to the soybeans is alleviated and less water is lost to deep drainage through the cracks. The soil also wets up faster and takes less time to irrigate; this increases irrigation efficiency. One grower reported that irrigation time on a cracking clay soil went from 4 to 5 days to 30 hours by irrigating earlier as recommended by GLYCIM/GUICS. Another grower, after noticing how much yield loss the model was predicting due to moisture stress, purchased an additional irrigation system realizing that it would pay for itself through increased yield.

An interesting side benefit of the DSS was that it provided incentive to the growers to go out to their fields and critically observe their soybean plants. Many growers, after viewing simulation results during the growing season, would go to their fields often to check their crop growth stage and compare to GLYCIM's predictions of phenology. As a result, they would be more aware of details of their fields and crops, and the crops' responses to the environment. There is also a learning component to using the DSS this way. After time, growers would recognize plant stress stages and critical soil moisture levels where irrigation would be necessary. A DSS might be less important for management at this stage.

RESEARCH EXPERIENCES

Phenology predictions for two farms in the Mississippi Valley are given in Figures 4.1 and 4.2. The data in Figure 4.1 are from 1997, 4 years after we calibrated the soybean cultivar files. Predictions of vegetative (Vstage) and reproductive (Rstage) stages are close to the measured values. The error bars suggest the wide range in variability in phenological development on the farms.

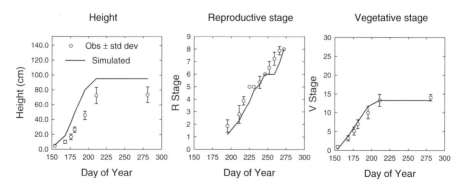

Figure 4.1   Predicted and measured phenology data for irrigated soybeans grown on the Hester farm in 1997. The cultivar is DPL3588 and soil is Sharkey clay loam (very-fine smectitic thermic Chromic Epiaquerts).

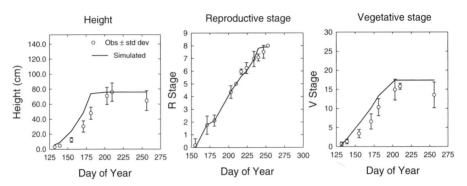

Figure 4.2   Predicted and measured phenology data for irrigated soybeans grown on the McCain farm in 1997. The cultivar is Asgrow 4922 and soil is Dubbs sandy loam (fine silty, mixed, active, thermic Typic Hapludalfs).

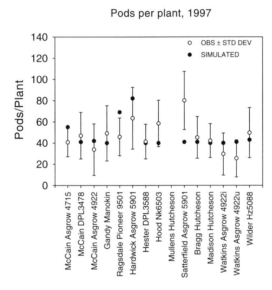

Figure 4.3   Simulated and measured pods per plant for seven different soybean varieties grown at various locations in the Mississippi Valley.

Plant height seems to be the most difficult variable to capture. In some cases, GLYCIM overestimates plant height. These data are typical of most of the measurements collected on the farms. The cultivar parameters also appear to be stable over years and sites. Predictions of pods per plant for a number of varieties and sites for 1997 are given in Figure 4.3. The variation for the GLYCIM calculated numbers appear to be less than for the measured values and some predicted values are outside the range of the measured. Nevertheless, GLYCIM does a reasonably good job of prediction especially given the range in varieties and farms (Reddy et al., 1995).

Predicted and measured soybean grain yields are shown in Figures 4.4 to 4.6. These data represent 3 growing seasons, six to 10 varieties, and 12 growers. In some cases, predicted yields are close to measured, in others they are off by as much as 50%. The relative variation in measured yields appears to be considerably less than that of the phenology data. On the whole, GLYCIM is more likely to overestimate soybean yield than underestimate it. In recent years weather was warmer than usual, and GLYCIM yield estimates were generally much higher than the measured yields. GLYCIM appears to estimate biomass correctly and seed number per pod (Figure 4.3) but has

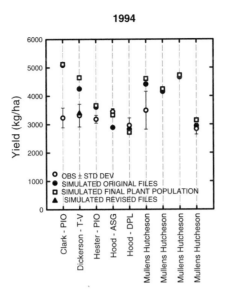

1994

Figure 4.4   Simulated and measured yields for soybean varieties grown in 1994 on several farms in the Mississippi Valley; and differences in predictions using measured and simulated plant populations.

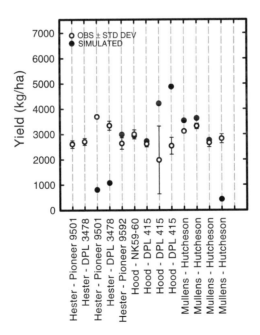

1995

Figure 4.5   Simulated and measured soybean yields for several varieties grown on three farms in the Mississippi Valley in 1995.

problems simulating the correct seed weights. This may indicate a problem with carbon allocation and is currently being investigated.

Currently, measured soil hydraulic and physical properties are used in GLYCIM. These properties include saturated hydraulic conductivity, parameters to describe the relationship between soil

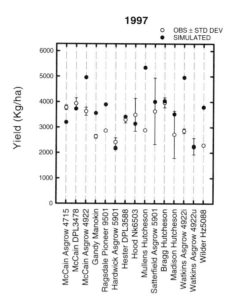

Figure 4.6   Simulated and measured soybean yields for the varieties tested on the cooperating farms in 1997.

matric potential and water content, and sand and silt percentages. Timlin et al., 1996 investigated the use of soil hydraulic properties that were estimated from soil texture. They reported large differences in yields predicted using measured and estimated soil hydraulic properties for single season simulations. Differences in long term averages of simulated yields were less. GLYCIM was most sensitive to the value used for available water content (currently the difference between the 30 kPa and 1500 kPa water contents) and less sensitive to saturated hydraulic conductivity.

Understanding the uncertainties involved with the model, the growers were not overly concerned with the quality of the predictions by GLYCIM/GUICS as long as the differences between simulated and observed yields were not too large. They were mainly concerned with how well the model predicted relative effects of water stress and the other variables to allow them to evaluate comparative management strategies. Irrigation timing predictions seemed to be correct, within 1 to 2 days, and were considered to be satisfactory. The maturity dates were correct about 80% of the time, though the worst error was 10 days. If parameters were not available for a particular cultivar, the use of parameters for a similar cultivar resulted in increased error.

ROLE OF CONSULTANTS

An early assumption of the research team was the expectation that the grower or a member of the family and/or associates would run the model. As farm operations are becoming larger and more complex however, the grower does not have time to make the many runs and analyses necessary to get the most benefit from the model. It would appear that, a consultant advisor could be a target group to service growers by delivering and interpreting the model's output. Cotton growers are accustomed to having such advisors for pest management and do not hesitate to use them for plant growth regulator and irrigation advice. That is not the case, however, for soybean and other small grain producers. In the mid-South, generally, a cotton grower also is a soybean and small grain producer and therefore is likely to use the services of consultants. North of the Cotton Belt, however, small grain producers are less likely to use consultants. At this point, costs cannot be realistically supported by user fees since the economic benefits are not always clear. About 50% of consultants and agricultural extension agents in the Great Plains believe that growers would pay for the service of running a simulation model and managing input (Ascough et al., 1999).

Another area of concern is reliance upon the Extension Service or agricultural consultants in each state to train model users. Computer use among agricultural consultants in the Great Plains area of the U.S. is high at 94% but only 79% use computers for their clients (Ascough et al., 1999). Models such as GLYCIM/GUICS are very different from WordPerfect, Excel, or most of the other computer tools that are rarely updated and even more rarely changed in format of input and output. Plant growth models, however, will probably be updated regularly and input and output requirements changed as new information from research on the target crop is integrated into the model. This seems to be a difficult concept to grasp for trainers who are used to the other types of computer tools.

Use of an Intermediary

Much of the progress and success of the GLYCIM/GUICS project could be traced to the presence of an intermediary to facilitate communications between the researchers and the growers. Whisler, one of the authors of this chapter, was such a facilitator. He had developed a trust relationship with the growers through the previous GOSSYM/COMAX project, and that trust was extended to GLYCIM/GUICS. He and his students were geographically close enough to carry out extensive data collection necessary to improve the model. Because they were close to the farms, they could maintain the consistent and frequent grower contact which helped the project avoid stagnation. The facilitator also often ran the model for the growers and many of the enhancements were made to ease the use of the model by the facilitator for a number of growers.

SUMMARY AND CONCLUSIONS

The USDA research team worked with soybean growers in the mid-south region of the U.S. for 10 years collecting data for model testing and evaluation, and developing a decision support tool for soybean management. The decision support tool included a simulation model (GLYCIM) and an interface (GUICS), an intuitive, easy–to–use tool to assemble the relative data and run the simulation model. The interface also served as an expert system to summarize important information from the simulations, allowing growers to make informed preplant decisions on cultivar selection and row spacing or postplant decisions on irrigation and harvest scheduling. Researchers found that growers preferred to view results of the simulations in terms that are most meaningful such as simple graphs and short, succinct tables.

The absolute estimates of yield by GLYCIM were sometimes very different from measured yields and often too high. The relative responses of GLYCIM to row spacing differences, varieties and irrigation, however, were reported to be realistic and useful. Growers reported a 14 to 24% increase in yields and up to 400% increase in water use efficiency by using GLYCIM/GUICS. Experience with the model encouraged the growers to visit their fields more often which provided important feedback to them.

On-farm research is critical to model development efforts and should not be delayed until the final stages of model development. The development of a working relationship with the growers early in the program helped the team develop a relevant interface. At the early stages the design of the interface was still plastic and modifications did not require major structural changes in the design. Most of the features and uses of GLYCIM/GUICS came not from developers, but from the growers, and were incorporated into the design of the DSS during development. In retrospect, the importance of involving extension agents and consultants in modeling work, especially at the design stage, is crucial. They could then become intermediaries between the growers and the DSS developers.

Decision support systems are very complex, and it is still difficult to promote their widespread use among growers. They are still not easy enough for growers, who have many other tasks competing for their time, to use. The growers most often involved in the use of models are those who are early adapters of technology and are willing to accept more risk. Agricultural extension agents and consultants still do not appear to be very supportive of simulation-based DSSs, as their use grows and agents become more familiar with them perhaps their acceptance will also grow. From the researcher's standpoint, it is difficult to find time for support, providing help with problems, and updating cultivar and soil files. At this point it is not clear if the costs of these models can be realistically supported by user fees.

Nevertheless, DSS systems, such as GLYCIM/GUICS, have been useful and therefore have definite potential to increase profits and reduce resource use. The future still holds promise as computers become faster and software technology allows developers to design "smart" systems that can reduce the complexity of the interfaces. At the same time, researchers are increasing their knowledge of plant growth and development and improving their models.

ACKNOWLEDGMENTS

The authors appreciatively acknowledge the assistance of the cooperating growers: Dennis Bragg, James Dickerson, O.J. and Gandy, Edward Hester, Kenneth Hood, David Madison, William McCain, Jay Mullens, William Ragsdale, Travis Satterfield, Steve Watkins, and Thomas Wilder. The authors also gratefully acknowledge the work of the late Dr. Anthony Trent, who designed the original interface, WINGLY.

REFERENCES

Acock, B., Y.A. Pachepsky, E.V. Mironenko, F.D. Whisler, and V.R. Reddy. 1999. GUICS: a generic user interface for on-farm crop simulations, *Agron. J.,* 91:657–665.

Acock, B., E.V. Mironenko, Ya.A. Pachepsky, and V.R. Reddy. 1998. *GUICS: Graphical User Interface for Crop Simulations with Windows 95. Version 1.7. User's Manual,* USDA-ARS, Remote Sensing and Modeling Lab., Beltsville, MD.

Acock, B., Y.A. Pachepsky, M.C. Acock, V.R. Reddy and F.D. Whisler. 1997. Modeling soybean cultivar development rates using field data from the Mississippi Valley, *Agron. J.,* 89:994–1002.

Acock, B. and A. Trent. 1991. *The Soybean Simulator, GLYCIM: Documentation for the Modular Version 91,* Agric. Exp. Stn., Univ. of Idaho, Moscow, ID.

Acock, B., V.R. Reddy, F.D. Whisler, D.N. Baker, H.F. Hodges and K.J. Boote. 1985. *The Soybean Crop Simulator GLYCIM, Model Documentation,* PB 851163/AS, U.S. Department of Agriculture, Washington D.C., Available from NTIS, Springfield, VA.

Ascough J.C. II, D.L. Hoag,, W.M. Frasier, and G.S. McMaster, 1999. Computer use in agriculture: an analysis of Great Plains producers, *Comp. Elect. Agric.* 23:189–204.

Baker, D.N., J.R. Lambert, and J.M. McKinion. 1983. GOSSYM: a simulator of cotton growth and yield, South Carolina Agricultural Experiment Station, Technical Bulletin 1089.

Gertsis, A.C. 1985. Validation of GLYCIM: a dynamic crop simulator for soybeans. M.S. Thesis, Agronomy Dept., Mississippi State University, Mississippi State, MS.

Greer, J.E. 1994. Explaining and justifying recommendations in an agriculture decision support system, *Comput. Elec. Agric.,* 11:195–214.

Hammer, G.L., T.R. Sinclair, K.J. Boote, G.C. Wright, H. Meinke, and M.J. Bell. 1995. A peanut simulation model: I. Model development and testing, *Agron. J.,* 87:1085–1093.

Jones, C.A. and J.R. Kiniry. 1986. *CERES-Maize: A Simulation Model of Maize Growth and Development,* Texas A&M Univ. Press, College Station, TX.

Landivar, J.A., G.W. Wall, J.H. Siefker, D.N. Baker, F.D. Whisler, and J.M. McKinion. 1989. Farm application of the model-based reasoning system GOSSYM/COMAX. In *Proc. 1989 Summer Computer Simulation Conf.,* J.K. Clema, Ed., Austin, TX, July 24–27, 1989. Soc. for Computer Simulation, San Diego, CA, 688–694.

Lemmon, H.E. 1986. COMAX: an expert system for cotton crop management, *Science,* 232:29–33.

Lightfoot, C. and R. Barker. 1988. On-farm trials: a survey of methods, *Agric. Adm. Ext.,* 30(1):15–23.

Manning, E. 1996. GLYCIM: crop model for soybeans, *Progressive Farmer,* 11:33.

Martin, A. and D. Eastman. 1996. *The User Interface Design Book for the Applications Programmer,* John Wiley & Sons, New York.

McKinion J.M. and H.E. Lemmon. 1985. Expert systems for agriculture, *Comput. Electron. Agric.,* 1:31–40.

Pachepsky, Ya. and D.J. Timlin. 1996. Infiltration into layered soil covered with a depositional seal, *J. Agrophysics,* 10:21–30

Reddy, V.R., Ya. Pachepsky, and F.D. Whisler. 1998. Allometric relationships in field-grown soybean, *Annals of Botany,* 82:125–131.

Reddy, V.R., B. Acock, and F.D. Whisler. 1995. Crop management and input optimization with GLYCIM: differing cultivars, *Comput. Electron. Agric.,* 13:37–50.

Reddy, V.R., Y.A. Pachepsky, F.D. Whisler, and B. Acock. 1997. A survey to develop decision support system for soybean crop management, vol, IV, *Proc. 1997 ACSM-ASPRS Annual Convention and Exhibition,* April 7–10, Seattle, WA, 11–18.

Reddy, V. R., D. J. Timlin, and Y. Pachepsky. 1999. Quantitative description of plant density effects on branching and light interception in soybean, *Biotronics,* 28:73–85.

Remy, K. 1994. *GLYCIM Soybean Model Proves its Worth, Research Highlights,* MAFES Publ., Mississippi State University, Mississippi State, MS, 57.

Rubin, J. 1994. Handbook on Usability Testing: How to Plan, Design and Conduct Effective Tests, John Wiley & Sons, New York.

Rudisill, M., C. Lewis, P.B. Polson, and T.D. McKay, Eds., 1996. *Human–Computer Interface Design: Success Stories, Emerging Methods, and Real-World Context,* Morgan Kaufmann, San Francisco, CA.

Timlin, D.J., Ya. Pachepsky, B. Acock, and F. Whisler. 1996. Indirect estimation of soil hydraulic properties to predict soybean yield using GLYCIM, *Agric. Sys.,* 52:331–353.

Whisler, F.D. 1982. *Soil Data Preparation for Crop Growth Models,* Agron. Sci. Ser. #1, Miss. State University, Mississippi State, MS.

CHAPTER 5

Benefits of Models in Research and Decision Support: The IBSNAT Experience

Gordon Y. Tsuji, Andre du Toit, Attachai Jintrawet, JamesW. Jones, Walter T. Bowen, Richard M. Ogoshi and Goro Uehara

CONTENTS

INTRODUCTION

The evolution of modern agricultural research into disciplines has resulted in major advances in understanding plant and animal processes at plant, organ, cellular, and biochemical levels of detail; however, this specialization resulted in scientists who know a lot about components but lacked an understanding of production systems as a whole. This led to a gap between research produced by disciplinary scientists and its application to improve production systems. More applied agricultural research efforts did not fully benefit from this disciplinary knowledge; they relied on costly trial and error tests and analysis methods to determine whether one management approach was better than another. One aberration of this situation was that production management trials were repeated continually to test technology because of differences in climate, soil, and pests, and other factors that vary over space and time. New methods were needed to help researchers transfer production technology to speed up the process and reduce the costs of repeating trials everywhere that new technology was needed.

During the 1980s and early 1990s, the IBSNAT project was created with the aim of accelerating the transfer of agrotechnology to increase food security and minimize environmental degradation. Systems analysis and simulation were the methods implemented and tested by this project. Principal products of the IBSNAT project were its global network of collaborators and the decision support system software referred to as DSSAT. Examples of its *post facto* applications at local levels in Albania, South Africa, Thailand, and Hawaii, and in studies of potential impacts of climate change on agriculture at the global level are presented. Output from the systems approach allowed scientists, educators, extension specialists, and other decision makers to analyze technology options and better enable them to match the biological requirements of crops to land characteristics.

Following the end of this project, global cooperation continues to increase, confirming the value of the approach and adding new tools for widespread use.

Methods for Agrotechnology Transfer

Agrotechnology transfer was a commonly used term by the international agricultural research community in the 1970s and 1980s. The term generally referred to taking technology and/or knowledge developed in one location and applying it in another location where it had not been used before. The transfer was generally effected from the industrialized countries in the upper or temperate latitudes to locations in the lower or tropical latitudes with varying degrees of success.

The principal issue in the transfer process is how to best match the biological requirements of a crop to the biophysical characteristics of the land. A successful match implies measurable production increase as a result of the use and adoption of transferred technology. Three methods are available to accomplish this transfer.

Trial-and-Error

The first is the traditional trial-and-error method. This method has provided agricultural scientists with a wealth of information and data on crop responses to application of varying rates of fertilizers from field trials conducted at their respective research sites; however, this method requires much time and money to implement, and the results are principally applicable only to the site where the experiment was installed. Results from such studies generate data and information for our understanding of processes such as photosynthesis and water and nutrient uptake. Much of the postWorld War II efforts could be categorized as outcomes from trial-and-error methods.

Analogy

A second method, transfer by analogy, was critically examined by the Benchmark Soils Project, a program of the U.S. Agency for International Development (USAID) implemented by both the University of Hawaii and the University of Puerto Rico, Mayaguez, in the mid-1970s. The purpose of that project was to determine the feasibility of transferring agrotechnology (management practices related to fertilizer N and P for maize and soybeans) from one location to another on the basis of similarly classified soils and climate (Silva, 1985). The soil family category of a hierarchical system, Soil Taxonomy (Soil Survey Staff, 1975), served as the level at which soils with similar chemical, physical and mineralogical characteristics were grouped. The nomenclature of this category also included characteristics on the specific ranges of both the soil temperature and soil moisture regimes for a given soil family.

Three soil families were determined to be the minimum necessary to test the concept of transferability (Silva and Beinroth, 1975). At that time, the three soil families had the following taxonomic nomenclature (Ikawa et al. 1985):

1. The thixotropic, isothermic Hydric Dystrandepts
2. The clayey, kaolinitic, isohyperthermic Tropeptic Eustrustox
3. The clayey, kaolinitic, isohyperthermic Typic Paleudults

With the addition of two soil orders, Andisols and Gelisols, in soil taxonomy (Soil Survey Staff, 1999), the nomenclature for these soil families has been modified.

Experimental sites were established only after characteristics of the soil pedon at a potential site were verified as having common taxonomic properties of a typical member of a soil family. Besides Hawaii and Puerto Rico, sites were identified in Brazil, Cameroon, Indonesia, and the Philippines. A total of 25 experimental sites representing three soil families became the network of soil family sites. A quantitative evaluation of transfer by analogy using the soil family concept was reported by Wood et al. (1985) and Cady et al. (1985).

Although the outcome was successful, the process to carry out tests verifying transferability was time-consuming. Furthermore, because of the lack of a standard, universal system of soil classification and methods of soil analysis, the task of locating and verifying the similarity of soil properties was, at best, difficult.

Transfer by analogy was shown to be technically possible, although the probability of its adoption as an acceptable practice was low. A more efficient method was needed.

Systems Analysis and Simulation

A third method is systems analysis and simulation. The systems approach is guided by the premise that it is more efficient to use models to find alternative ways to improve agroecosystem performance than to only experiment with the system itself. Improvements to one component of a system cannot be assumed to lead to an improvement in the performance of the whole system without an understanding of how system components interact (Uehara and Tsuji, 1993). The systems approach to agrotechnology transfer required a balanced development of two interactive components, crop models and databases (Nix, 1984). The systems approach is frequently used to complement field research by reducing the number of options tested in the field to those that have been shown to be well suited for the soil and weather conditions of that location.

IBSNAT

The IBSNAT (International Benchmark Site Network for Agrotechnology Transfer) project was formally established as a program of USAID and implemented by the University of Hawaii through a cooperative agreement in 1983. The project was formulated as a program to develop a methodology to accelerate the transfer of agroproduction technology developed in one location to any other location in the world. An efficient method of transfer of agrotechnology would likely result in an effective mechanism to provide options that would enable these countries attain self-sufficiency in food production.

The systems approach was considered by the designers of IBSNAT (ICRISAT, 1984) as the most efficient and effective means for agrotechnology transfer. Instead of focusing on similar soil properties, the systems approach takes into account the soil–plant–atmosphere–management inter-actions. It relies on what we already know of individual components of the continuum and helps identify gaps in our knowledge.

For example, trial-and-error research provided a general understanding of crop responses to management variables for specific situations. Analog transfer provided an understanding of the role of soil and climate as defined in soil taxonomy (Soil Survey Staff, 1975) on crop responses to technology. A systems approach allows us to now capture, condense, and organize knowledge generated from both transfer approaches and from disciplinary knowledge to make informed decisions for achieving desired outcomes relative to crop production and land use. Figure 5.1 (Nix, 1984) illustrates the pathway or framework for knowledge generation for understanding of bio-physical processes and knowledge utilization for prediction to control outcomes.

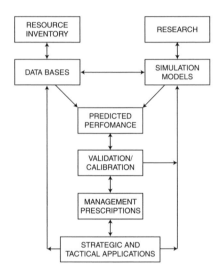

Figure 5.1   Diagram outlining the systems approach to developing and testing models. (From Nix, H.A., *Proceedings of the International Symposium on Minimum Data Sets for Agrotechnology Transfer,* ICRISAT Center, Patancheru, India, 1984. With permission.)

The purpose of the IBSNAT project was to assemble and distribute a portable, user-friendly, computerized decision support system, which enabled users to match the biological requirements of crops to the physical characteristics of land to attain objectives specified by the user (Uehara and Tsuji, 1993). Two products of the IBSNAT project were the global network of collaborating scientists and organizations, and the decision support system software referred to as DSSAT for Decision Support System for Agrotechnology Transfer. This chapter presents *post facto* case exam-ples of user applications of DSSAT. Before doing so, we present a brief background and description of DSSAT and its components.

IBSNAT PRODUCTS

The Global Network

Members of the global network had a shared vision and understanding of their roles in the application of a systems approach to achieve project objectives. The network collaborators con-tributed resources, both human and capital, and information in their roles as model developers, data generators, and systems users. In many instances, collaborators had roles in more than one of these areas. More than 75 countries were represented in the IBSNAT network.

Table 5.1   Members of the IBSNAT Technical Advisory Committee Represented the Global Community and Many Disciplines

Name	Country of Residence	Discipline
Juan A. Comerma	Venezuela	Pedology
J. Barry Dent	Scotland	Agricultural economics
L. Anthony Hunt	Canada	Plant breeding
Henry A. Nix	Australia	Agroecology
Joe T. Ritchie	U.S.	Soil and water sciences
Paul S. Teng	U.S.	Plant pathology

The diversity of scientific disciplines involved in the IBSNAT network can be best expressed by those who served as members of the Technical Advisory Committee or TAC. Table 5.1 lists the individual, home country, and discipline.

Members of the TAC met annually to provide advice and counsel to the network and to the IBSNAT principal investigator at the University of Hawaii. Members of the network understood their roles in the overall scheme for the development and subsequent testing of components of what was eventually to become DSSAT.

A systems approach was agreed upon conceptually as the methodology to effect efficient transfer of agrotechnology. To implement the concept, general agreement was reached on the use of process-based models to simulate crop growth and development for 12 food crops — wheat, maize, rice, sorghum, millet, barley, soybeans, peanut, dry beans, potato, cassava, and aroids (ICRISAT, 1984). For members of the IBSNAT network, model developers were to establish a standard programming structure to develop additional simulation models for other crops not listed and both data generators and model developers had to agree on the data requirements to test the efficacy of the models. Finally potential and prospective users of the technology needed to have a "menu" of data necessary to the run the models for their respective purposes. This set of data was referred to as a minimum data set (MDS) (Nix, 1984).

The minimum data set was defined as the minimum input necessary to operate each of the crop simulation models. Daily maximum and minimum air temperatures, rainfall, and solar radiation were the minimum set of weather data required. For soils, the minimum set included the following by horizon or depth: texture, bulk density, organic carbon, soil pH, and surface albedo. Crop growth and development data included information such as the number of days from planting to flowering or to tassel initiation or to physiological maturity under optimum conditions, seed yield, above-ground biomass, and grain or tuber size. Management data included information such as date of planting, timing, kind, and amounts of N fertilizer applied, and irrigation amounts and dates (IBSNAT, 1986; IBSNAT, 1990; Jones et al., 1998).

The Decision Support System for Agrotechnology Transfer (DSSAT)

The DSSAT is a software system that serves as a shell to integrate three principal components: the database management system, the crop models, and application programs. (Jones et al. 1998) The shell was written in the C programming language, the database management system in dBaseIV, the models in FORTRAN, and application programs in BASIC and FORTRAN. The first version was released in 1989 as DSSAT v2.1 (IBSNAT, 1989). For the first time, scientists, educators, and other users had access to a system where any one of the models used a common database system. This allowed users to conduct *ex ante* experiments to assess a range of "What if?" questions that would take the lifetime of a research agronomist to do in the field.

Initially, there were simulation models for 10 food crops in DSSAT v2.1. In the later version of DSSAT, v3 (Tsuji et al., 1994; Jones et al. 1998), application programs for single season (Thornton and Hoogenboom, 1994) and multiple season crops (Thornton et al., 1995) were included.

Table 5.2 Listing of the Crops That Can Be Simulated in DSSAT v3.5 and the Model Used to Simulate Each

Crop	Model	Crop	Model
Maize	CERES	Soybean	CROPGRO
Wheat	CERES	Peanut	CROPGRO
Rice	CERES	Dry bean	CROPGRO
Sorghum	CERES	Chickpea	CROPGRO
Millet	CERES	Tomato	CROPGRO
Barley	CERES	Bahia grass	CROPGRO
Potato	SUBSTOR	Fallow	CROPGRO
Sunflower	OILCROP	Cassava	CROPSIM
Sugarcane	CANEGRO		

Source: From Hoogenboom, G. et al., *DSSAT v.3.5, DSSAT version 3, vol. 4,* ICASA, Honolulu, HI, 284.

The current version, DSSAT v3.5, has 17 crop simulation models (Hoogenboom et al., 1999) as listed in Table 5.2.

The crop models in DSSAT are characterized by their portability, i.e., they are applicable globally. They are site-specific and local when the minimum data set for a specific location is used to operate the models. To be able to use DSSAT as a matching tool, the user will need to assemble the minimum data set of biophysical information on soils and weather (Hunt and Boote, 1998). Each of the crop models in DSSAT incorporate coefficients that account for the way in which genotypes differ from each other in the duration of development phases of growth, in their response to specific environmental factors, or in morphological characteristics (Hunt et al., 1989; Ritchie, 1991).

ACCEPTANCE AND ADOPTION

Research methodologies have high probability of acceptance if they can adequately meet the three purposes of research. The first purpose is to advance our **understand**ing of the processes of a system. For example, knowledge of the processes of photosynthesis is important to our understanding of growth and yield of most crops. The second purpose is **predict**ion. If the knowledge and information generated permits us to understand the processes of a system, we should be able identify or diagnose probable causes of problems within the system and prescribe a range of options as remedies. Hence, if we understand the processes of a system and can prescribe remedies or solutions to problems or deficiencies within the system, we should be able meet the third purpose of research, to **control** outcomes.

The systems approach allows integration of our state of knowledge and understanding of processes to make predictions that enables the user to choose options to control outcomes. The DSSAT software continues to be an open-ended system that attempts to capture the essence of the systems approach.

The decade of the IBSNAT project, from 1983 to 1993, coincided with the evolutionary growth of the computer industry. The acceptance and adoption of DSSAT as a research and planning tool by users can be traced to the ready access to computers ranging from the mainframes to the desktops and eventually to the laptop and notebooks. To assess the acceptance and adoption of the systems approach in 1993 would have been premature. Today, nearly a decade later, we may be able to address the question of adoption and impact with a number of case examples.

From 1989 (v2.1) to 2000 (v3.5), nearly 1000 copies of DSSAT have been distributed to individuals/organizations in 83 countries and listed in Table 5.3.

Table 5.3   List by Countries of Registered Users of DSSAT in the IBSNAT Network of Model Developers, Data Generators, and Users

Africa and Middle East	Asia and Near East	Europe	The Americas	The Pacific
Benin	Bangladesh	Albania	Antigua	Australia
Botswana	China	Austria	Argentina	Guam
Burkina Faso	India	Belgium	Brazil	New Zealand
Cameroon	Indonesia	Bulgaria	Canada	
Cote D'Ivorie	Japan	Czech Republic	Chile	
Egypt	Malaysia	Denmark	Colombia	
Ghana	Nepal	France	Costa Rica	
Iran	Pakistan	Germany	Dominican Republic	
Israel	Philippines	Hungary	Ecuador	
Kenya	South Korea	Ireland	Guatemala	
Lesotho	Taiwan	Italy	Guyana	
Malawi	Thailand	Netherlands	Lesser Antilles	
Mali	Vietnam	Norway	Mexico	
Mauritius		Poland	Nicaragua	
Mozambique		Romania	Panama	
Namibia		Spain	Peru	
Niger		Switzerland	Trinidad and Tobago	
Nigeria		Turkey	Uruguay	
Senegal		United Kingdom	United States	
South Africa			Venezuela	
Syria				
Tanzania				
Uganda				
Zambia				
Zimbabwe				

Acceptance and adoption of DSSAT by a range of users for application at the global level relative for climate change studies to local application and adaptation of crop models in four widely separated geographic regions, South Africa, Thailand, Albania, and the U.S. are reported in the following sections.

Global Application

Global climate change and its impact on food production and global trade were, and continue to raise concern about food security and environmental sustainability of the planet. Global circulation models (GCMs) were commonly used tools of scientists to demonstrate the impact of rising atmospheric carbon dioxide and other so-called greenhouse gases on possible increases in global temperatures and changes in rainfall patterns over time and space; however, in order to relate global climate change to crop production, models to simulate the impact of rising temperature and increased CO_2 on crop production were required.

The crop models in DSSAT were selected for use in a global study supported by the U.S. Environmental Protection Agency titled "Climate Change and World Food Supply" (Rosenzweig and Parry, 1994). A description of the linkage of DSSAT to the GCM with a special version of DSSAT (version 2.5) was reported by Hoogenboom et al., (1995). The later version, DSSAT v3, was then used by Rosenzweig and Iglesias (1998) to assess the potential impacts of climate change on world food crop production using GCM scenarios affecting water availability and with and without direct physiological effects of carbon dioxide on crop growth. For details of this global application study, readers are referred to authors cited in this chapter.

Local Examples

Although DSSAT found application for policy decisions at the global level relative to studies on the potential impact of global climate change, the DSSAT system became widely available to researchers with the rapid advances in computer technology. What were once considered "computer toys," crop models became computerized tools, helping researchers organize data and information into electronic databases for easier access. Furthermore, it soon became apparent that the ability of the models to predict outcomes was as good as the knowledge of the infinite number of combinations that affect plant growth and development. Hence, the models became tools to help identify gaps in the knowledge base and allow improved design of more efficient and effective experiments.

The first such case study is from South Africa where the CERES-Maize model was modified to suit local farmer practices and cultivars. The second is from Thailand where researchers collaborated with modelers from South Africa and the U.S. in the development, testing, and application of a crop model for a nonfood crop, sugarcane. The third is from Albania. The CERES-Wheat model was used to provide information on the impact of delaying nitrogen fertilizer on grain yield for a crop that was already planted. A final example, derived from research activities in Hawaii, demonstrates the linkage of crop model outputs from DSSAT to GIS software for spatial analyses and the potential application of DSSAT on the Internet.

Potchefstroom, South Africa

In the early 1990s, the Agricultural Research Council-Grain Crops Institute (ARC-GCI) maize modeling group at Potchefstroom considered using a modified version of CERES-Maize (De Vos and Mallet, 1987) and PUTU for maize (De Jager, 1989) to study genotype and management practices used in the region. CERES-Maize (Jones and Kiniry, 1986; Ritchie et al., 1998) was developed for crop production practices distinctly different from those in the western Highveld of South Africa, where the combination of very low plant populations (2.0 plants m^{-2}) and wide rows (2.1 m) are standard. Local cultivars provide adequate grain yields to compensate for a low plant population and the practice of having wide rows contributes to sustainable yields under both drought and nondrought conditions (Du Toit, 1991).

Suggested modifications to improve the simulation of silking date for CERES-Maize v2.1 were included in a later version, CERES-Maize v3 in DSSAT v3 (Tsuji et al. 1994). The simulation of the silking date allowed a better estimate of the level of yield compensation that will likely occur with wide row spacing.

The ARC-GCI maize modeling team used a correlation matrix to quantify the error (observed — simulated values). Linear regression ($y = a + bx$) was determined between climatic and stress data during different growth stages (x) with the error (y). Using this method, Du Toit et al. (1994) reported that 44% of the error in yield could be explained by water stress during silking, and 30% of the error by water stress before silking. This analysis was conducted on a cultivar x planting date trial at Potchefstroom, South Africa. Hence, the ARC-GCI team was able to identify gaps in our understanding of maize growth and development under conditions considered extreme relative to those used in the development of CERES-Maize. They were then able to demonstrate causes of the poor match between observed and simulated grain and kernel numbers and modify the DSSAT maize model to accommodate their local needs. The feasibility of using fitted genetic coefficients for the simulation of yield compensation was determined for different cultivars. Differences in yield compensation among cultivars were quantified; results were then used to develop regression equations for simulating differences in yield compensation among cultivars.

An important characteristic of the local maize cultivar is the production of tillers at low plant populations. Functions for tiller simulation were developed and included in the modified version of

CERES-Maize v3 by the ARC-GCI team. This modification resulted in an improved capacity of the model to simulate tiller production and, hence, improved accuracy in simulating yield compensation.

Modifications made to CERES-Maize v3 were tested using a historical field trial, as well as commercial data in order to determine whether the modifications were site specific. Outcomes from these tests ascertained that the modified CERES-Maize retained its portability for application elsewhere. The ARC-GCI maize modeling team, in 1996, prepared an interface between the modified CERES-Maize v3.0 and the Free State Department of Agriculture geographic information system (GIS). This made it possible to add a spatial component to outputs derived from DSSAT v3 and make them widely available to users.

Chiang Mai, Thailand

Systems modeling and simulation are now commonly accepted methods for assessing the match between the biological requirements of crops and the biophysical characteristics of the land in Thailand. This acceptance was demonstrated by the institutionalization of the systems approach by the Thailand Department of Agriculture, the research unit of the Ministry of Agriculture and Cooperatives. The process of introducing the approach and proving its utility to researchers and policy makers was made simpler by the parallel success of computer and software technology over the past 15 years. Furthermore, advancement of computer technology made it possible to introduce courses on crop models and system simulation into the core curriculum of the Multiple Cropping Centre at Chiang Mai University in the early 1990s.

Recognition for increased capacity building in systems research led to support for a project titled "Estimating Sugarcane Yield of a Large Scale Production Area Using Information Technologies" by the Thailand Research Fund (TRF) to the Faculty of Agriculture of Chiang Mai University in 1993. Model development and testing, spatial and attribute database development, and interface shell development for sugarcane were the principal components to this project. Training was an essential output of the project.

From 1993 through 1998, Thai scientists collaborated with developers of the CANEGRO model in South Africa (Inman-Bamber, 1991) to test it under localized conditions and with local cultivars. An improved version of CANEGRO (Inman-Bamber and Kiker, 1999), patterned after the CROPGRO models (Jones et al., 1998), was installed in DSSAT v3.5 (Hoogenboom et al. 1999). This linkage was made possible through members of the IBSNAT network and serves as an example of the growing partnerships among network members.

Thai scientists used the earlier versions of the model to assess the effects of planting dates on cane and sugar yields in the Northeast region of Thailand (Jintrawet et al. 2001). Even with the preDSSAT version of CANEGRO, outputs from the exercise were encouraging. Results helped Thai scientists identify gaps in their knowledge and understanding of the effects of inter-nodal leaf development from different planting dates on stalk and sugar yields.

The crop simulation model for sugarcane in Thailand was developed through a participatory effort involving collaborators from the industry and scientists affiliated with research institutions. CANEGRO is a good example of a model initially programmed and calibrated for application in South Africa that is now portable and applicable globally. With the limited amount of data available to test the model at any one geographical area (relative to food crops), the availability of CANEGRO globally as part of DSSAT should accelerate its testing.

Lushnja, Albania

Immediately after the collapse of the Soviet Union, several Eastern European countries that relied on economic trade with them faced a food shortage crisis. In 1991, after 45 years of Communist rule, Albania began the transition to democratic pluralism and a market economy.

During this early period, the U.S. Agency for International Development forwarded a request received from the Albanian government for assistance in determining to what extent wheat imports might be offset with emergency nitrogen fertilizer imports. The request asked if the timely importation and distribution of nitrogen fertilizer would result in improved grain yield forecasts. If so, then the demand for emergency wheat aid imports would be substantially reduced. The Bureau for Europe and the Near East (ENE) of USAID asked the IBSNAT project for assistance in late October 1991.

Using its network of collaborators, a systems scientist member of the IBSNAT network from the International Fertilizer Development Center (IFDC) traveled to Albania in November 1991. A meeting with Albanian scientists and officials of the Ministry of Agriculture and AID/ENE was arranged to demonstrate the capacity of a systems approach using DSSAT and the CERES-Wheat model.

After that initial meeting, Albanian scientists organized results from earlier studies and provided soil and weather data sets for the Lushnja, an important winter wheat growing area in Western Albania. Initially, the CERES-Wheat model was calibrated with data from an earlier 3-year study (Figure 5.2). Then, simulated outputs from DSSAT using a single top dressing of nitrogen on different dates showed the outcomes of applying two rates of nitrogen at 2, 3, 4, 5, and 6 months after planting (Figures 5.3 and 5.4).

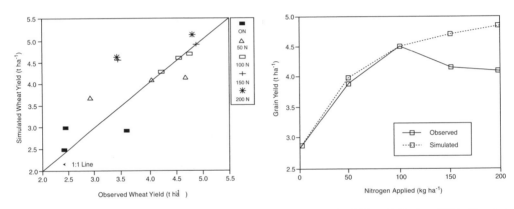

Figure 5.2   (a) Comparison of observed and simulated winter wheat grain yields with 5 rates of N fertilizer in kg ha^{-1} averaged over a 3-year study period. (From Bowen, W. and Papajorgji, P., *Agrotechnology Transfer*, 16, 9–12, 1992. With permission.) (b) Observed and simulated winter wheat grain yields with 5 rates of N fertilizer treatments. Observed values are average values from the 3-year study. (From Bowen, W. and Papajorgji, P., *Agrotechnology Transfer*, 16, 9–12, 1992. With permission.)

The Ministry of Agriculture, in consultation with USAID/ENE, reviewed the simulated outcomes and decided to import N-fertilizer instead of wheat. If a similar request were received from USAID 10 years earlier, a similar response would have been unlikely. Through the systems analysis and simulation, data input and information from a specific site were used to calibrate the crop model with field data and observations of the local wheat cultivar. Trial-and-error field trials to determine outcomes of fertilizer trials would require more human and fiscal resources, and the outcomes would not have been available to support decisions at the national level.

The long-term impact of this action was not easily assessed then, although we are aware of a subsequent request from USAID to the International Fertilizer Development Center (IFDC) to assist in developing small business enterprises to handle the marketing and distribution of fertilizers in Albania. Since then, IFDC has reported how local small enterprises evolved successfully in a free market environment.

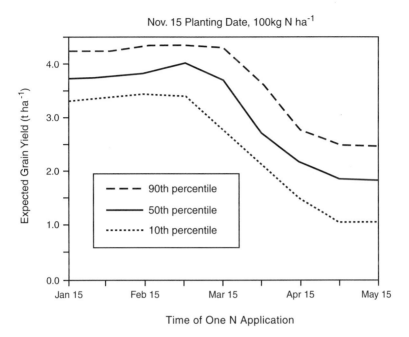

Figure 5.3a   Simulated expected winter wheat grain yields for a crop planted on November 15 with a single top-dressing of 100 kg N ha⁻¹ fertilizer at 2, 3, 4, 5, and 6 months after planting. (From Bowen, W. and Papajorgji, P., *Agrotechnology Transfer,* 16, 9–12, 1992. With permission.)

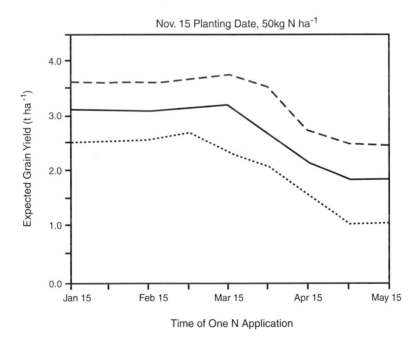

Figure 5.3b   Simulated expected winter wheat grain yields for a crop planted on November 15 with a single top-dressing of 50 kg N ha⁻¹ fertilizer at 2, 3, 4, 5, and 6 months after planting. (From Bowen, W. and Papajorgji, P., *Agrotechnology Transfer,* 16, 9–12, 1992. With permission.)

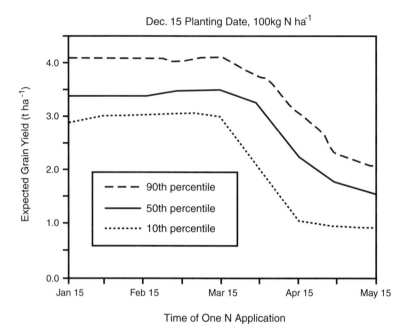

Figure 5.3c   Simulated expected winter wheat grain yields for a crop planted on December 15 with a single top-dressing of 100 kg N ha^{-1} fertilizer at 2, 3, 4, 5, and 6 months after planting. (From Bowen, W. and Papajorgji, P., *Agrotechnology Transfer,* 16, 9–12, 1992. With permission.)

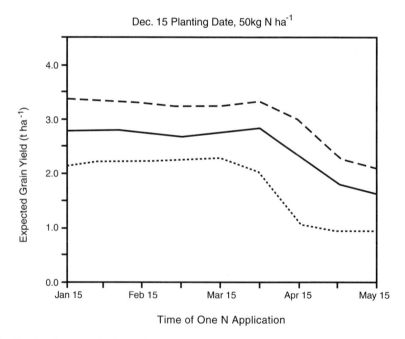

Figure 5.3d   Simulated expected winter wheat grain yields for a crop planted on December 15 with a single top-dressing of 50 kg N ha^{-1} fertilizer at 2, 3, 4, 5, and 6 months after planting. (From Bowen, W. and Papajorgji, P., *Agrotechnology Transfer,* 16, 9–12, 1992. With permission.)

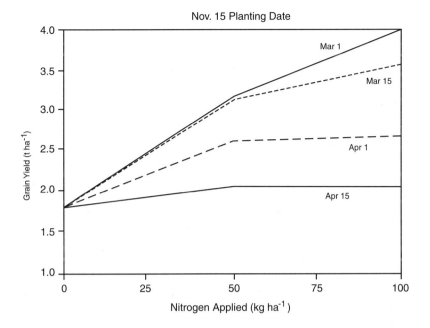

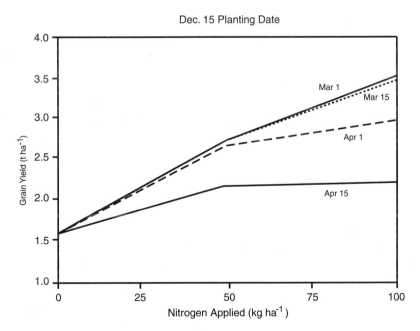

Figure 5.4a and 5.4b   Simulated winter wheat grain yield for two dates of planting as affected by dates and rates of N fertilizer application. (From Bowen, W. and Papajorgji, P., *Agrotechnology Transfer,* 16, 9–12, 1992. With permission.)

Hawaii, U.S.

Two case examples are presented involving the application of the systems approach and DSSAT. The first involves alternative crops for lands formerly planted with sugarcane, and the second involves site selection for cabbage production to minimize insect damage.

Options to Replace Sugarcane

Global economic changes resulted in changes in the agricultural landscape in Hawaii from the mid-1980s to the mid-1990s. Sugar was the leading agricultural commodity and occupied a major portion of the arable land in the state. By 1995, the number of sugar plantations declined from 23 to 2 within a period of a decade. Sugarcane was grown in a range of environmental conditions on the islands of Kauai, Oahu, Maui, and Hawaii. Areas planted with sugarcane included dry and warm coastal areas with mean annual rainfall of 1600 mm and mean annual temperature of 23°C to the cool and wet mountainous locations (up to 600 m elevation) with mean annual rainfall of 2800 mm and a mean annual temperature of 17°C

With the demise of sugarcane, government decision makers faced a major dilemma or a major opportunity for the economic and environmental well-being of the state if the right choices were made on land use. Information and knowledge to make those decisions to explore options, however, were not readily available or limited.

Taro (*Colocasia esculenta* L.), an aroid, was considered a prime candidate as a replacement crop for sugarcane in the Hamakua district of the island of Hawaii. Scientists linked to the former IBSNAT network collaborated on a proposed project to test the efficacy of a prototype aroid model (Singh et al., 1998) to match crop requirements with land characteristics in the Hamakua district. Network members represented the Universities of Hawaii and Puerto Rico, the Agricultural Research Service, U.S. Department of Agriculture, Tropical Agricultural Research Station, Mayaguez, Puerto Rico, and the International Fertilizer Development Center.

A comparison between simulated and observed taro corm yields from the lower to higher elevation experimental sites along an elevation transect on the slopes of Mt. Haleakala was carried out. The model was able to mimic the longer maturation period (more than 18 months) for plants grown at higher elevations versus the nine months at the warmer lower elevations.

Currently, taro is grown in the lower elevations of the Hamakua District with eucalyptus trees commonly planted in the cooler higher elevations in a diversified agricultural plan for the county of Hawaii. We are not aware if decision makers at the government level or at the farm level used any of the outputs or recommendations from the research project. Creating awareness and/or conditions for adoption of research outputs and products are issues researchers will have to confront in response to new GRPA or government reporting and performance act.

Site Selection for Cabbage Production

Cabbage is the principal vegetable crop planted all year round on the slopes of Mt. Haleakala on the island of Maui. Repeated spraying to control the diamond back moth gradually lost its impact with the increasing resistance of the moth to the insecticide Bt. The result of this growing resistance was major crop failures in the early 1990s. One of the options considered was to move cabbage production to another location where the potential for crop failure was low.

An efficient and low cost method was needed to assess cabbage performance over a range of environments. The typical cabbage growing area on the slopes of Mt. Haleakala has a mean annual rainfall that ranges from 508 mm to 6096 mm with mean temperatures of 8°C to 24°C.

Scientists from the Universities of Hawaii and Florida, who were part of the IBSNAT network, collaborated in the development of a simulation model for cabbage (Ogoshi et al., 1997) using the programming framework established for the CROPGRO models in DSSAT (Boote et al., 1998).

The model was then linked to a geographical information system (GIS) database that included attribute soils and weather data from the island of Maui. With management input data such as variety, planting date, irrigation regime, nitrogen fertilizer rate and schedule, and plant density for the cabbage model, outputs from DSSAT v3 were linked to a GIS database. This linkage offers "point-and-click" access to information on potential performance at any location on the island. The user can expect to obtain simulated outcomes of crop performance over a five year period and be provided with information on crop yield, crop duration, irrigation required, and nitrate leached. Expansion of the program sometimes referred to as "DSSAT on the Web" to the other islands would be possible if similar attribute datasets are readily available.

Prior to conclusion of this effort, it was learned cabbage growers were adopting a new pesticide to control diamond-back moths. Economics of acquiring new land versus purchasing a new pesticide or a new more resistant cabbage variety made the simulated outcomes meaningless to the grower or to the extension agent. The utility of the spatial component to DSSAT will be to a hypothetical new cabbage grower who is interested in determining which land area he currently owns is best suited to grow cabbage.

Outcomes from the collaborative research were presented in the final report (Ogoshi and Uehara, 1997) of the project to sponsoring agency, USDA/T-STAR (Tropical-SubTropical Agricultural Research). Training on utilization of products for stakeholders were not included in this research project.

LESSONS LEARNED

There is widespread agreement that a systems approach based on interdisciplinary effort is needed to address agricultural and environmental issues, however, large scale implementation of this approach is constrained by the inability of the development community to field the necessary interdisciplinary teams. One important lesson learned during the IBSNAT project was that better integration of effort offers the easiest and most cost-effective way to increase research efficiency. Unfortunately, although many researchers have embraced the application of the systems approach over the past decade, most research institutions are organized and administered in a way that fosters continued reliance on disciplinary research for prestige and scholarly excellence. The reward systems and the research and publications standards set by disciplinary scientists also contribute to perpetuation of the existing situation (IBSNAT, 1993).

The IBSNAT experience provides evidence that establishment of multidisciplinary, international collaborative research networks composed of individuals with the following characteristics is possible and essential. The individuals should be:

- Mission- and goal-oriented
- Committed to systems-based interdisciplinary research
- Able to set research priorities based on client needs
- Prepared to develop tools to empower clients to diagnose and solve problems
- Product-oriented
- Process-oriented and understand the value of basic research
- Able to share a common vision of the purpose of research
- Eager to form networks that enable them to attain higher goals that are otherwise unattainable

With these characteristics, it is not only possible and worthwhile, it is essential to have a global participatory network for dealing with systems problems. The four examples briefly described here were selected to demonstrate the utility of a research network. Accomplishments at the local levels would not have been possible without the network and a portable computerized tool, DSSAT. By portability, we mean the software is globally applicable for any location on Earth.

Collectively, the network of collaborators contributed information and data for a global assessment of the impact of increased carbon dioxide levels on food production and trade. The data sets and outputs from global circulation models (GCM) were used as inputs to crop simulation models in DSSAT for a range of crops (Rosenzweig et al., 1995).

ICASA (International Consortium for Agricultural Systems Applications)

The acceptance of systems-oriented methodologies in agriculture and natural resources is the focus of a network of like-thinking systems scientists referred to as ICASA for International Consortium for Agricultural Systems Applications. Former members of the IBSNAT project and systems scientists from the Wageningen Agricultural University in the Netherlands joined in formulating the network in 1991. In 1994, ICASA was formally established as a nonprofit corporation. In 1998, systems scientists from the APSRU (Agricultural Production Systems Research Unit) group in Towoomba, Australia, joined the network.

One of the early accomplishments of ICASA was the collaborative organization of three international symposia on Systems Approaches to Agricultural Development (SAAD) in Bangkok, Thailand in 1991 (Penning de Vries et al., 1993), in Los Banos, the Philippines, in 1995 (Teng et al., 1997), and in Lima, Peru, in 1999 (Bowen et al, 2001). These symposia provided systems scientists with an international forum to share results and information on advances in systems analysis and simulation. By the end of the third symposium, an international set of data standards for a range of systems tools was established. Such standards should result in a more cost-effective generation, recording, and storage of data sets for universal application (Hunt et al. 1999). In a memorandum of understanding with Global Climate Change and Terrestrial Ecosystems (GCTE) of International Geosphere-Biosphere Project (IGBP) (Ingram, 2000), scientists involved in that program adopted the ICASA data standards for their global program. These data standards are intended to be specific guides for data collection and handling to narrow the gap between the growing paucity of quality data relative to the number of models.

SUMMARY

What has been accomplished? The case examples presented here are brief reports of efforts in Albania, South Africa, Thailand, and the U.S. to utilize systems analysis and simulation as more efficient and cost-effective methods to improve our capacity to match crop requirements with land characteristics. Trial-and-error and analogy are still commonly used and acceptable, although both methods are unlikely to provide management options and conditions necessary to make decisions in a timely manner.

When the IBSNAT project ended in 1993, it was premature then to make a statement on impact of its network of collaborators and DSSAT. Now, almost eight years later, we are able to report on four examples easily accessed through the Internet. Many more exist. The momentum in application of the systems analysis and simulation approach to accelerate the transfer of agrotechnology started with the IBSNAT project and continues to build.

The continuing growth of communications networks has complemented this momentum. Agricultural scientists require access to modern information tools to address problems and issues confronting farmers at the farm and household levels as well as those confronting policymakers at the national and global levels. Newer information tools will eventually allow scientists and many other users of information to seamlessly integrate outputs from crop simulation models and decision support systems with remote sensing satellite imageries and geographical information systems to render educated decisions on land use policies rather than on "best guesses." Further, outputs from biophysical models will be inputs to economic systems models to help farmers assess impacts at the farm level and to assist policymakers to assess agricultural and environmental policies to the economic and social well-being of their communities.

Finally, the IBSNAT project served as a catalyst in promoting the development of user-oriented and functional systems tools. Through IBSNAT, simulation models previously confined to laboratories were transformed into practical tools for use by a wide range of users. These tools can now be operated with data sets obtained directly from farmer's fields. Although the IBSNAT project ended in 1993 as a program supported by USAID and by partner universities in 1993, the IBSNAT concept and its network continued to expand and grow. The network is now a truly collaborative partnership of system developers, data generators, and systems users. This global network and the systems approach provide researchers, educators, and decision makers with the most efficient means to better use our understanding of biophysical processes to match crop requirements with land characteristics to enable stakeholders to better understand options for agricultural production and environmental protection.

REFERENCES

Boote, K.J., J.W. Jones, G. Hoogenboom and N.B. Pickering. 1998. The CROPGRO model for grain legumes. In *Understanding Options for Agricultural Production,* G.Y. Tsuji, G. Hoogenboom, and P.K. Thornton, Eds., Kluwer Academic Publishers, Dortrecht, The Netherlands, 99–128.

Bowen, W. and P. Papajorgji. 1992. Using DSSAT to predict wheat productivity in Albania, In *Agrotechnology Transfer,* IBSNAT, Dept. of Agron. and Soil Sci., College of Trop. Agric. and Human Resour., Univ. of Hawaii, Honolulu, HI, 9–12.

Bowen, W.T., P. Malagamba, R. Quioz, M. Holle, J. White, C. Leon-Velarde, H.H. von Laar, Eds. 2001.*Proc. 3rd Int. Symp. on Systems Approaches for Agricultural Development,* Lima, Peru, Nov. 8–10, 1999 (CD-Rom computer file). International Potato Center (CIP), Lima, Peru.

Cady, F.B., C.P.Y. Chan, J.A. Silva, and C.L.Wood. 1985. Transfer yield response to phosphorus and nitrogen fertilizer. In J.A. Silva, Ed., *Soil-Based Agrotechnology Transfer,* Benchmark Soils Project, Dept. of Agron. and Soil Sci., College of Trop. Agric. and Human Resour., University of Hawaii, Honolulu, HI, 55–73.

De Jager, J.M. 1989. *PUTU 90 Maize Crop Growth Model, User Instructions and Model Description,* Dept. of Agromet., Univ. of the Orange Free State, Summer Grain Centre, Dept. of Agric. and Water Supply, Bloemfontein, South Africa.

De Vos, R.N. and J.B. Mallet. 1987. Preliminary evaluation of two maize (*Zea Mays* L.) growth simulation models, *South Afr. J. Plant Sci.,* 4:131–135.

Du Toit, A.S. 1991. Die invloed van rywydte op die voorspellingswaarde van die CERES-Maize mieliegroeisimulasiemodel in die Wes-Transvaal. M.Sc. Agric. Thesis, U.O.V.S. Bloemfontein, Suid Africa.

Du Toit, A.S., P.J. van Rooyen, and J.J. Human. 1994. Evaluation and calibration of CERES-Maize: 1. Nonlinear regression as an alternative method to determine genetic coefficients, *South Afr. J. Plant and Soil,* 11:96–100.

Hoogenboom G., G.Y. Tsuji, N.B. Pickering, R.B. Curry, J.W. Jones, U. Singh and D.C. Godwin. 1995. Decision support system to study climate change impacts on crop production. In *Climate Change and Agriculture: Analysis of Potential International Impacts,* Amer Soc. Agron. Special Pub. No. 59, Madison, WI. 51–75.

Hoogenboom, G., P.W. Wilkens, and G.Y. Tsuji, Eds., 1999. *DSSAT v3.5, DSSAT version 3, volume 4,* International Consortium for Agricultural Systems Applications (ICASA), Honolulu, HI, 284.

Hunt, L.A. and K.J. Boote. 1998. Data for model operation, calibration, and evaluation. In *Understanding Options for Agricultural Production,* G.Y. Tsuji, G. Hoogenboom, P.K. Thornton, Eds., Kluwer Academic Publishers, Dortrecht, The Netherlands, 9–39.

Hunt, L.A., G. Hoogenboom, J.W. Jones, and J. White. 1999. ICASA files for experimental and modelling work, International Consortium for Agricultural Systems Applications (ICASA), Honolulu, HI.

Hunt, L.A., J.W. Jones, J.T. Ritchie, and P.S. Teng. 1989. Genetic coefficients for the IBSNAT crop models. In *(IBSNAT) Decision Support System for Agrotechnology Transfer,* Part I, IBSNAT Symposium Proc., Las Vegas, Nevada Oct. 1989, IBSNAT, Dept. Agron. and Soil Sci., College of Trop. Agr. and Human Resour., Univ. of Hawaii, Honolulu, HI, 15–29.

IBSNAT. 1986. *Technical Report 5: Documentation for the IBSNAT Crop Model Input and Output Files, Version 1.0,* Dept. of Agron. and Soil Sci., College of Trop. Agric. and Human Resour., University of Hawaii, Honolulu, HI.

IBSNAT. 1989. *Decision Support System for Agrotechnology Transfer, Version 2.1 (DSSAT Version 2.1),* Dept. of Agron. and Soil Sci., College of Trop. Agric. and Human Resour., University of Hawaii, Honolulu, HI.

IBSNAT. 1990. *Technical Report 5: Documentation for the IBSNAT Crop Model Input and Output Files, Version 1.1,* Dept. of Agron. and Soil Sci., College of Trop. Agric. and Human Resour., University of Hawaii, Honolulu, HI.

IBSNAT. 1993. *The IBSNAT Decade. Final Report of the IBSNAT Project,* Dept. of Agron. and Soil Sci., College of Trop. Agric. and Human Resour., University of Hawaii, Honolulu, HI.

ICRISAT. 1984. *Proc. Int. Symp. on Minimum Data Sets for Agrotechnology Transfer,* March 21–26, 1983. ICRISAT Center, Patancheru, India.

Ikawa, H., G.N. Alcasid,Jr., F.H. Beinroth, W.H. Hudnall, S.N. Lyonga, D. Muljadi, G. Uehara, A.T. Valmidiano, and G.W. van Barneveld. 1985. Soil family network. In *Soil-Based Agrotechnology Transfer,* J.A. Silva, Ed., Benchmark Soils Project, Department of Agronomy and Soil Science, College of Trop. Agr. and Human Resour., University of Hawaii, Honolulu, HI.

Ingram, J. 2000. GCTE and ICASA reach agreement, *ICASA News,* 5:1–2, (ICASA, Honolulu, HI).

Inman-Bamber, N.G. 1991. A growth model for sugarcane based on a simple carbon balance and the CERES-Maize water balance, *S. African J. Plant Soil,* 8:93–99.

Inman-Bamber, N.G. and G.A. Kiker. 1999. CANEGRO 3.10 In *DSSAT v3,* G. Hoogenboom, P.W. Wilkens, and G.Y. Tsuji, Eds., vol. 4, ICASA, Honolulu, HI.

Jintrawet, A., S. Laohasiriwong, and C. Lairuengroeng. 2001. Predicting the effects of planting dates on sugarcane performance in Thailand. In *Proc. Int. Workshop on the CANEGRO Sugarcane Model,* G.J. O'Leary and G.A. Kiker, Eds., South African Sugar Association Exp. Station, KwaZulu-Natal, South Africa, August 4–7, 2000.

Jones, C.A. and J.R. Kiniry. 1986. *CERES-Maize: A Simulation Model of Maize Growth and Development,* Texas A&M University Press, College Station, TX.

Jones, J.W., G.Y. Tsuji, G. Hoogenboom, L.A. Hunt, P.K. Thornton, P.W. Wilkens, D.T. Imamura, W.T. Bowen, and U. Singh. 1998. Decision support system for agrotechnology transfer: DSSAT v3. In *Understanding Options for Agricultural Production,* G.Y. Tsuji, G. Hoogenboom, and P.K. Thornton, Eds., Kluwer Academic Publishers, Dortrecht, The Netherlands, 157–178.

Nix, H.A. 1984. Minimum data sets for agrotechnology transfer. In *Proc. Int. symp. on Minimum Data Sets for Agrotechnology Transfer.* ICRISAT, Ed., March 21–26, 1983. ICRISAT Center, Patancheru, India.

Ogoshi, R.M., K.J. Boote, J.W. Jones, H. Ikawa, and G.Y. Tsuji. 1997. Modeling cabbage growth. In *1997 Agronomy Abstracts,* Amer. Soc. Agron., Madison, WI, 19.

Ogoshi, R.M. and G. Uehara. 1997. Decision support systems for vegetable crop production. Final report to USDA/T-STAR. Dept. Agron. and Soil Sci., College of Trop. Agr. and Human Resour., University of Hawaii, Honolulu, Hawaii.

Penning de Vries, F.W.T., P.S. Teng, and K. Metselaar, Eds. 1993. Systems approaches for agricultural development. *Proc. Int. Symp.,* Bangkok, Thailand, December 1991. Kluwer Academic Publishers, Dortrecht, The Netherlands.

Ritchie, J.T. 1991. Genetic specific data for crop modeling. In *Systems Approaches for Agricultural Development,* F. Penning de Vries, P. Teng, K. Metselaar, Eds., Kluwer Academic Publishers, Dortrecht, The Netherlands, 77–93.

Ritchie, J.T., U. Singh, D.C. Godwin, and W.T. Bowen. 1998. Cereal growth, development and yield. In *Understanding Options for Agricultural Production,* G.Y. Tsuji, G. Hoogenboom, and P.K. Thornton, Eds., Kluwer Academic Publishers, Dortrecht, The Netherlands, 79–98.

Rosenzwieg, C. and A. Iglesias. 1998. The use of crop models for international climate change impact assessment. In *Understanding Options for Agricultural Production,* G.Y. Tsuji, G. Hoogenboom, and P.K. Thornton, Eds., Kluwer Academic Publishers, Dortrecht, The Netherlands, 267–292.

Rosenzweig, C. and M.L. Parry. 1994. Potential impact of climate change on world food supply, *Nature* (London), 367:133–138.

Rosenzweig, C., L.H. Allen, Jr., L.A. Harper, S.E. Hollinger, and J.W. Jones, Eds., 1995. Proc. of a symposium on Climate Change and Agriculture: Analysis of Potential International Impacts, ASA Special Pub. No. 59, Amer. Soc. of Agron, Madison, WI.

Silva, J.A., Ed. 1985. *Soil-Based Agrotechnology Transfer, Benchmark Soils Project,* Dept. of Agron. and Soil Sci., College of Trop. Agric. and Human Resour., University of Hawaii, Honolulu, HI, 269.

Silva, J.A. and F.H. Beinroth. 1975. Report of the workshop on experimental design for predicting crop productivity with environmental and economic inputs. Dept. Paper 26, Dept. Agron. and Soil Sci., College of Trop. Agric. and Human Resour., University of Hawaii, Honolulu, HI.

Singh, U., R.B. Matthews, T.S. Griffin, J.T. Ritchie, L.A. Hunt and R. Goenaga. 1998. Modeling growth and development of root and tuber crops. In *Understanding Options for Agricultural Production,* G.Y. Tsuji, G. Hoogenboom, and P.K. Thornton, Eds., Kluwer Academic Publishers, Dortrecht, The Netherlands, 129–156.

Soil Survey Staff. 1975. *Soil Taxonomy: A Basic System of Soil Classification for Making and Interpreting Soil Surveys,* Soil Conservation Service, U.S. Dept. of Agric. Handbook 436, U.S. Gov. Printing Office, Washington, D.C.

Soil Survey Staff. 1999. *Soil Taxonomy: A Basic System of Soil Classification for Making and Interpreting Soil Surveys,* 2nd edition, Natural Resource Conservation Service, U.S. Dept. of Agric. Handbook 436, U.S. Gov. Printing Office, Washington, D.C.

Teng, P.S., M.J. Kropff, H.F.M. ten Berge, J.B. Dent, F.P. Lansigan, and H.H. van Laar , Eds. 1997. Applications of Systems Approaches at the Farm and Regional Levels, Vols. 1 and 2, *Proc. Second Int. Symp. on Systems Approaches for Agricultural Development,* Los Banos, Philippines, Kluwer Academic Publishers, Dordrecht, The Netherlands.

Thornton, P.K. and G. Hoogenboom. 1994. A computer program to analyze single-season crop model outputs, *Agron. J.,* 86(5):860–868.

Thornton, P.K., G. Hoogenboom, P.W. Wilkens, and W.T. Bowen. 1995. A computer program to analyze multiple-season crop model outputs, *Agron. J.,* 87(1):131–136.

Tsuji, G.Y., G. Uehara, and S. Balas, Eds., 1994. *Decision Support System for Agrotechnology Transfer version 3 (DSSAT v3),* Dept. of Agron. and Soil Sci., College of Trop. Agric. and Human Resour., University of Hawaii, Honolulu, HI.

Uehara, G. and G.Y. Tsuji. 1993. The IBSNAT project. In *Systems Approaches for Agricultural Development,* F. Penning de Vries, P. Teng, and K. Metselaar, Eds., Kluwer Academic Publishers, Dortrecht, The Netherlands, 505–513.

Wood, C.L., F.B. Cady, and C.P.Y. Chan. 1985. Development of transfer evaluation methodology. In *Soil-based Agrotechnology Transfer. Benchmark Soils Project,* J.A. Silva, Ed., Dept. Agron. and Soil Sci., College of Trop. Agric. and Human Resour., University of Hawaii, Honolulu, HI, 45–52.

CHAPTER 6

Decision Support Tools for Improved Resource Management and Agricultural Sustainability

Upendra Singh, Paul W. Wilkens, Walter E. Baethgen, and T.S. Bontkes

CONTENTS

1-56670-0563/02/$0.00+$1.50
© 2002 by CRC Press LLC

INTRODUCTION

The need for increasing agricultural productivity on a sustainable basis is the primary concern of the agricultural research and development community. The International Fertilizer Development Center's (IFDC) interest in natural resource and environmental management, specifically, efficient and improved use of inorganic, organic, and biological sources of nutrients, led to the realization that many of the problems in the nutrient efficiency domain can be adequately handled only by a multidisciplinary research approach. To achieve multidisciplinarity, a switch from reductionist scientific approach to systems approach was necessary (Figure 6.1). A systems approach is essential in addressing agrotechnology transfer and sustainability and environmental concerns for the following reasons:

1. Systems approaches serve to identify the agroecological production systems that characterize different environments, including the social, cultural, and economic components of those systems.
2. Agroecological systems theory distinguishes a hierarchy of system levels and therefore serves to clearly define the geographic scale and the temporal dimensions of the problem to be solved.
3. Systems approaches call for greater attention to the relationships between production and the environment, with an inventory of the environmental resources and better understanding of how they are used.

Thus, fertilizer evaluation needed a more dynamic approach supported with information technology development.

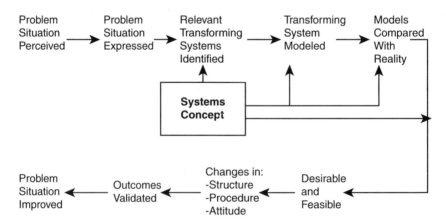

Figure 6.1 Systems approach in agricultural research.

When agricultural research is undertaken, it tends to be highly location specific. It is impossible to cover an entire country or region with field trials in an effort to derive appropriate cropping practices for the range of soil types and climatic conditions that exist in the country or region. Thus, the method of technology transfer by trial and error is ineffective. The analogue approach whereby the experience for a particular recipient agroenvironment has been generated elsewhere is useful, but it is qualitative and can obviously be applied only in limited situations. Statistical methods are suited primarily for summarizing results and interpolation rather than extrapolation. The use of systems simulation and decision support systems for the transfer of agricultural technology has been adopted and promoted because the trial and error and the analogy approach are not only time consuming and costly, but often they ignore the multidisciplinary impact of the technology. The country- and location-specificity of information availability (or lack of it) and accessibility also calls for a systems-based approach.

Developments in computers and telecommunications is revolutionizing agricultural research. Information and systems tools have the potential to improve the quality of agricultural research through effective sharing and use of information and knowledge among researchers, policymakers and farmers; however, the poor adoption of many agrotechnological innovations by the farmers in developing countries is a serious concern that the research and development community faces. We consider the limited information flow that typically exists between farmers, researchers, extension workers, policy makers, and agribusiness personnel to be one of the key factors for the poor adoption and use of innovations in developing countries. In most countries, the use of systems tools has been negligible beyond the completion of projects. Thus, improving the availability and the accessibility of information remains one of the key activities of research and technology transfer.

This chapter describes:

1. The development of soil-crop simulation models and decision support systems
2. Use of systems tools in research to improve basic understanding
3. Applications of models at the farm and global level
4. The training and confidence-building efforts that have taken place to promote the use and adoption of the modeling approach

The chapter also discusses the importance of the information technology in agricultural research and proposes some strategies for integrating information technology in research and decision making.

FIELD RESEARCH AND IMPROVED MODELING CAPABILITIES

In agriculture, systems analysis through modeling evolved in the late 1960s as a means of integrating knowledge about plant physiological processes to explain the functioning of crops as a whole (De Wit, 1965). In the early years, the models were used to gain insights to basic crop physiological and soil hydrological processes. These comprehensive models served their purpose as research tools — their use was limited to advanced country research institutes due to comprehensive and complex input data requirements.

From 1980 to 1990, the general emphasis and funding of agricultural research started to shift from understanding and explaining to practical application of results. The U.S. Department of Agriculture-Agricultural Research Service (USDA-ARS), Temple, Texas; the U.S. Agency for International Development (USAID)-funded International Benchmark Sites Network for Agrotechnology Transfer (IBSNAT) project, University of Hawaii; and IFDC played a crucial roles in making crop models practical and user-oriented (Tsuji et al., 1994). In 1984 IFDC established a Fertilizer Evaluation Program and Information System (FEPIS) to promote the use of proper methodologies for fertilizer experimentation, to provide agroeconomic evaluation of fertilizer products and practices, and to facilitate the exchange of data on fertilizer research results between IFDC and other international agencies and national organizations. In this section, some of the model development work and the important linkage with field experimentation will be highlighted.

Simulating Nitrogen Dynamics

Soil and fertilizer nitrogen undergoes many transformations involving numerous pathways and states, all of which are influenced by the weather, soil properties, and management practices. The need for systems approach involving simulation models to appropriately manage N in a cropping system was evident given the complexity of soil N cycle, the myriad of pathways for N transformations, and the interactions with weather. In collaboration with USDA-ARS the development of an N model for maize and wheat was initiated (Godwin and Jones, 1991). The model simulated the effect of genotype, weather, water availability, and soil N and fertilizer N dynamics on crop growth, yield, N uptake, and N losses.

The N submodel for rice was inevitable because N limitation is the key yield-determining factor for irrigated rice. The presence of floodwater in rice paddies leads to large differences in nitrogen behavior in rice cropping systems compared to upland systems. In these systems the presence of a shallow layer of floodwater limits oxygen transfer to deeper layers of soil. At the same time it provides a very biologically active environment for many organisms at the soil/floodwater–atmosphere interface. The model thus had to handle both reduced soil conditions and, on the disappearance of floodwater, simulate aerobic soil conditions. The following N processes are simulated for both upland and lowland rice: mineralization and immobilization, including the effect of crop residues and pools of soil organic matter; fertilizer placement, sources, and incorporation methods; urea hydrolysis; ammonia volatilization; nitrification; denitrification; nitrate and urea movement (leaching); vegetative and grain N concentration; N uptake; plant N stress indices; and existence of floodwater N pool, oxidized layer, and reduced soil N pools (Godwin and Singh, 1998).

The development of a lowland N dynamics model synthesized the information and process level understanding generated by the Nitrogen Research Program at IFDC, the IFDC–International Rice Research Institute (IRRI) Collaborative Program, and the Agronomy and Soil Divisions at IRRI. The model development effort led to identification of knowledge gaps in the existing soil N processes, residue incorporation, and nitrification–denitrification loss mechanisms. The model findings also led to field verification for effect of transplanting shock on growth and development, effect of nitrate accumulation during fallow period (Buresh et al., 1989; George et al., 1994), and quantification of fertilizer mixing on the presence of N in floodwater (Padilla et al., 1990). The following examples further illustrate the linkages between modeling and field research and the need for effective transfer of information and knowledge to improve the models on one hand and identify knowledge gaps for future research on the other hand.

Biological Inhibitors

Information on the use of biological inhibitors for improving nitrogen fertilizer efficiency for rice has been transferred to users through publications and presentations at international meetings. This information has also been synthesized and incorporated into the CERES-Rice model as one of the outputs of the Special Purpose Grant from the Australian Center for International Agricultural Research (ACIAR) (IFDC, 1997).

Transfer of these results to the agricultural industry depends on commercial production of urease inhibitors such as cyclohexyl phosphorictriamide (CHPT) and nitrification inhibitors such as encapsulated calcium carbide. Future research should include testing of the inhibitor model with existing and new data in different rice-growing soils and climate. This effort would lead to improved prediction and identification of favorable regimes for inhibitor applications.

Deep Point-Placement of USG

Deep point placement of urea supergranules (USG) has generally resulted in increased rice yields and improved nutrient efficiency, although field results have shown that this technology is soil and site dependent. The CERES-Rice model was used to verify and identify key processes and factors leading to improved N use efficiency with deep point placement (Singh and Thornton, 1992). This effort was later expanded by Mohanty et al. (1999) to identify appropriate niches for USG application on rainfed lowland rice in India. Results from ongoing projects on Participatory Evaluation, Adaptation and Adoption of Environmentally Friendly Management Technologies for Resource-Poor Farmers are being used to evaluate and validate niches for USG application in Bangladesh. Economic benefits to farmers would differ under different soil-climate-management regimes; thus, it is critical that such information is available to decision makers before recommendations are passed on to farmers.

Nitrate Leaching in Soils with Variable-Charged Colloids

Simulation of nitrate leaching as described in the CERES models (Godwin and Singh, 1998) assumes complete reservoir mixing with soil water in a given layer, and the flux of nitrate in solution is equal to the soil water flux; however, in soils with variable-charged surfaces and anion-retention capacity, flux of nitrate is less than soil water flux. The retention factor (ratio of nitrate flux to soil water flux) is dependent on bulk density of the soil, soil water content above the drained upper limit and the retention or adsorption coefficient of the nitrate ion. Bowen and Wilkens (1998) allowed the retention coefficient to increase from 0 (no retention) to 1×10^{-3} m^3 kg^{-1}. Since the retention coefficient of the nitrate ion is not a readily available input for simulation models, it was estimated for subsoils with variable-charged surfaces as a function of organic matter content and difference in pH measured in KCl and in water (Singh and Uehara, 1998).

Effect of N Stress on Phenology

Crop growth simulation models reliably predict effects of temperature and photoperiod on crop growth stages and duration, a primary determinant of yield; but they usually ignore possible effects of extreme high/low temperature, drought stress, or nutrient deficiencies on duration. Drought stress and N and P deficiencies during the vegetative phase can delay floral initiation and anthesis. Similarly, stresses during the ripening stage can cause early senescence and maturity.

Based on field results from sub-Saharan Africa (SSA), India, Hawaii, and Florida, the CERES-Maize model was modified to accommodate the effect of N stress on phenological development in maize (Singh et al., 1999). The average N stress effect over the reproductive period was used to modify the anthesis to silking interval (ASI), which, in turn, affects the number of grains per ear. The phyllochron, or leaf appearance rate, is least affected by N deficiency; as a result, the final leaf number changes only slightly. Under N limiting conditions, the model captures the effect on growth, grain number, and the shortened duration of the grain-filling stage to contribute to lower grain yield.

Simulating N Supply from Organic Sources

N release from organic sources depends on their nutrient content, quality, and the environmental and management factors. Combating Nutrient Depletion Consortium (CNDC), in partnership with National Agricultural Research Systems (NARS), is synthesizing information generated from integrated nutrient management trials in SSA and Latin America. Field results linked with the Organic Resources Database and Agricultural Production Systems Simulator (APSIM) and Decision Support System for Agrotechnology Transfer (DSSAT) models are being used to develop and test a model that captures N dynamics in an integrated nutrient management system. The standard DSSAT and APSIM models simulate the effects of N supply from different organic sources based on their N (or C:N ratio), carbohydrate, cellulose, and lignin content. The models reliably simulate N mineralization and immobilization for organic sources with varying C:N ratios. On the other hand, the decomposition rates of the organic materials were not as sensitive to the carbohydrate, cellulose, and lignin content.

Field and laboratory results have shown that high polyphenolic content reduces decomposition rate of organic materials even for those having similar N content. Based on these data the N model in DSSAT has been modified. Work is in progress to develop decomposition rate modifiers for use in the APSIM and DSSAT based on the polyphenolic content of the organic materials as available from the Organic Resource Database. The modified decision support tools would offer researchers and farmers in the region the options to choose appropriate local materials to increase crop yields and improve soil fertility.

New Methodologies for Use with Models

Although accuracy is a desirable attribute of decision support systems (with initial emphasis on model validation and refinement), the authors emphasize that the primary objective is to use model outputs to make better choices. This is reflected in some of the methodologies developed to make simulation models and databases more practical for decision making.

Climate Analysis and Synthetic Weather Generation

The critical analysis of climatic variability over time and its consequences for management requires that we have tools to statistically evaluate historical weather and to generate synthetic weather sequences. Typically, when addressing the impact of temporal variability within the context of the impact assessment of new methodologies, the availability of appropriate climatic data is limited. Several tools have been developed to address these data gaps.

MarkSim (Jones and Thornton, 2000a) is a software package designed to generate daily weather data for much of the arable surface of the earth, given inputs of latitude, longitude, and elevation of a given point. The methodology for the development of the system has been well documented (Jones and Thornton, 1993; Jones and Thornton, 1997; Jones and Thornton, 1999). The third-order Markov rainfall model, which was utilized, has been extensively tested and works well for given climate stations. Errors in interpolation in the climate surfaces can be natural to such a system, but it has been proven to be a useful tool in modeling climatic risk (Jones and Thornton, 2000b).

A second climate characterization tool utilized in crop modeling is WeatherMan (Pickering et al., 1994), a component of the DSSAT (Tsuji et al., 1994). International Consortium for Agricultural Systems Applications (ICASA) has specified a minimum daily weather data set and format for use with the crop models. The required daily variables are solar radiation (MJ $m^{-2}d^{-1}$) and maximum and minimum temperature (°C), and rainfall (mm). The extended climate data set includes optional variables of photosynthetically active radiation (PAR, mol $m^{-2}d^{-1}$), dew point (°C), and wind speed (m s^{-1}). The WeatherMan program is designed to simplify or automate many of the tasks associated with handling, analyzing, and preparing weather data for use with crop models or other simulation software. WeatherMan can also generate complete sets of weather data using historic data or synthetically generated records (Geng et al., 1988; Richardson and Wright, 1984). Research priorities into large-scale atmospheric circulation patterns such as the El Niño-Southern Oscillation (ENSO) have led to the development of methodologies in WeatherMan to generate synthetic weather sequences to match monthly target goals along with the incorporation of improved weather generation capabilities (Hansen, 1999; Hansen and Mavromatis, in press; Mavromatis and Hansen, in press).

Risk Analysis Associated with Weather

Nutrient losses, fertilizer recovery, grain yield, and the processes affecting these vary greatly from year to year in any location. To develop optimal management strategies in any location, it would be desirable to have experiments conducted over many years (although costly and time consuming). Field experiments are rarely conducted for more than two seasons, and thus long-term data providing insights into the nature of temporal variability are usually not available. Procedures were developed for DSSAT models (Thornton and Hoogenboom, 1994; Thornton et al., 1994a) with long-term weather data or using stochastic generation of weather data for quantification of variations in yield, fertilizer recovery, and nutrient losses associated with weather. In addition to the biophysical outputs, the system also generates an analysis of gross margin of production based on expected product values and production costs. The risk analysis procedure allows for the selection of strategies under conditions of uncertainty and provides due recognition to farmers' attitudes to risk and mean outcome (Thornton and Wilkens, 1998).

Sustainability of Cropping Sequences

Improving our understanding of long-term, sustainable productivity of cropping systems requires an integration of field experimentation, resource monitoring, and the capability to simulate and analyze long-term cropping sequences. The crop models distributed in DSSAT have the capability to be linked together in a rotation or a continuous sequence of crops and for the results of these simulations to be analyzed in both graphical and tabular fashion (Thornton et al., 1994b, 1995a). Using different weather years or stochastically generated weather sequences, the cropping sequence effect may also be optimized and risk quantified (Bowen et al., 1998). The linkage of multiple cost–price scenarios and a series of stochastic weather sequences permits the user to explore many possible scenarios beyond yield sustainability, and allows the assessment of economic risk with cropping systems.

Information and Decision Support Tools (IDST)

The influence of agricultural research and development is usually measured via testable, quantifiable impacts on the biophysical characteristics of a system; however, it is often difficult to assess the economic impacts of agricultural development that arise from technology-induced changes in yield potential and/or production costs at the farm level. These broader economic effects depend upon a range of biophysical, social, and market factors. To address this problem, we developed an information and decision support tool (IDST) linking a geographic information system (GIS), a crop simulation modeling system (Singh et al., 1993a; Baethgen et al., 1999; Wilkens et al., 2000), and the DREAM (Dynamic Research EvaluAtion for Management) economic model (Alston et al., 1998; Wood et al., 2001). Long-term, sequential cropping simulations at different technology levels can be compared with the analysis of the biophysical sustainability of a system coupled with the generation of relevant and structured economic information to support decision makers implementing agricultural policy, assigning priorities, and allocating limited resources over a large area.

In a test case in Carimagua, Colombia, system development included the calibration and application of a phosphorus-enabled version of the DSSAT crop models for the evaluation of an improved maize (*Sikuani*)/soybean (*Soyica altillanura*) rotation designed to provide a higher and more sustainable level of agricultural production on the highly weathered Oxisol and Ultisol soils of the Eastern Plains (Llanos Orientales) of Colombia. These soils exhibit Al-toxicity and high levels of P fixation. The DSSAT CERES-Maize and CROPGRO soybean models were calibrated using field trial data from the CIAT/CORPICA experimental station at Carimagua. The calibrated DSSAT crop models were then used to test alternative crop residue, fertilizer, and liming management practices to find options that may increase long-term stability in yield and income. The analysis was conducted within the framework of the IDST spatial interface to crop modeling.

The IDST interface is based on soil and climate coverages as the basis of crop simulation. Unique combinations of soils and climate produce unique model outcomes with given technology inputs. All unique polygons are then simulated based on criteria established in the selected ICASA experimental file, either in seasonal mode to examine year-to-year variability of a system or in sequential mode, where long-term sustainability of a given cropping rotation can be analyzed within the context of a spatial database.

The generation of DREAM input files involves the selection of a baseline technology scenario and a prototype simulation where model parameters have been adjusted to reflect known or likely effects of new technology or some intervention. A DSSAT-DREAM linkage file documenting the differences between the baseline scenario and the prototype technology is then generated. The data generated by the IDSS is being used to compare divergent technologies (baseline vs. prototype) to test the biophysical and economic sustainability of alternate systems on both a temporal and spatial basis. The scope, flexibility, and reliability of technology evaluation analysis is enhanced several ways:

- Providing the capacity to evaluate a broad range of technology and natural resource management options
- Explicit modeling of the soil water, P, N, and OM to provide improved assessment of natural resource impacts of technological change
- Linkage of plot-scale crop simulation to market-scale analysis
- Support of nested analyses across a broad range of geographic scales

The initial success of the IDST effort is evident from application of simulation models-GIS for projects in Malawi (Thornton et al., 1995b), Albania (Tsuji et al., 1994), Burkina Faso (Thornton et al., 1997), and Uruguay (Baethgen et al., 1999). Additional case studies of IDST applications for Chhattisgarh, India, and Colombia are presented later in the chapter.

TECHNOLOGY TRANSFER FOR SMALLHOLDING FARMERS

It is essential to identify and quantify the relative importance of key sources of current yield gaps in farmers' fields in relation to potential production (nonlimiting water and nutrients and no production losses due to pests and diseases) and rainfed potential production (limited only by rainfall and water availability). Such analyses would help implementation of appropriate policies regarding infrastructure, credit, extension services, and others that may be required to increase and rationalize input use and the adoption of productivity-enhancing technologies. The question, "Could risks associated with rainfall be reduced by identifying low-risk planting windows, appropriate crops and genotypes that would be less prone to drought stress or simply complete their life cycle before the onset of a drought spell?" was explored in the three case studies presented. The technical problems of agriculture in much of SSA with respect to population growth, resource base depletion, and weather risks are such that all available tools should be brought to bear in attempts to find solutions.

Agrotechnology Transfer Using Biological Modeling in Malawi

The primary objective of the case study conducted from 1990 to 1993 was to validate the CERES-Maize model through a series of field trials carried over three seasons at a number of locations in the mid-altitude and lower-altitude maize ecologies of Malawi. The rationale was to determine if simulation techniques had potential to enhance the efficiency of the research and development process by helping relieve the pressure on scarce research resources using a computer model to screen large numbers of production alternatives. Promising alternatives identified in this way could then enter field testing for eventual transference to the farmer, in the search of increased smallholder maize production to enhance food security for a rapidly growing population.

The CERES-Maize model appeared to work reasonably well by simulating a range of yields from 0.5 to 6.5 t ha^{-1} over three seasons at three research stations and several farmers' fields (Singh et al., 1993b; Thornton et al., 1995b). A large amount of model experimentation was carried out to investigate effects of weather, soil type, and planting date on planting windows and fertilizer response. For example, the complex interaction between soil moisture and nutrient rates over 25 seasons is presented in Figure 6.2. An effort was also made to link the model to the spatial

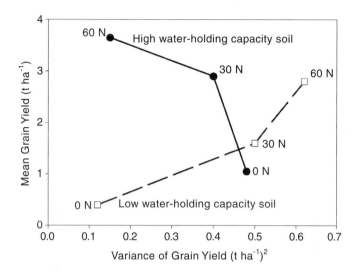

Figure 6.2   Water and nitrogen interaction effects on maize grain yield (mean 0 and variance.

climate and soils databases of a GIS for a small area in Kasungu Agricultural Development Division, primarily to illustrate the potential for regional analysis using these tools (Thornton et al., 1995b).

The small start made in modeling activities in Malawi provided a base on which to build for the future. The rapid expansion in the availability and use of computers and expertise of staff in using them over the period of the project and later is clear evidence that there are little inherent barriers to the use of what are often perceived as relatively sophisticated technology to attack research problems in Malawi. In the last 10 years, the soil resources of Malawi have been digitized, and the models and GIS have been applied to address issues relating to agricultural production and the environment. The Department of Meteorology now e-mails weekly weather summaries for the entire country on a regular basis in contrast to the 1990s when none of the meteorological data was in an accessible format.

Soil Fertility and Climatic Interactions

Rainfall variability plays a dominant role in the use of inorganic fertilizers for small-scale maize production in Kenya. Soils are being mined of the essential nutrients (Stoorvogel and Smaling, 1990). An integrated nutrient management approach requires the use of crop simulation models that allow the extrapolation of biophysical performance of the maize crop across time and space in response to climatic and management inputs. During the past 10 years, various projects have used different maize models to address the above issues (Keating et al., 1991, 1993; Rotter, 1993; Wafula, 1995).

The models have:

- Highlighted and quantified the risks associated with smallholder maize production in Kenya
- Improved understanding of the interactions between soil, management, and weather for more productive systems
- Showed the importance of soil fertility and fertilizer applications for stable and increased production
- Identified constraints associated with water logging, aluminum toxicity, and striga for different areas.

These constraints also led to the improvement and inclusion of water and nutrient stress effects on phenology, simulation of water table depth, and the effect of oxygen stress due to water logging, and in case of severe water stress that forces the model to kill crops (Carberry and Abrecht, 1991).

Combating Nutrient Depletion Consortium (CNDC)

Inefficient use of nonrenewable resources, destruction of ecosystem and overexploitation of renewable resources have led to the general agreement that the current human-environment relationship may be untenable. In confronting this dilemma, CNDC is working with farmers, researchers, and the ecoregional programs in West Africa and East African Highlands to reverse the degradation of tropical soils through identification of sustainable practices for managing soil, water, and nutrients. The challenge of CNDC is to develop methodologies that allow nutrient management technology to be transferred effectively from one area to another, given the limitations of time, money, and resources that all research and donor organizations face. Although the primary research thrust of CNDC is on soil fertility improvement through an integrated use of organic/inorganic fertilizers and amendments, the approach is not so much to initiate new research as to achieve greater impact from present knowledge. The primary focus is therefore to use tools such as simulation models and decision support systems that synthesize available information and make it accessible to a wide range of clients.

Yield-Gap under Water and Nutrient Limitation

Long-term historical weather data for 12 years from Koukombo, Togo, was used with the CERES-Maize model to determine the rainfed potential yield — with rainfall and soil water-holding capacities influencing maize yield. As evident from the mean-variance (E-V) plot, the ideal planting timeframe for rainfed-maize at Koukombo is the period May–August (Figure 6.3a). During this period more than 85% of the potential yield is reached under rainfed conditions (Figure 6.3b). A sharp drop in yield is associated with late planting. Maize planting before May would result in increased risk associated with lower mean yields and increased variance. With additional limitation of N and P (only 20 kg N ha^{-1} and 15 kg P_2O_5 ha^{-1}), simulated mean yields under May-August plantings ranged from 1.5 to 3.2 t ha^{-1}. These amounted to less than 30% of the potential grain yield for each of the planting dates (Figure 6.3b). The simulated results show that in contrast to what is often stated, nutrients may be more limiting than water in SSA. With external nutrient input and appropriate planting date and genotype, maize grain yields could be increased by 2–4 times the current farmers' yields.

Nitrogen Fertilizer Economics

Because fertilizer response is dependent on seasonal weather variation, investigators used the past 12 years of weather data to capture the mean N response and the standard deviation at Koukombo. Even in the P-deficient soils of SSA, adequate N application is necessary to achieve the full benefits of P application. The P rate was set at 45 kg P_2O_5 ha^{-1} and other nutrients were assumed nonlimiting. Based on field observations, the soil hospitality factor that allows for root growth in the model was set to zero (no root growth) below 45 cm. The apparent N recovery simulated by the model ranged from 30-35%. Based on the current prices of inputs and products and costs of production, the mean-Gini efficient N rate was determined at 155 kg N ha^{-1} with the monetary returns of $566 ha^{-1}. With improved recoveries of applied fertilizer (50%), the mean-Gini efficient N requirement rate will drop to 110 kg N ha^{-1} with the monetary returns of $644 ha^{-1}. Figure 6.4 further illustrates the dependence of optimum N rates on fertilizer-related costs and maize grain price.

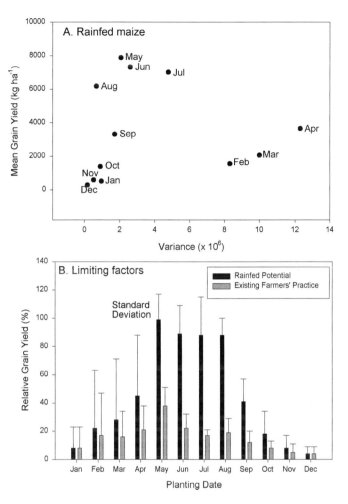

Figure 6.3 Optimum planting window (A) and yield gaps due to water and nitrogen limitation (B) at Koukombo, Togo, as simulated by long-term simulation with CERES-Maize model.

It is envisaged that the model applications described previously would lead toward soil fertility improvement and overcoming the key constraints that result in nutrient mining. The major constraints to the use of inorganic fertilizers are:

1. Limited accessibility
2. High prices
3. Poor nutrient recoveries
4. Lack of market and price stability for excess production beyond the farmer's need

SUSTAINABILITY OF RICE–WHEAT SYSTEMS

Rice followed by wheat is a dominant cropping sequence under a wide range of management regimes across some 26 million hectares in South and East Asia with variable productivity. The rice–wheat rotation provides food and livelihood for millions of people. Both rice and wheat are exhaustive feeders, and the double cropping system is heavily depleting the soil of its nutrient content. The continuing degradation of resources threatens food security for an ever-expanding population.

Mean Gini Efficient N Rate

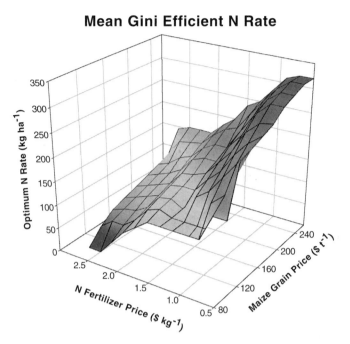

Figure 6.4   Effect of grain price, N fertilizer cost, and seasonal differences (12 years) on optimum N rate for maize.

The rice–wheat rotation system has developed by the introduction of rice into traditional wheat areas (Punjab, India, and Pakistan) and wheat into traditional rice areas (Bangladesh, Eastern India). Consequently, wide variation in growing conditions due to soil and climatic factors prevails. In the rice–wheat system each crop is affected by a multitude of environmental and management factors. The dominant feature is the repeated transitions from the anaerobic conditions with rice to aerobic regime in wheat. In addition cultural practices of one crop affect the other: the effect of soil puddling in rice influences the establishment of the wheat crop; the changes in planting date of wheat due to method of rice planting (transplanted versus direct-seeded), and tillage practice on wheat (minimum versus conventional); and the varietal differences.

It is therefore imperative to apply a systems approach to understand processes in such systems (Singh and Timsina, 1994). Simulation models provide the best option with which to quantify sustainability of a rice–wheat system and technology transfer across a diverse region (Timsina et al., 1995, 1997, 1998). Validation of both the CERES-Rice and CERES-Wheat models has been carried out in collaboration with the Wheat Research Center, Bangladesh and G.B. Pant University of Agriculture and Technology, India (Timsina et al., 1995, 1998). In some cases, simulated results were not in close agreement with observed field results (Figure 6.5) because the model did not take into account many biotic factors that influenced the crop's performance, e.g., pest damage.

Using results from the long-term rice–wheat experiments at G.B. Pant University, an attempt was made to validate the rice–wheat sequence model (Timsina et al., 1997). In the sequence model the soil water, residual N, and organic matter status at harvesting of one crop becomes the starting conditions for the next. Figure 6.6 shows the results for continuous rice–wheat cropping without N fertilizer application at Pantnagar, India from 1979 to 1993. The long-term yield trends based on the observed data indicated nonsignificant changes in rice yield; however, wheat yields showed a significant increase over the time of study. The model was able to capture the trends in both the crops. With the help of a cropping sequence model, compounding effects of weather variability, varietal changes, planting date shifts, and failures of the irrigation system on yield, trends in long-term trials were eliminated by running the model in sequence over multiple years with the same

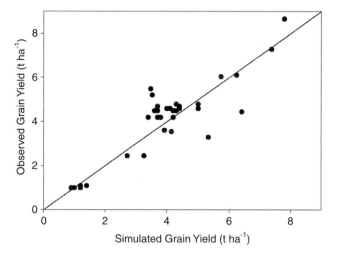

Figure 6.5   Rice and wheat grain yields, Nashipur, Bangladesh.

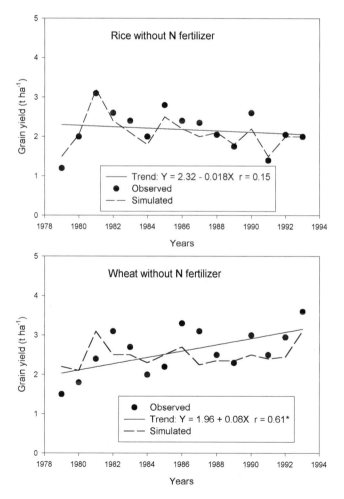

Figure 6.6   Observed and simulated yield trends for rice-wheat sequence at Pantnagar, India.

variety and planting date and under rainfed and fully irrigated conditions. Much more effort is required to comprehend the processes leading to the current yield trends. The rice–wheat sequence model has contributed toward identifying the knowledge gap by focusing the attention of the researchers on key processes. Further refinements of the existing models and simulation of pest and disease effects, tillage options, and nutrients other than N and P are required to assist research and technology transfer in the rice–wheat cropping systems.

CLIMATIC VARIABILITY AND GLOBAL CLIMATE CHANGE

Climate variability measured in the interannual and interseasonal scales is one of the key factors affecting agricultural production. Crop and pasture productivity greatly depend on the environmental conditions during the growing season including temperature regime and extremes, total rainfall, onset of the rainy season, rainfall during critical growth stages, etc. Variability in these environmental conditions inevitably results in variable yields and economic returns. Although often public or media attention is focused on extreme events, climate variability can also have subtler consequences but with significant economic impacts. Under the most commonly expected scenarios of climate change — increased temperatures and increased rainfall variability — such systems are likely to become even more vulnerable.

Model Applications with Seasonal Climate Forecasts

Two strategies are possible to diminish the climatic variability: reduce risks by controlling the limiting factors (e.g., introducing irrigation in water-limited environments), and adjust production management practices that consider information about the most likely environmental conditions expected for the upcoming growing season. Early warning of impeding poor crop harvests in variable environments can allow policy makers the time they need to take appropriate actions to ameliorate the effects of regional food shortages on vulnerable rural and urban populations. Historically the main limitation to define *a priori* adjusted management practices has been the lack of means to predict climate conditions (e.g., precipitation, temperature) with sufficient skill and lead-time. In some regions of the world, this situation has started to change due to recent advances in the capacity to predict climate anomalies linked to the onset and intensity of a warm or cold event as part of the El Niño/Southern Oscillation (ENSO) phenomenon. ENSO is the main source of interannual climate variability in many parts of the world.

In recent years, the scientific community has started using this knowledge to issue probabilistic climate forecasts, i.e., define the likelihood of expected rainfall and temperature scenarios (normal, below normal or above normal) for the following 3 to 6 months. Studying the effects of climatic variability and identifying adequate management responses would require many decades of experimentation, particularly in areas where such variability is high. Alternatively, well-tested crop simulation models are being used to assess the effect of varying temperatures and precipitation in different crop growth stages, to better identify the most vulnerable periods of the growing season and to establish climatic thresholds. The models are also being linked to weather generators conditioned to ENSO phases and used to determine crop management practices that will minimize negative expected weather conditions and take full advantage of expected favorable conditions. Simulation modeling is particularly useful to quantitatively compare alternative management options in areas where seasonal climatic variability is high, such as Australia, Southeast Asia, Africa, and Latin America (Meinke et al., 2001; Travasso et al., 1999; Keating and Meinke, 1998; O'Meagher et al., 1998; White et al., 1998).

An Application in Southeastern Latin America

The research approach used in southeastern Latin America for developing applications of climate forecasts in the agricultural sector started with the identification and description of the ENSO impacts on rainfall and temperature anomalies. Then the researchers identified crops with the highest sensitivity to these ENSO-related anomalies and quantified the impact of the anomalies on the obtained crop yields. Finally, simulation models and decision support systems were used to evaluate the ability of agronomic practices to reduce risks and/or increase farmers' profits (Baethgen and Magrin, 2000; Podestá et al., 1998).

The first detailed studies conducted in southeastern Latin America revealed the existence of a near symmetry between impacts of El Niño and La Niña on precipitation and on crop productivity. Positive rainfall anomalies prevail in El Niño years, and negative rainfall anomalies prevail in La Niña years, during the austral spring and summer months. Some research results also suggested that the impacts of La Niña were stronger and less variable in both rainfall and crop yields than the impacts of El Niño.

Once the effects of ENSO-related anomalies on crop production were characterized, simulation models were used to explore the best-adapted management practices. One of the limitations that were often found for this analysis was the relatively small number of years with available daily weather data that is required by the simulation models. The analyses require the separate consideration of El Niño and La Niña years, which occur every 5 to 7 years, and therefore, the number of years for each ENSO phase is typically reduced. To overcome these limitations, scientists improved existing or developed new weather generators conditioned to ENSO (or SOI) phases (e.g., Grondona et al., 1999). These weather generators produce synthetic weather data sets with similar statistical properties of the observed data for each ENSO (or SOI) phase.

For example, Baethgen (1998a) found that the probability for obtaining low maize yields in Uruguay is about twice as high in La Niña years than in normal years. A large number of crop management practices were then explored using the CERES-Maize simulation models and a conditioned weather generator. As evident from Figure 6.7, using short-season maize hybrids and delaying the planting date by 2 months could minimize the yield losses in La Niña years.

In Argentina, several activities were carried out to evaluate the acceptance and value of ENSO-based climate forecasts for agricultural decision making. In a preliminary study, Hansen et al. (1996) found that predicted benefits of tailoring soybean planting dates to forecasted ENSO phases ranged from U.S. $2.40 to $32.40, according to location and soybean prices. Messina (1999) found optimal crop combinations for each ENSO phase in different Pampas locations that depend on location, risk aversion, and initial wealth. On the other hand, Magrin et al. (1999) concluded that the best management option in La Niña years for both maize and soybean was to delay the planting date. Inversely, early sowings were more likely to optimize yields and incomes in El Niño years. These researchers also found that nitrogen rates in maize should be higher during El Niño events to maximize expected maize yield and profit; however, due to important intraphase climate variability (e.g., rainfall during flowering is not always lower in La Niña years), making crop management recommendations based on mean values may still lead to negative results (Magrin et al., 1999).

Effective application of climate forecasts must frame the climate information in broader decision support tools that also include data on prices, land use feasibility, evaluation of technologies, etc. With this challenge in mind, researchers in the region have been developing an information and decision support system (IDSS) for the agricultural sector (Baethgen et al., 1999).

Expected Impact of Climate Change

The vulnerability of the agricultural sector in any region to future possible climate change scenarios is determined largely by the vulnerability of the sector to current climatic, economic, and

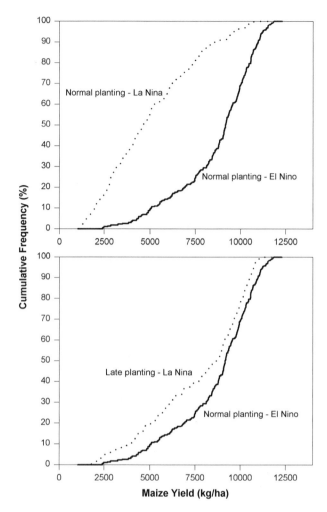

Figure 6.7   Simulated maize yield distribution for El Niño and La Niña years in Uruguay for normal planting on September 15 with medium-duration maize and late planting with short-duration maize.

policy scenarios. Most agricultural systems are currently subject to severe climatic interannual variability (e.g., drought, flood, storms, etc.). Such systems are likely to become even more vulnerable under the most commonly expected scenarios of climate change (i.e., increased temperatures, increased rainfall variability). Similarly, agricultural systems that are currently subject to drastic changes in economic and policy scenarios are also prone to become more vulnerable under climate change conditions.

The few theoretical studies conducted to specifically assess the impact of future climatic change on agriculture revealed expected reductions and increased variability in crop productivity. Possible scenarios of increased atmospheric CO_2 concentration were generated with different General Circulation Models (GCMs) including National Aeronautics and Space Agency/Goddard Institute of Space Studies (NASA/GISS), United Kingdom Meteorological Office (UKMO) Model, and General Fluid Dynamics Laboratory (GFDL). Additional scenarios were generated modifying observed long-term weather data by increasing temperatures or varying precipitation (sensitivity analysis). These modified climate data were then used with the DSSAT models to estimate the expected impact of climate change on the production of key food crops: rice, wheat, sorghum, maize, millet, and soybean (Hoogenboom et al., 1995; Rosenzweig and Iglesias, 1998; Singh and Ritchie, 1993).

The impact based on simulated outcomes for winter crop production in Argentina and Uruguay was 10 to 30% crop yield reductions (Baethgen and Magrin, 1995; Baethgen 1994). Climatic impact studies in Southeast Asia on rice (Singh and Padilla, 1995) and the Oceania–Fiji Islands (Singh et al., 1990) also revealed yield reductions of 10 to 35% and a lower response to nitrogen fertilizer application with the consequent negative effects on economic returns for farmers with existing varieties. Based on these studies the potential impacts and possible adaptations of climatic change on world food supply, demand and trade were presented (Reilley et al., 1996). A common conclusion was that globally, the expected impact of climate change on crop production was small; however, under all simulated scenarios possible negative impacts were mostly observed in low latitudes where the majority of developing countries are located, thus tending to increase the disparity between developed and developing countries.

The Global Change and Terrestrial Ecosystems (GCTE) Project under Focus 3 — Agroecology and Production Systems — also uses soil-crop simulation models to assess the impact of climate change and identify possible adaptive measures. Focus 3 includes a network for wheat, rice, rice–wheat, pests and diseases, pastures, soil organic matter roots, roots and tubers, grain legumes, and tropical cereals. One of the objectives of the network is to provide relevant models for global change studies by validating the models with good quality data under existing climatic conditions.

REGIONAL APPLICATION OF INFORMATION AND DECISION SUPPORT TOOLS

IFDC has been finding increased interest in the public and private sectors of different countries to develop and establish information and decision support tools (IDST). In addition to taking advantage of the large volume of information generated in the past by the research institutes, IDSTs provide excellent support for the decision-making process in public institutions, private companies, and research centers.

IDST for Resource Management in Chhattisgarh

An IDST, which interfaces DSSAT crop models with GIS, is illustrated for Chhattisgarh State, India, where large amounts of relevant information have been generated (Patil et al., 2001). The IDST is based on prototype developed by IFDC for sorghum in the semiarid tropics of India (Singh et al., 1993a) and for wheat in Uruguay (Baethgen, 1998b). The complexity of agroecosystems, the need for taking a long-term view of biophysical processes to assess sustainability, and the limited availability of research resources also support the notion of an IDST.

The Chhattisgarh state represents the rainfed, lowland rice–growing area of eastern India. This predominantly tribal area (14.4 million ha) has highly variable soil types and climatic conditions. The region provides complex microclimatic and soil conditions and offers a unique opportunity of applying simulation GIS–integrated IDST in decision making to ensure sustainable and environmentally safe resource management. The agricultural intensification activities are increasing due to population pressure and to making the area double-cropped. These have caused changes in the socioeconomic status of the farmers.

Identification and screening of suitable management strategies and their adoption by farmers requires consideration of soil, climate and socioeconomic factors. An IDST helps in organizing the information in a database that can be used by a simulation model or for answering queries as specified by the user. A soil, weather and agricultural characteristics database for the eight districts of Chhattisgarh was created in the form of soil maps using Arc View (v3.0). The weather coefficients for each of the weather stations were interpolated to form grid maps of weather parameters using spatial analysis of Arc View 3.1 (ESRI, 1996).

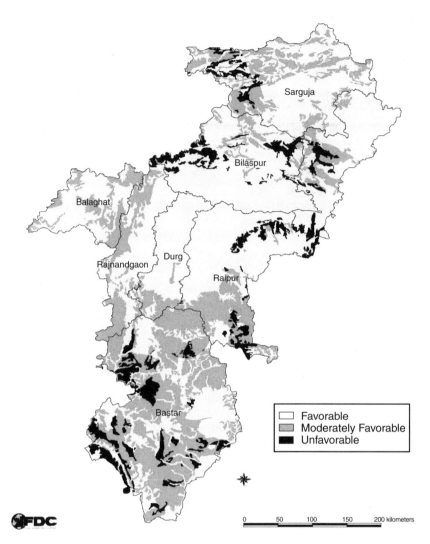

Figure 6.8 Crop suitability map for wet-season (Kharif) crops in Chhattisgarh, India.

Using soil and weather parameters from Chhattisgarh, suitability of areas for wet- and dry-season crops (Figure 6.8) were determined. Such delineation helped identify areas that should be targeted for increasing production. The criteria chosen for delineating the area into classes of suitability were soil depth, texture, slope, and rainfall (June to September). The rainfall map prepared from recent weather data (1988 to 1998) was used for the purpose. The criteria for dry season (October to January) included soil depth, texture, slope, dry season rains, and groundwater table depth in dry season. The rainfall map for October to January prepared from weather data of 1988 to 1998 was used. The water table map for the month of November was considered while identifying suitable areas for dry season crops. The maps so derived may be very useful in crop planning and identifying the priority areas that can be targeted for increasing production and crop diversification. Such analyses can be extended to individual crops by considering their specific soil and climatic requirements.

The CERES-Rice model was used to derive the relationships between soil-climate parameters and rice crop yield. These relationships were used to prepare rainfed potential yield maps and

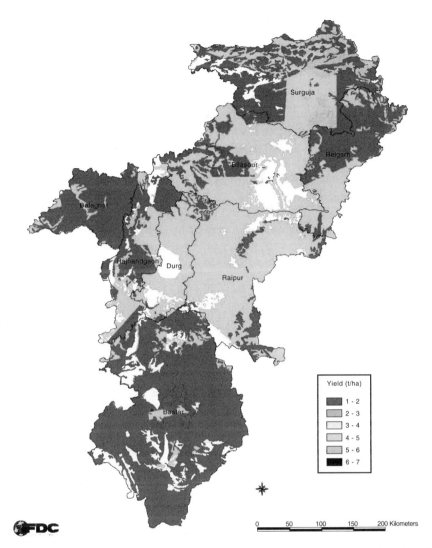

Figure 6.9 Existing yield gaps in dry-seeded rice as determined from potential rainfed yield (simulated) and yields limited by nutrient stress and management (based on observed and simulated data) at Chhattisgarh, India.

identification of the yield-gaps for rainfed rice (Figure 6.9). The IDST is a very effective aid for researchers and extension workers in identifying proper crop planning and better soil management practices. It will also be very useful in agrotechnology transfer.

The slow rate of technology transfer and subsequently poor adoption by the farmers is one of the most important bottlenecks for the development of agriculture in Chhattisgarh. Therefore, the agriculture of the region is relatively less developed as compared to most of the Indian states. One of the most important considerations for adoption of research results is identifying areas where the technology can be most effective and appropriate modifications for changing local needs can be implemented so that the risks associated with soil and climatic variability, pest and disease incidence, and price fluctuations can be minimized. Using the IDST, technology packages can be designed and assessed quickly and efficiently using computers on a district-wide basis. This ensures that only the most promising research results enter into the field-testing program. Further, the suitability of the same management appropriate for one district can be pretested in other districts by simply changing the model inputs (soil and weather) and the technology can be fine-tuned accordingly. The stability of the technology in a particular area can be identified using historical or generated weather. A decision support system can therefore help in identification and transfer of the most effective nutrient management technologies to the farmers.

METHODOLOGIES FOR HARMONIZING AND PROMOTING USE OF SYSTEMS TOOLS FOR DECISION MAKING

For more than 20 years, researchers have attempted to introduce modeling as a tool for decision making related to a sustainable agriculture in sub-Saharan Africa (SSA). During that period, several systems tools were developed from econometric and mathematical models at regional levels to crop growth simulation and nutrient management models. Although these efforts have resulted in a large number of publications, these models have had limited uses in agricultural decision making (Breman, 1995; Matthews et al., 2000; Newman et al., 1999; Parker, 1999).

Causes of Limited Adoption in Sub-Saharan Africa (SSA)

First, a set of hypotheses regarding the lack of adoption was developed based on literature and discussions with knowledgeable persons and with members of the AGMODELS discussion group on the Internet. Then, a number of persons and organizations, who were known to have been exposed to models, were visited and interviewed. Countries visited were Mali, Burkina Faso, Nigeria, Niger, Kenya, Malawi, and Zimbabwe.

Several modeling projects were discontinued during the phase of data collection. This related to the problem that data collection required more time than anticipated and that the planning of such projects had been too optimistic in terms of time and money. Availability of reliable data is a general problem in many developing countries; by the time the data are collected, the situation may have changed in such a way that the models are not relevant anymore. In the case of large interdisciplinary models, completion was sometimes hampered by the incompatibility of the methodologies that were used by the different disciplines. It also appeared that most modeling projects did not have a follow-up. An important reason for this was the fact that model development was very often initiated in donor countries that would like to try out a model or a modeling approach for a concrete case — searching a problem case for a model instead of searching for a model that can help solve the problem. In such a case, the potential user was hardly involved in the development of the model. This may reduce the confidence of the local user in the appropriateness of the model to address the problems that are deemed essential by him or her. The recent initiatives in Mali to orchestrate the modeling efforts better and to screen the projects as to their utility for decision making may serve as an important step forward.

Local scientists were sometimes used to collect data but were rarely involved in model development. Once the project ended, local scientists were left behind with a model they might be able to use as a black box but with limited understanding on adjusting the model sensibly. On the other hand, several projects did involve local scientists in model development, often enabling them to obtain a Ph.D. within the framework of the project. Although this certainly contributed to an improvement in modeling skills of local scientists, these skills were not always implemented because there was often no opportunity to use the models developed, either because of lack of funds or because of a lack of interest by potential users.

The reluctance to use models is also caused by lack of knowledge, as modeling is only included in the curricula of the universities to a very limited extent. In addition, the complexity of some models constitutes a barrier; it is important that models are kept simple and that their results are presented in a comprehensible way to the decision makers. On the other hand, there were also complaints that the models were not realistic, which would call for more complex models.

Opportunities

What are the lessons to be learned from this inquiry, and what should be done to enhance the use of models in decision making in agriculture? It seems imperative to include the envisaged users to a much larger extent in the development and selection of models; in other words, a much more participatory approach is required.

Though large interdisciplinary models are interesting from a holistic point of view in the sense they allow insight into the relations between the various aspects of the problem studied, they also have a number of drawbacks, as indicated above. These drawbacks, along with the frequently heard call for simple models, suggest it is preferable to use models that do not cover a wide range of disciplines. In this respect, the ideas of Newman et al. (1999) appear to be very useful. They propose a "hybrid" approach to DSS development that recognizes the need for simultaneous application of hard and soft systems approaches. In West Africa, a similar approach could be followed, using models as a learning and discussion tool rather than for direct recommendations. Because simple models address only a limited number of aspects, it is also possible to use a number of complementary models.

The inquiry also suggested a lack of knowledge and awareness regarding models and their use. Ph.D. scholarships, related to a particular project, are unlikely to be sufficient to create a critical mass for the use of models. It is therefore necessary to include systems approaches and modeling in university curricula and to train national research and extension staff in the use of models. Such training should not be limited to how to handle the models but, more important, how to use them to help solve problems.

The COSTBOX Approach

The Client-Oriented Systems Tool Box for Agrotechnology Transfer (COSTBOX) project has embarked on a number of activities to promote the use of models in agricultural decision making in a number of west African countries, while trying to avoid the pitfalls described earlier (IFDC, 1999). The project focuses on the introduction of relatively simple models that address problems at the field level. To be able to cater to the clients' needs, a toolbox of various models is developed comprising models such as DSSAT (Tsuji et al., 1994), Cotton model–COTONS (Jallas, 1998), Qualitative Evaluation of the Fertility of the Tropical Soils–QUEFTS (Janssen et al., 1990), and Nutrient Management Support System–NuMaSS (Osmond et al., 2000). DSSAT and COTONS are crop growth models. QUEFTS and NuMaSS are models that provide decision support in nutrient management. It is possible to use combinations of these models, e.g., DSSAT and QUEFTS–DSSAT takes climate, water, N, and P into consideration, whereas QUEFTS considers N, P, and K, and their interactions. Other models may also be considered for inclusion in the toolbox, e.g., models predicting the evolution of soil organic matter.

To acquire an understanding for the possibilities and the limitations of the different models in "real-life" conditions, a village was selected where a number of farmers were monitored in collaboration with the Togolese Agricultural Research Institute and the Togolese Agricultural Extension Service. Two plots per crop were monitored per farmer — a plot where the advice of the extension service was followed and a plot where the farmer used the local practice. Crop, soil, and meteorological data were collected. The meteorological data were collected by means of a solar- powered weather station, and the crop and soil samples were analyzed at the soil laboratory of the Togolese Agricultural Research Institute.

Although the models correctly predicted the trends, precision was low, partly due to factors that were executed such as plot variability caused by the presence of trees, less than perfect weeding, insects and diseases. Another problem was the reliability of the soil data. Soil laboratories in these countries are often poorly equipped, and the results of analyses are not always reliable, a fact confirmed by comparing results with those obtained by other laboratories. This suggested that researchers should use more reliable laboratories and improve models to fit the conditions better; however, this would require a lot of time and funds beyond the resources available in these countries. It is therefore more useful to explore how models can be used knowing that model results are less then perfect.

In addition to field activities, workshops were organized for research and extension staff to become acquainted with models and to discuss the possibilities and the limitations of these models, to determine usefulness to their research programs and integration of models in their work. Workshop participants revealed a great interest in the use of models and learned of many opportunities to apply these models in their work. Results of the workshop pointed to the necessity of developing a unit that is acquiring and maintaining models of interest, promoting the use of models by the research staff, providing support to researchers who want to make use of models, as well as developing and maintaining a database on climate, soils, and crop varieties. It was suggested that the unit could be organized in a similar way as the existing units within the agricultural research institutes that provide support to the researchers regarding statistics. An additional conclusion was that more training is required and that modeling ought to be part of the curriculum of the university to create a critical mass of scientists in a country that is familiar with models. Similar workshops were later organized in Benin and Nigeria.

These conclusions resulted in the following activities:

- Short introductory courses were organized at the universities of Togo, Benin, and Ghana to acquaint students and staff with simulation modeling and to explore the possibility of including the subject in the regular course program.
- Modeling units will be established within the national agricultural research institutes of Togo and Benin, and further training will be provided in GIS and the application of models. Similar activities are envisaged in Ghana and Nigeria.
- The national research institutes were invited to identify *existing* research topics that may benefit from the use of models, which resulted in some interesting examples in cotton and maize production.

Balanced Fertilizer for Cotton

In Togo, the Society Togolese for Cotton Growers (SOTOCO) is responsible for the input supply for cotton. Until recently, the SOTOCO advised the farmers to apply compound fertilizer as a basal dressing followed by topdressing of urea, although farmers often fail to apply the top dressing, resulting in lower yields. The SOTOCO therefore decided to increase the N content of the compound fertilizer and to cancel the topdressing; however, during the tests only one formulation was used, implying a fixed relationship between N, P, and K. To evaluate the appropriateness of this relationship the QUEFTS model (Janssen et al., 1990) was used, using existing soil data of various areas in Togo and the minimum and maximum concentrations of N, P, and K found in cotton. The results suggested that the compound fertilizer is not well balanced and that a relative increase in N would benefit the areas in the center and the northern part of Togo. The new option is being tested in a few areas.

Maize Varieties and Sowing Time

The southern part of Togo experiences two rainy seasons; thus farmers can grow two crops of maize per year. Average rainfall varies from 700 to 1000 mm in the southern region and strongly varies from year to year; the rainfall in the second (short) season is especially unreliable. Most of the farmers cultivate local varieties although improved varieties are becoming more popular. The growth duration of these varieties is approximately 100 days. A problem here is that late sowing or an early onset of the (short or long) dry season may seriously affect production. In such cases, the availability of an earlier maturing variety could be helpful.

Recently, new maize varieties with a growth duration of 85 to 95 days were released. The question then is when farmers in a particular area should sow which variety. In case of early sowing, they should use the late-maturing variety for a high yield, but after a certain date it may be wise to use an earlier maturing variety to reduce the risk of a low yield caused by an early onset of the dry season. To assist in advising farmers in a particular area with planting date and varietal selection, researchers used DSSAT to calculate the yields for the different varieties over a large number of years in different areas. At the same time, trials were being conducted in two different villages (high and low rainfall), comparing the performance of the varieties at different sowing dates.

It should be emphasized that this is different from other research that aims at determining optimum sowing dates for maize (e.g., Wafula, 1995). That research seems less relevant since labor and other constraints often force farmers to sow on suboptimal dates. The present research does not advise the farmer when to sow but rather which variety to sow at a particular date.

SUMMARY

The IFDC focuses on increasing and sustaining food and agricultural production in developing countries through the establishment and transfer of effective and environmentally sound plant nutrition technology and agribusiness. To that end, the research priorities at IFDC have focused on the creation of sustainable technologies for the developing world and the transfer of that technology. A multidisciplinary research team approach was implemented at IFDC, utilizing agricultural economists, biometricians, crop physiology, and soil fertility modelers, and GIS specialists. The transfer of the work product technology from research activities was largely through systems analysis and modeling, via the IBSNAT project and associated global network of collaborators within the context of a decision support system, DSSAT. In this chapter, researchers have outlined efforts involved in the development of soil–crop simulation models and decision support systems, the use of these system tools in research to improve basic understanding, their applications at farm and global levels, and the training events that have taken place to promote the use and adoption of these approaches.

The research initiatives and case examples presented in this chapter represent only a sampling of the different types of applications that have been implemented by IFDC in conjunction with research partners. The significant effort involved in the development of soil-crop simulation models and decision support systems, the use of these system tools in research to improve basic understanding and the application of models at the farm and global level have been focused on training events that have promoted the use and adoption of the modeling approach. Although the use of models for decision making in agriculture has been limited in sub-Saharan Africa and South Asia, there is increasing scope and interest for further application of systems research; however, the introduction of this technology should consider the prevailing conditions in the countries such as limited knowledge of models, limited data availability, limited quality of data, limited resources, and the fact that environmental factors in agriculture are less well controlled.

The authors therefore postulate that the emphasis should be placed on the inclusion of system modeling in the regular curricula of the universities, coupled with the use of simple models, oriented toward an awareness of the limitations of the models, and using the models as a generator of ideas and insights, instead of uncritically using the results. It is important to use a variety of complementary models in order to allow users to appropriately address specific problems. It is essential that systems units be created at the research institutes that develop and maintain the toolbox, and that these units promote the use of models by the research staff and provide support to them.

DSSAT v3.5 with a CD User's Guide can be ordered from http://www.icasanet.org/dssat/getdssat.html. DREAM may be downloaded at http://www.ifpri.org/dream.htm. Information and support services on nutrient modeling, as well as information and decision support tools, can be obtained from research@ifdc.org. For information on modeling training programs, please contact hrdu@ifdc.org or visit http://www.ifdc.org/.

REFERENCES

Alston, J.M., G.W. Norton, and P.G. Pardey. 1998. *Science under Scarcity,* CAB International (CABI), Wallingford, U.K.

Baethgen, W.E. 1998a. El Niño and La Niña Impacts in Southeastern South America. In *Review on the Causes and Consequences of Cold Events: A La Niña Summit.* M. Glantz, Ed., NCAR, Boulder, CO, http://www.dir.ucar.edu/esig/lanina/.

Baethgen, W.E. 1998b. Applying scientific results in the agricultural sector: information and decision support systems, Proc. Inter-American Institute for Global Change Research (IAI) Science Forum: Global Change in the Americas, Arlington, VA, June 1998.

Baethgen, W.E. 1994. Impacts of climate change on barley in Uruguay: yield changes and analysis of nitrogen management systems. In *Implications of Climate Change for International Agriculture: Crop Modeling Study,* C. Rosenzweig and A. Iglesias, Ed., USEPA 230-B-94-003, Washington, D.C.

Baethgen, W.E., and G.O. Magrin. 2000. Applications of climate forecasts in the agricultural sector of South East South America. In *Climate Prediction and Agriculture,* M.V.K. Sivakumar, Ed., Proc. START/WMO Int. Workshop, Geneva, International START Secretariat, Washington, D.C.

Baethgen, W.E., and G.O. Magrin. 1995. Assessing the impacts of climate change on winter crop production in Uruguay and Argentina using crop simulation models. In *Climate Change and Agriculture: Analysis of Potential International Impacts.* C. Rosenzweig et al., Eds., American Society of Agronomy Special Publication 59, Madison WI, 207–228.

Baethgen, W.E., R. Faria, A. Giménez, and P. Wilkens. 1999. Information and decision support systems for the agricultural sector, *Third Int. Symp. on Systems Approaches for Agricultural Development,* Lima, Peru, November 8–10, 1999.

Bowen, W.T., P.K. Thornton, and G. Hoogenboom. 1998. The simulation of cropping sequences using DSSAT. In *Understanding Options for Agricultural Production,* G.Y. Tsuji, G. Hoogenboom, and P. K. Thornton, Eds., Kluwer Academic Publishers, Dordrecht, The Netherlands, 313–327.

Bowen, W.T. and P.W. Wilkens. 1998. Application of a decision support system (DSSAT) at the field level: nitrogen management in variable-charged soils. In *Information Technology as a Tool to Assess Land Use Options in Space and Time: Proc. Int. Workshop,* J.J. Stoorvogel, J. Bouma, and W.T. Bowen, Eds., September 28–October 4, 1997, CIP, Lima, Peru, Quantitative Approaches in Systems Analysis No. 16, Wageningen Agricultural University and AB-DLO, 23–31.

Breman, H. 1995. Modélisation et simulation dans l'élaboration de systèmes de production durables. In *Livestock and Sustainable Nutrient Cycling in Mixed Farming Systems of Sub-Saharan Africa,* Powell, J.M., S. Fernandez-Rivera, T.O. Williams and C. Renard, Eds., Volume II, Technical Papers, ILCA (International Livestock Centre for Africa), Addis Ababa, Ethiopia, 473–492.

Buresh, R.J., T. Woodhead, K.D. Shepherd, E. Flordelis, and R.C. Cabangon. 1989. Nitrate accumulation and loss in a mungbean/lowland rice cropping system, *Soil Sci. Soc. Am. J.,* 53:477–482.

Carberry, P.S. and D.G. Abrecht. 1991. Tailoring crop models to the semiarid tropics. In *Climate Risk in Crop Production: Models and Management for Semiarid Tropics and Subtropics,* R.C. Muchow and J.A. Bellamy, Eds., CAB International, Wallingford, U.K., 157–182.

De Wit, C.T. 1965. *Photosynthesis of Leaf Canopies, Agricultural Research Report 663,* PUDOC, Wageningen, The Netherlands.

ESRI. 1996. *Using the ArcView Spatial Analyst,* Environmental Systems Research Institute (ESRI), Redlands, CA.

Geng, S. Auburn, E. Brandsetter and B. Li. 1988. A program to simulate meterological variables: Documentation for SIMMETEO. Agronomy Progress Report No. 204. Dept. of Agronomy and Range Sci., Univ. of California, Davis, CA.

George, T., J.K. Ladha, D.P. Garrity and R.J. Buresh. 1994. Legumes as nitrate catch crops during the dry-to-wet season transition in lowland rice-based cropping systems, *Agron. J.,* 86:267–273.

Godwin, D.C. and U. Singh. 1998. Nitrogen balance and crop response to nitrogen in upland and lowland cropping systems. In *Understanding Options for Agricultural Production,* G.Y. Tsuji, G. Hoogenboom, and P.K. Thornton, Eds., Kluwer Academic Publishers, Dordrecht, The Netherlands, 55–78.

Godwin, D.C. and C.A. Jones. 1991. Nitrogen dynamics in soil-crop systems. In *Modeling Soil and Plant Systems. Agronomy Monograph 31,* R.J. Hanks and J.T. Ritchie, Ed., American Society of Agronomy, Madison, WI, 287–321.

Grondona, M.O., G.P. Podestá, M. Bidegain, M. Marino, and H. Hordij. 1999. A stochastic precipitation generator conditioned on ENSO phase: a case study in southeastern South America, *J. Climate,* 13: 2973–2986.

Hansen, J.W. 1999. Stochastic daily solar irradiance for biological modeling applications, *Agricultural and Forest Meteorol.,* 94:53–63.

Hansen, J.W. and T. Mavromatis, T. Correcting low-frequency bias in stochastic weather generators. *Agricultural and Forest Meteorol.* In press.

Hoogenboom, G., G.Y. Tsuji, J.W. Jones, U. Singh, D.C. Godwin, N.B. Pickering, and R.B. Curry. 1995. 1995. Decision support system to study climate change impacts on crop production. In *Climate Change and Agriculture: Analysis of Potential International Impacts,* C. Rosenzweig et al., Eds., ASA Spec. Publ. 59. ASA, Madison, WI, 51–75.

IFDC. 1997. Improving the efficiency and predictability of biological inhibitors to reduce nitrogen losses and enhance flooded rice productivity, *Termination Report for ACIAR Special Purpose Grant,* International Fertilizer Development Center, Muscle Shoals, AL.

IFDC. 1999. A client-oriented systems tool box for technology transfer related to soil fertility improvement and sustainable agriculture in West Africa, An Ecoregional-Fund Project, International Fertilizer Development Center, Muscle Shoals, AL.

Jallas, E. 1998. Improved Model-Based Decision Support by Modeling Cotton Variability and Using Evolutionary Algorithmes, Ph.D. Dissertation, Mississippi State University, Mississippi State, MS.

Janssen, B.H., F.C.T. Guiking, D. van der Eijk, E.M.A. Smaling, J. Wolf and H. van Reuler. 1990. A system for quantitative evaluation of the fertility of tropical soils (QUEFTS), *Geoderma,* 46:299–318.

Jones, P.G. and P.K. Thornton. 1993. A rainfall generator for agricultural applications in the tropics, *Agric. Forest Meteorol.,* 63:1–19.

Jones, P.G. and P.K. Thornton. 1997. Spatial and temporal variability of rainfall related to a third-order Markov model, *Agric. Forest Meteorol.,* 86:127–138.

Jones, P.G. and P.K. Thornton. 1999. Fitting a third-order Markov rainfall model to interpolated climate surfaces, *Agric. Forest Meteorol.,* 97:213–231.

Jones, P.G. and P.K. Thornton. 2000a. MarkSim: Software to generate daily weather data for Latin America and Africa, *Agron. J.,* 92:445–453.

Jones, P.G. and P.K. Thornton. 2000b. Spatial modeling of risk in natural resource management applying plot-level, plant-growth modeling to regional analysis. Paper presented at the Workshop on Integrated Natural Resource Management, Penang, Malaysia. August 2000. Submitted to Conservation Ecology (in review).

Keating B.A., and H. Meinke. 1998. Assessing exceptional drought with a cropping systems simulator: A case study for grain production in north-east Australia, *Agric. Syst.,* 57:315–332.

Keating, B.A., R.L. McCown, and B.M. Wafula. 1993. Adjustment of nitrogen inputs in response to a seasonal forecast in a region of high climatic risk. In *Systems Approach for Agricultural Development. Proc. Int. Symp. on Systems Approach for Agricultural Development,* F.P.de Vries et al., Eds., December 2–6, 1991, Bangkok, Thailand, 233–252.

Keating, B.A., D.C. Godwin, and J.M. Watiki. 1991. Optimising nitrogen inputs in response to climatic risk. In *Climate Risk in Crop Production: Models and Management for Semi-arid Tropics and Subtropics.* R.C. Muchow and J.A. Bellamy, Eds., CAB International, Wallingford, U.K., 329–358.

Magrin G.O., M.I. Travasso J.W. Jones, G.R. Rodriguez , and D.R. Boullón. 1999. Using climate forecast in agriculture: a pilot application in Argentina, *Third Int. Symp. on Systems Approaches for Agricultural Development (SAADIII),* November 8–10, Lima, Peru.

Matthews, R., W. Stephens, T. Hess, T. Mason and A. Graves. 2000. Applications of crop/soil simulation models in developing countries. Final Report. Institute of Water and Environment, Cranfield University, Silsoe, Bedfordshire MK45 4DT, U.K.

Mavromatis, T., and J.W. Hansen. Interannual variability characteristics and simulated crop response of four stochastic weather generators, *Agricultural and Forest Meteorol.* In press.

Meinke, H., W.E. Baethgen, P.S. Carberry, M. Donatelli, G.L. Hammer, R. Selvaraju, and C.O. Stockle. 2001. Increasing profits and reducing risks in crop production using participatory systems simulation approaches, *Agric. Syst,* 70:493–513.

Messina, C.D. 1999. El Niño–Southern oscillation y la productividada de cultivos en la zona Pampeana: Evaluación de estragias para mitigar el riesgo climático. Tesis Magister Scientiae, Producción Vegetal. Escuela para Graduados, Facultad de Agronomia, University of Buenos Aires, Argentina. 132 pp.

Mohanty, S.K., U. Singh, V. Balasubramanian, and K. P. Jha. 1999. Nitrogen deep-placement technologies for productivity, profitability, and environmental quality of rainfed lowland rice systems, *Nutrient Cycling in Agroecosystems,* 53:43–57.

Newman, S., T. Lynch, and A.A. Plummer. 1999. Aspects of success and failure of decision support systems in agriculture. In *Inaugural Australian Workshop on the Application of Artificial Intelligence, Optimisation and Bayesian Methods in Agriculture,* Abbass, H.A. and M. Towsey, Eds., Queensland University of Technology Press, Brisbane, Australia, 17–30.

O'Meagher, B., L.G. du Pisani, and D.H. White. 1998. Evolution of drought policy and related science in Australia and South Africa, *Agric. Syst.,* 57:231–258.

Osmond, D.L., T.J. Smyth, R.S. Yost, W.S. Reid, W. Branch and X. Wang. 2000. *Nutrient Management Support System (NuMaSS) Version 1.5. Software Installation and User's Guide,* Soil Management Collaborative Research Support System Technical Bulletin No. 2000–02. North Carolina State University, Raleigh, NC.

Padilla, J.L., R.J. Buresh, S.K. De Datta, and E.U. Bautista. 1990. Incorporation of urea in puddled rice soils as affected by tillage implements, *Fertilizer Res.,* 26:169–178.

Parker, C. 1999. Decision support systems: lessons from past failures, *Farm. Manage.,* 10:273–289.

Patil, S.K., U. Singh, C.J. Neidert, and P.W. Wilkens. 2002. Integrated information and decision support toolbox for nutrient management in Chhattisgarh, India: Soil, climate, and crop database. International Fertilizer Development Center, Muscle Shoals, AL. 120 p.

Pickering, N.B. et al. 1994. WeatherMan: a utility for managing and generating daily weather data, *Agron. J.,* 86:332–337.

Podestá, G.P., C.D. Messina, M.O. Grondona, and G.O. Magrin. 1998. Associations between grain crop yields in central-eastern Argentina and El Niño-Southern Oscillation, *J. Applied Meteorol.,* 38:1488–1498.

Reilly, J., W.E. Baethgen, F.E. Chege, S.C. van de Geijn, Lin Erda, A. Iglesias, G. Kenny, D.Patterson, J. Rogasik, R. Ritter, C. Rosenzweig, W. Sombroek and J. Westbrook. 1996. Agriculture in a changing climate: impacts and adaptation. In *Changing Climate 1995: Impacts, Adaptation and Mitigation of Climate Change: Scientific-Technical Analyses,* Report of Working Group II of the Intergovernmental Panel on Climate Change. Cambridge University Press, London. 427–467.

Richardson, C.W. and D.A. Wright. 1984. *WGEN: A Model for Generating Daily Weather Variables,* United States Department of Agriculture, Agricultural Research Service, ARS-8, Washington, D.C.

Rosenzwieg, C., and A. Iglesias. 1998. The use of crop models for international climate change impact assessment. In *Understanding Options for Agricultural Production,* G.Y. Tsuji, G. Hoogenboom, and P.K. Thornton, Eds., Kluwer Academic Publishers, Dortrecht, The Netherlands, 267–292.

Rotter, R. 1993. Simulation of the biophysical limitations to maize production under rainfed conditions in Kenya, Ph.D. dissertation, University of Trier and Wageningen Agricultural University. The Netherlands.

Singh, U., P.W. Wilkens, V. Chude, and S. Oikeh. 1999. Predicting the effect of nitrogen deficiency on crop growth duration and yield. In *Proc. Fourth Int. Conf. on Precision Agriculture,* ASA-CSSA-SSSA, Madison, WI, 1379–1393.

Singh, U., and G. Uehara. 1998. Electrochemistry of the double-layer: principles and applications to the soils. In *Soil Physical Chemistry,* 2nd edition, D.L. Sparks, Ed., CRC Press, Boca Raton, FL, 1–46.

Singh, U., and J.L. Padilla. 1995. Simulating rice response to climate change. In *Climate Change and Agriculture: Analysis of Potential International Impacts,* C. Rosenzweig et al., Eds., ASA Spec. Publ. 59. ASA, Madison, WI, 99–121.

Singh, U., and T. Timsina. 1994. Rice-wheat systems: problems, constraints and modelling issues. In *Agroecological Zonation, Characterization and Optimization of Rice-Based Cropping Systems, SARP Research Proc.,* F.P. Lansigan, B.A.M. Bouman, and H.H. Laar, Eds., DLO, Wageningen, The Netherlands, 47–57.

Singh, U., and J.T. Ritchie. 1993. Simulating the impact of climate change on crop growth and nutrient dynamics using the CERES-Rice Model, *J. Agric. Meteorol.,* 48(5):819–822.

Singh, U., J.E. Brink, P.K. Thornton, and C.B. Christianson. 1993a. Linking crop models with a geographic information system to assist decisionmaking: a prototype for the semiarid tropics, IFDC Paper Series, P-19, International Fertilizer Development Center, Muscle Shoals, AL.

Singh, U., P.K. Thornton, A.R. Saka, and J.B. Dent. 1993b. Maize modeling in Malawi: a tool for soil fertility research and development. In *Systems Approaches for Agricultural Development,* F.W.T. Penning de Vries et al., Eds., Kluwer Academic Publishers, Dordrecht, The Netherlands, 253–273.

Singh, U., and P.K. Thornton. 1992. Using crop models for sustainability and environmental quality assessment, *Outlook Agric.,* 21:209–218.

Singh, U., D.C. Godwin, and R.J. Morrison. 1990. Modelling the Impact of Climate Change on Agricultural Production in the South Pacific. In *Global Warming-Related Effects on Agriculture and Human Health and Comfort in the South Pacific,* Philip J. Hughes and Glenn McGregor, Eds., University of Papua New Guinea, Port Moresby, 24–40.

Stoorvogel, J.J., and E.M.A. Smaling. 1990. Assessment of soil nutrient depletion in Sub-Saharan Africa: 1983–2000, Winand Staring Centre, Report 28, Vol. 1, Wageningen, The Netherlands.

Thornton, P.K., and P.W. Wilkens. 1998. Risk assessment and food security. In *Understanding Options for Agricultural Production,* G.Y. Tsuji, G. Hoogenboom, and P.K. Thornton, Eds., Kluwer Academic Publishers, Dordrecht, The Netherlands, 339–345.

Thornton, P.K., W.T. Bowen, A.C. Ravelo, P.W. Wilkens, G. Farmer, J. Brock, and J.E. Brink. 1997. Estimating millet production for famine early warning: an application of crop simulation modeling in Burkina Faso, *Agric. and Forest Meteorol.,* 83:95–112.

Thornton, P.K., G. Hoogenboom, P.W. Wilkens, and W.T. Bowen. 1995a. A computer program to analyze multiple-season crop model outputs, *Agron. J.,* 87:131–136.

Thornton, P.K., A.R. Saka, U. Singh, J.D.T. Kumwenda, J.E. Brink, and J.B. Dent. 1995b. Application of a maize crop simulation model in the central region of Malawi, *Experimental Agric.,* 213–226.

Thornton, P.K. and G. Hoogenboom. 1994. A computer program to analyze single-season crop model outputs, *Agron. J.,* 86:860–868.

Thornton, P.K., G. Hoogenboom, P.W. Wilkens, and J.W. Jones. 1994a. Seasonal analysis. In *DSSAT v3,* Vol. 3. G.Y. Tsuji, G. Uehara and S. Balas, Eds., University of Hawaii, Honolulu, HI, 1–66.

Thornton, P.K., P.W. Wilkens, G. Hoogenboom, and J.W. Jones. 1994b. Sequence analysis. In *DSSAT v3,* Vol. 3, G.Y. Tsuji, G. Uehara and S. Balas, Eds., University of Hawaii, Honolulu, HI, 67–136.

Timsina, J. et al. 1998. Cultivar, nitrogen, and moisture effects on a rice-wheat sequence: experimentation and simulation, *Agron. J.,* 90:119–130.

Timsina, J., U. Singh, and Y. Singh. 1997. Addressing sustainability of rice-wheat systems: analysis of long-term experimentation and simulation. In *Applications of Systems Approaches at the Field Level,* M.J. Kropff et al., Eds., Kluwer Academic Publishers, Dordrecht, The Netherlands, 383–397.

Timsina, J., U. Singh, Y. Singh, and F.P. Lansigan. 1995. Addressing sustainability of rice-wheat systems: testing and application of CERES and SUCROS models. In *Fragile Lives in Fragile Ecosystems, Proc. Int. Rice Research Conference,* International Rice Research Institute, Philippines, 633–656.

Travasso, M.I., G.O. Magrin, and M.O. Grondona. 1999. Relations between climatic variability related to ENSO and maize production in Argentina. In *Proc. 10th AMS Symp. on Global Change Studies,* January 10–15, 1999, Dallas, TX. 67–68.

Tsuji, G.Y., G. Uehara, and S. Balas. 1994. *Decision Support System for Agrotechnology Transfer version 3 (DSSAT v3),* Dept. of Agron. and Soil Sci., College of Trop. Agric. and Human Resour., University of Hawaii, Honolulu, HI.

Wafula, B.M. 1995. Applications of crop simulation in agricultural extension and research in Kenya, *Agric. Syst.,* 49, 399–412.

White, D.H., S.M. Howden, J.J. Walcott, and R.M. Cannon. 1998. A framework for estimating the extent and severity of drought, based on a grazing system in south-eastern Australia, *Agric. Syst.,* 57:259–270.

Wilkens, P, S. Wood, M. Rivera, and S. Daroub. 2000. Linking biophysical and economic models to evaluate the impact of agricultural research, 92nd Annual Meeting Abstracts, Minneapolis, MN, 71.

Wood, S., L. You and W. Baitx. 2001. *DREAM User Manual,* International Food Policy Research Institute, Washington, D.C.

CHAPTER 7

An Evaluation of RZWQM, CROPGRO, and CERES-Maize for Responses to Water Stress in the Central Great Plains of the U.S.

Liwang Ma, David C. Nielsen, Lajpat R. Ahuja, James R. Kiniry, Jonathan D. Hanson, and Gerrit Hoogenboom

CONTENTS

INTRODUCTION

Simulation of crop production is one of the most difficult tasks in agricultural system models because of its dependence on simulations of soil water, soil nutrient, diseases, and weeds. When there is a disagreement between simulated and observed crop production, it can be difficult to identify the source of the problem(s). Discrepancies may be due to:

1. Inaccurate soil water and nutrient simulation
2. Lack of model sensitivity to plant environmental stresses
3. Unrecorded damage from natural disasters, extreme weather events, pests, diseases, and weeds
4. Variability in field measurements
5. Lack of accuracy in model parameters or processes simulation

Based on the understanding of a biological process, data availability, and experimental conditions, there can be more than one modeling approach to describe an experimental phenomenon. One approach may work better than another for certain experimental conditions. Crop modeling has not matured to a point where one model can be used for all combinations of environmental and experimental conditions.

Because plants are a central component of an agricultural production system, all models have plant growth components that are either simple or complex, and either generic or crop specific. For example, the USDA-ARS Root Zone Water Quality Model (RZWQM), which is available through the Water Resources Publications (*www.wrpllc.com*), was developed primarily for water quality applications in the 1990s. It has a generic crop growth component so that management effects on both water quality and crop yield can be simulated. Although this crop growth component does not predict detailed phenology and yield components, it does divide plant growth into seven stages: dormant seeds, germinating seeds, emerged plants, established plants, plants in vegetative growth, plants in reproductive stage, and senescent plants (Hanson, 2000; Hanson et al., 1999). Another unique feature of the generic plant growth model is that it assumes nonuniform plant population development. In other words, not all the plants in a field are at the same growth stage. Another group of agricultural system models is more focused on crop production, such as the CROPGRO and CERES family of models. These models have detailed plant growth processes and simulate each yield component. CROPGRO is for legume crops and CERES is for cereal crops (Boote et al., 1998, Ritchie et al., 1998). The CROPGRO and CERES models are distributed through the DSSAT (Decision Support System for Agrotechnology Transfer) package (*www.icasanet.org/dssat/getdssat.html*).

The objective of this chapter is to compare RZWQM, CERES, and CROPGRO for simulating crop production in the Central Great Plains of the U.S.A. The second objective is to determine how much benefit we gain from added details in crop growth and development components (e.g., CROPGRO and CERES-Maize) or soil water balance components (e.g., RZWQM). Applications of the three models for other aspects of agricultural systems are available from Ma et al. (2000), Ma et al. (2001), Singh et al. (2002), Kiniry et al. (2002), and Tsuji et al. (1998, 2002). CROPGRO has the most detail for simulating biological processes. RZWQM has the most detailed components for simulating soil water, nutrients, pesticides, and management practices (Ahuja et al. 2000). CROP-GRO and CERES share the same soil water and nitrogen components (Ritchie, 1998). For the processes considered in all of the three models, methods for simulating the processes differ (e.g., photosynthesis, water uptake, and N uptake). Also, the response of plants to environmental stresses is simulated differently among the models, both in the way of quantifying stresses and in the way the processes are affected by the stresses (Ritchie et al., 1998; Boote et al., 1998; Hanson, 2000).

Data sets for corn and soybean from the USDA-ARS, Central Great Plains Research Station in Akron, CO, were used to evaluate the three models. These data were collected from 1984 to 1986 under different irrigation treatments (e.g., gradient line source, drip, and rain shelter system). Because RZWQM has a generic crop growth component and is parameterized for both corn and soybean, it was compared to CROPGRO for soybean production and CERES-Maize for corn production. All the models were run under an assumed nonlimited nitrogen condition. An evaluation for different N conditions is reported in Ma et al. (2002). The purposes of the evaluations presented in this chapter are to:

1. Compare plant responses to water stress simulated in the three models.
2. Demonstrate the applications of agricultural system models for field research.

EXPERIMENT DESCRIPTION

Maize Experiment Design

Gradient Line-Source Irrigation System

Studies were conducted during the 1984, 1985, and 1986 growing seasons at the USDA Central Great Plains Research Station, 6.4 km east of Akron, CO (40° 9' N, 103° 9' W, 1384 m elevation). The soil type is a Rago silt loam (fine smectitic, mesic Pachic Argiustolls). Soil texture was analyzed with the hydrometer method (Gee and Bauder, 1986) (Table 7.1). Although it is called a Rago silt loam, measured soil texture indicates that the soil is closer to a loam than a silt loam. Therefore, when using soil texture to estimate hydraulic properties, we used the measured soil texture (loam) rather than the soil mapped texture.

Corn was planted on May 14, 1984, May 3 1985, and May 1, 1986, with corresponding seeding densities of 72,400; 76,100; and 76,100 seeds/ha. Prior to each planting, the plot area was fertilized with ammonium nitrate at a rate of 168 kg N/ha. Corn (Pioneer Hybrid 3732) was grown under a line-source gradient irrigation system, with full irrigation next to the irrigation line, and linearly declining water application as distance increased away from the line. Details regarding the irrigation system can be found in Nielsen (1997). Four replications of four irrigation levels (only three levels in 1984) existed along the line-source system, with a soil water measurement site and irrigation catch gauge at each of the 16 locations (12 locations in 1984). Irrigations were initiated just prior to tasseling (stage VT, Ritchie et al., 1986) in each year. A total numbers of five irrigation events were run from July 20 to September 2, 1984; 11 from June 29 to August 22, 1985; and 10 from July 21 to August 26, 1986. Total irrigation water applied ranged from 2.3 to 10.6 cm in 1984, 7.1 to 18.9 cm in 1985, and 14.6 to 30.0 cm in 1986 (Table 7.2). Corn was harvested on October 1, 1984; September 27, 1985; and October 15, 1986.

Table 7.1 Measured Soil Texture of the Rago Silt Loam

Soil Depth (cm)	Bulk Density (g/cm³)	Sand	Silt (%)	Clay	Drainage Limit[a] (cm³/cm³) Upper	Lower	Saturated Hydraulic Conductivity (cm/hr) Soil Texture[b]	Effective Porosity[c]
0–30	1.33	39.0	41.7	19.3	0.224	0.092	1.32	10.67
30–60	1.33	32.3	44.3	23.3	0.236	0.104	1.32	9.32
60–90	1.36	37.0	40.7	22.3	0.230	0.098	1.32	10.04
90–120	1.40	45.7	36.7	17.7	0.221	0.090	1.32	9.80
120–150	1.42	45.7	42.3	12.0	0.215	0.084	1.32	8.75
150–180	1.42	48.0	41.7	10.3	0.212	0.081	1.32	8.20

[a] Calculated from Ritchie et al. (1999), *Trans ASAE* 42:1609–1614, the upper limit was assumed as soil water content at 33 kPa and the lower limit as soil water content at 1500 kPa in RZWQM.
[b] Estimated from Rawls et al. (1982), *Trans ASAE* 25:1316–1320, and used for soybean field.
[c] Estimated from Ahuja et al. (1989), *Soil Sci.* 148:404–411, and used for corn field.

Table 7.2 Irrigation Timing and Amount (cm) for the Line-Source (LS) Gradient Irrigation System in Corn Production from 1984 to 1986

Date (1984)	Irrigation Level 1	2	3	Date (1985)	Irrigation Level 1	2	3	4	Date (1986)	Irrigation Level 1	2	3	4
7/20	0.61	1.98	3.05	6/29	0.33	0.28	0.36	0.48	7/21	1.19	1.83	2.36	2.79
7/30	0.43	1.22	2.03	7/30	3.40	3.35	5.84	7.62	7/23	1.27	1.96	2.44	2.84
8/20	0.43	1.07	1.68	8/8	0.28	0.46	0.71	0.86	7/25	1.19	1.68	2.34	3.10
8/25	0.25	0.81	1.37	8/9	0.08	0.13	0.18	0.23	7/29	1.22	1.55	1.85	2.08
9/21	0.58	1.73	2.49	8/12	0.79	1.35	2.08	2.54	8/4	1.80	2.26	2.59	2.54
				8/14	0.30	0.53	0.84	1.02	8/6	1.75	2.49	3.02	3.02
				8/15	0.53	0.89	1.40	1.68	8/12	1.17	1.52	1.96	2.26
				8/17	0.28	0.48	0.74	0.91	8/19	1.32	1.88	2.62	3.33
				8/19	0.38	0.64	0.99	1.19	8/20	1.55	1.91	2.79	4.01
				8/20	0.36	0.61	0.97	1.14	8/26	2.18	3.23	3.81	4.04
				8/20	0.38	0.64	0.99	1.19					
Total	2.30	6.81	10.62		7.11	9.38	15.04	18.85		14.64	20.31	25.78	30.00

Soil water content was measured at planting and harvest and at several intermediate dates during the growing season in 1985. These measurements were made at 15, 45, 75, 105, 135, and 165 cm depths below the soil surface with a neutron probe calibrated against soil water samples taken at the time of access tube installation. Crop water use was calculated as the difference between successive soil water measurements plus precipitation and irrigation during the sampling interval. Deep percolation and runoff were assumed to be negligible.

Leaf area measurements were made periodically during the growing season by destructively sampling a 1-m row, separating leaves from the stalks, and measuring the leaf area with a leaf area meter (LI-Cor LI-3100, Lincoln, NE). An automated weather station recorded air temperature, wind run, solar radiation, rainfall, and humidity. Total annual rainfall was 47.2, 45.4, and 33.0 cm for 1984, 1985, and 1986, respectively. Total growing season rainfall (May through September) was 29.9, 31.7, and 20.5 cm for the 3 years.

Drip Irrigation System

This study was conducted only in the 1985 growing season. The plot area was fertilized prior to planting with 184 kg/ha N. Corn was planted similarly to the gradient line-source irrigation area. Corn was planted on May 9, 1985, and the final population was 74,100 plants/ha. Irrigations were applied according to four levels of the Crop Water Stress Index (CWSI) as determined by canopy temperature (Nielsen and Gardner, 1987). Irrigations were applied through drip tubing at a rate of 0.32 cm hr^{-1} when CWSI exceeded levels of 0.1 0.2, 0.4, or 0.6 (where 0.0 = no water stress, 1.0 = maximum water stress) (Table 7.3). Plants were harvested on September 27, 1985.

Soybean Experiment Design

Studies were conducted during the 1985 and 1986 growing seasons at the same station in Akron, CO. Three experiments were conducted to provide a range of available water conditions in which to evaluate water stress effects on soybean productivity. The experiments varied in the method of water application, and will be referred to as the gradient line-source irrigation experiment (LS), the rain shelter experiment (RS), and the drip irrigation experiment (Drip). Details of some cultural practices are given in Table 7.4 and irrigation amounts are shown in Table 7.5. Other details for each experiment are provided below. In all experiments the soybean variety was Pioneer Brand 9291 (late-maturity group II).

Table 7.3 Irrigation Timing and Amount (cm) for the Drip Irrigation Study for Corn Production in 1985

Date	Irrigation Level			
	1	2	3	4
7/9	—	0.48	0.48	0.48
7/10	—	2.26	2.26	2.26
7/13	—	—	—	1.55
7/14	—	—	—	1.12
7/15	—	—	1.30	—
7/16	—	1.30	2.46	1.30
7/25	—	—	0.91	0.91
8/6	—	1.07	1.07	1.07
8/8	1.57	—	1.57	1.57
8/14	—	—	1.60	1.60
8/15	0.94	0.94	—	0.94
8/16	1.47	1.47	—	1.47
8/21	—	—	2.06	2.06
8/27	2.46	2.46	2.46	2.46
8/29	—	—	2.31	2.31
8/30	—	1.19	—	—
Total	6.44	11.13	18.48	20.70

Table 7.4 Cultural Practices for Soybean Experiments

Experiment	Year	Planting Date	Harvest Date	Row Spacing (m)	Plot Dimensions (m)	Population (plants/ha)	Irrigation Method
Solid Set	1985	23 May	03 Oct	0.76	4.1 × 12.2	375,600	Overhead impact sprinklers
Solid Set	1986	20 May	25 Sep	0.76	4.1 × 12.2	262,200	Overhead impact sprinklers
Rain Shelter	1985	28 May	31 Sep	0.53	2.7 × 2.7	331,100	Flood
Rain Shelter	1986	20 May	25 Sep	0.53	2.7 × 2.7	397,600	Flood
Drip	1986	20 May	25 Sep	0.76	4.6 × 9.0	271,100	Drip

Gradient Line-Source Irrigation System

This experiment was conducted as a limited irrigation study, with irrigations applied from June 23 to August 28 in 1985, and from June 26 to August 25 in 1986. Most of the irrigations were applied in the last half of the growing season (flowering and grain-filling). Irrigations were applied with a line-source gradient irrigation system, with full irrigation next to the irrigation line, and linearly declining water application as distance increased away from the line. Details regarding the irrigation system can be found in Nielsen (1997). Four irrigation levels existed along the line-source system. These four levels were replicated twice in 1985 and four times in 1986. A soil water measurement site and irrigation catch gauge was located at the center of each plot. There were seven irrigations in 1985 and nine irrigations in 1986 (Table 7.5).

Rain Shelter and Drip Irrigated Experiments

Details for these experiments are found in Nielsen (1990). Briefly, both experiments had four levels of irrigation determined by four threshold levels of the CWSI, which was computed from

Table 7.5   Irrigation Seasonal Amounts (cm) in the Akron, Colorado Soybean Study in 1985 and 1986

	Irrigation Time	Irrigation Amount (cm)			
		1	2	3	4
Gradient line-source	6/22	0.00	1.64	3.47	4.20
irrigation system (1985)	8/17	0.00	0.61	1.54	3.64
	8/21	0.17	0.60	1.07	1.50
	8/23	0.10	0.05	0.90	1.27
	8/26	0.00	0.06	1.04	1.31
	8/28	0.01	0.42	0.85	1.00
	Total	0.28	3.38	8.86	12.92
Gradient line-source	6/25	0.07	0.83	2.56	3.33
irrigation system (1986)	6/27	0.07	0.83	2.56	3.33
	7/01	0.08	0.83	2.56	3.34
	7/30	0.23	0.77	1.71	2.41
	8/07	0.06	0.56	1.16	1.85
	8/11	0.04	0.35	0.71	1.13
	8/15	0.08	0.77	1.74	3.05
	8/21	0.45	1.15	2.07	3.30
	8/25	0.45	1.13	2.04	3.25
	Total	1.55	7.22	17.11	24.98
Rain shelter irrigation	5/29	5.08	5.08	5.08	5.08
system (1985)	6/24	1.69	3.39	5.08	1.69
	6/25	3.39	1.69	0.00	3.39
	7/10	5.08	5.08	5.08	5.08
	7/24	0.00	0.00	2.54	5.08
	7/30	2.54	2.54	2.54	5.08
	8/06	1.69	1.69	2.54	0.85
	8/08	0.00	0.85	0.00	1.69
	8/09	3.39	2.54	4.23	3.39
	8/12	0.85	0.85	0.85	0.85
	8/14	0.00	0.00	0.85	0.00
	8/15	1.69	1.69	0.85	2.54
	8/16	0.00	0.00	0.85	0.00
	8/20	1.69	1.69	1.69	2.54
	8/21	0.00	0.85	0.00	0.85
	8/23	2.54	1.69	3.39	1.69
	8/26	0.00	0.85	0.85	0.85
	8/27	0.00	0.85	0.00	0.85
	8/28	0.85	0.00	0.85	0.00
	8/29	1.69	0.85	0.00	0.00
	8/30	0.00	0.00	2.54	5.08
	9/05	2.54	2.54	1.69	2.54
	9/06	0.00	0.00	0.85	0.85
	Total	34.71	34.71	42.33	49.95
Rain shelter irrigation	6/19	5.08	5.08	5.08	5.08
system	7/02	5.08	5.08	5.08	5.08
	7/11	5.08	5.08	5.08	5.08
	7/15	0.00	0.00	5.08	5.08
	7/18	5.08	5.08	0.00	0.00
	7/23	0.00	0.00	5.08	5.08
	7/25	5.08	5.08	0.00	0.00
	7/29	0.00	0.00	5.08	5.08
	8/01	5.08	5.08	0.00	0.00
	8/04	0.00	0.00	0.00	5.08
	8/06	0.00	0.00	5.08	0.00
	8/11	5.08	5.08	0.00	5.08

Table 7.5 (continued)   Irrigation Seasonal Amounts (cm) in the Akron, Colorado Soybean Study in 1985 and 1986

	Irrigation Time	Irrigation Amount (cm)			
		1	2	3	4
	8/14	0.00	0.00	5.08	5.08
	8/21	5.08	5.08	0.00	0.00
	8/25	0.00	0.00	5.08	0.00
	8/27	0.00	0.00	0.00	5.08
	8/28	5.08	5.08	0.00	0.00
	9/03	0.00	0.00	5.08	5.08
	9/04	0.00	5.08	0.00	0.00
	Total	45.72	50.8	50.8	55.88
Drip irrigation system	7/18	0.00	0.00	0.00	1.32
(1986)	7/21	0.00	0.00	0.00	1.17
	7/23	2.26	2.26	2.26	2.26
	7/25	0.00	3.33	3.32	0.00
	7/30	0.00	0.00	0.00	2.90
	8/01	1.96	1.96	1.96	0.00
	8/08	2.64	0.00	2.64	2.64
	8/12	0.00	4.65	0.00	0.00
	8/18	2.62	0.00	2.62	2.62
	8/26	2.54	2.54	2.54	2.54
	9/04	0.00	2.69	2.69	2.69
	9/05	2.51	0.00	0.00	0.00
	Total	14.52	17.42	18.02	18.13

Table 7.6   Calibrated Plant Model Parameter Values of RZWQM for Corn and Soybean. (Parameters with asterisk are suggested calibration parameters by the model developers.)

Parameter Name	Corn	Soybean
Minimum leaf stomatal resistance (s/m)[a]	100	100
Proportion of photosynthate lost to respiration (dimensionless)[a]	0.28	0.17
Photosynthesis rate at reproductive stage compared with vegetative stage[a]	61	69
Photosynthesis rate at seeding stage compared with vegetative stage[a]	61	69
Coefficient to convert leaf biomass to leaf area index, CONVLA (g/LAI)[a]	15.5	1.9
Plant population on which CONVLA is based (plants/ha)[a]	68,992	370,137
Maximum rooting depth (cm)[a]	300	300
Maximum plant height (cm)	210	70
Aboveground biomass at ½ maximum height (gm)	60	4
Aboveground biomass of a mature plant (gm)	152	13
Minimum time needed from planting to germination (days)	5	3
Minimum time needed from planting to emergence (days)	20	7
Minimum time needed from planting to 4-leaf stage (days)	35	22
Minimum time needed from planting to end of vegetative growth (days)	75	62
Minimum time needed from planting to physiological maturity (days)	115	92
Growth stage advanced from planting to germination (dimensionless)	0.0356	0.0356
Growth stage advanced from planting to emergence (dimensionless)	0.065	0.065
Growth stage advanced from planting to 4-leaf stage (dimensionless)	0.20	0.20
Growth stage advanced from planting to end of vegetative growth (dimensionless)	0.75	0.75
Growth stage advanced from planting to physiological maturity (dimensionless)	0.90	0.90

[a] Model developers' suggested calibration parameter.

crop canopy temperatures measured daily with an infrared thermometer. In both experiments, the irrigation treatments were laid out in a randomized complete block, with three replications in the rain shelter and five replications in the drip irrigated experiment. Irrigations were flood-applied in the rain shelter. In the drip-irrigated plots, irrigations were applied through drip irrigation tubing laid on the surface of every other interrow space. Timing and amount of each irrigation event are shown in Table 7.5.

Data Needed for the Three Models

RZWQM

RZWQM requires daily weather data for minimum and maximum daily temperature, wind run, solar radiation, relative humidity, and rainfall. These data were available from an on-site weather station. Soil texture and bulk density were determined from soil samples taken in the field (Table 7.1). The model requires a minimum input of soil water content at 33 kPa suction, which is estimated from Ritchie et al. (1999) by assuming that it is the drained upper limit. Saturated soil hydraulic conductivity was calculated from effective porosity (Ahuja et al., 1989) for the corn fields or soil texture mean values (Rawls et al., 1982; 1998) for the soybean fields. RZWQM uses the Brooks–Corey equations to describe the soil water retention curve, and the required parameters were estimated from soil texture classes (Rawls et al., 1982) and scaling with respect to bulk density and 33 kPa water content (Ahuja et al., 2000). Corn growth parameters were based on Farahani et al. (1999), with slight modification (Ma et al. 2002). Soybean parameters were based on model testing in Ohio (Landa et al., 1999), Missouri (Ghidey et al., 1999), and Iowa (Jaynes and Miller, 1999). Initial soil water content in the profile was assumed to be at field capacity, and the models were run from January 1 to December 31 every year.

CERES-Maize and CROPGRO-Soybean

Researchers used the versions included in the Decision Support System for Agrotechnology Transfer (DSSAT) family models (version 3.5) (Hoogenboom et al., 1999; Tsuji et al., 1994). CERES-Maize and CROPGRO-Soybean use the same soil water balance component, which requires weather data of minimum and maximum daily temperature, solar radiation, and rainfall. Measured soil texture and bulk density were used (Table 7.1). Drained upper and lower limits were calculated as suggested by Ritchie et al. (1999). Required corn and soybean growth parameters were calibrated as suggested in the DSSAT manual (Boote, 1999, Hoogenboom et al., 1994). Initial soil water content in the profile was assumed to be at field capacity and the models were run from January 1 to December 31 every year.

APPLICATIONS OF RZWQM AND CERES-MAIZE FOR CORN

Calibration of RZWQM and CERES-Maize

Calibration of RZWQM for Corn

Calibration of RZWQM followed the methods suggested by Hanson et al. (1999) and Rojas et al. (2000). Data from the 1985 line-source irrigation system were used for calibration because of its frequent soil water measurements. As suggested by Boote (1999), the authors selected data from the highest irrigation level (or least stress) as the calibration dataset (level 4 in Table 7.2), although all irrigation levels were not irrigated for full crop water use. In addition, they assumed that corn was not under N stress. Goodness-of-fit for the model calibration was based on a comparison of measured and simulated soil water content, estimated evapotranspiration (ET), leaf area index (LAI), plant height, plant biomass, and harvest grain yield. Root mean square errors (RMSE) were also calculated as an indication for model accuracy.

Table 7.1 lists the measured soil texture, bulk density, and estimated 33 kPa soil water contents from Ritchie et al. (1999). Estimated 33-kPa soil water contents were very close to the 33 kPa value of 0.233 cm^3 cm^{-3} given by Rawls et al. (1982) for a loam soil. In addition, the RMSE indicated that soil water contents were better predicted with saturated soil hydraulic conductivity estimated from effective porosity (Ahuja et al., 1989). Plant growth parameters were calibrated from previous work in Colorado (Farahani et al. 1999), and calibrated values are listed in Table 7.6. Minimum leaf stomatal resistance was set from 250 s/m to 100 s/m based on literature reports (Fiscus et al., 1991, Bennett et al., 1987). Aboveground biomass for a mature plant changed from 70 to 152 g, based on experimental measurements. Maximum rooting depth extended from 180 cm to 300 cm to accelerate root growth without changing other model parameters. Maximum plant heights were 210 cm instead of the 250 cm as calibrated by Farahani et al. (1999). Minimum days from planting to physical maturity was set to 115 in the model under optimal growth conditions, which was reasonable compared to observed actual life spans of 127 to 158 days under semiarid Colorado conditions.

Predicted soil water content and soil water storage are shown in Figure 7.1 with RMSE of 0.023 cm^3 cm^{-3} and 2.82 cm, respectively. In general, RZWQM over predicted soil water contents, but, as shown in Ma et al. (2002), soil water contents were more accurately predicted if the calibrated 33 kPa soil water contents were used. Predicted ET from June 13 to September 25, 1985 was 51.4 cm, which is very close to the estimated ET of 50.6 cm, based on soil moisture contents (changes in soil water storage + rainfall + irrigation water). The model simulated a 0.7 cm of surface runoff and a 0.4-cm deep percolation in 1985.

LAI was adequately simulated whereas plant height was underpredicted in the early growth stage and aboveground biomass was overpredicted in later growth stage (Figure 7.2). Simulated corn grain yield was 9813 kg/ha, which was similar to measured yield of 9854 kg/ha. At the observed silking date of July 26, 1985, when reproductive growth was initiated, RZWQM also simulated 20% of the plant population in the field entering reproductive growth. During model calibration, we put more weight on ET, LAI, and grain yield. We tolerated small errors in plant height and biomass simulations as long as they were within reasonable ranges of measured values. In addition, we tried to use as many default values as possible and used the same values in both RZWQM and CERES-Maize models without in-depth calibration so that bias on model calibration was minimized. Alternatively, we could have calibrated the 33 kPa soil water contents in both models, but we would end up with different calibrated 33 kPa soil water contents in the two models for the same soil.

Calibration of CERES-Maize for Corn

CERES-Maize was calibrated using the same irrigation treatment (line-source, irrigation level 4 in Table 7.2) in 1985 as RZWQM. The same soil hydraulic properties were used in the CERES-Maize model. CERES-Maize requires the drained upper and lower limits (Table 7.1). The model produced the same results with saturated hydraulic conductivity estimated from either Ahuja et al. (1989) or Rawls et al. (1982). Figure 7.3 shows simulated soil water storage and soil moisture contents. Simulated soil water contents are more scattered compared to that of RZWQM with RMSE of 0.036 cm^3 cm^{-3}; however, RMSE for CERES-Maize simulated water storage was 2.39 cm, which was better than that from RZWQM. Simulated ET from June 13 to September 25, 1985 was 48.0 cm, which is slightly lower than the observed 50.6 cm. CERES-Maize also simulated a 3.1 cm runoff and 2.0-cm deep percolation of water in 1985, which were higher than we would expect under the semiarid Colorado conditions with minimum irrigation, but neither were measured in the field plots.

Six cultivar-related parameters can be defined by the model user and the values used for this experiment are listed in Table 7.7. Species-specific parameters were not calibrated as suggested by the model developers. As shown in Figure 7.4, CERES-Maize provided reasonable simulations of

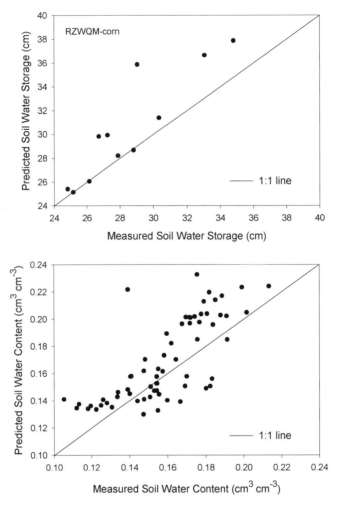

Figure 7.1 Predicted soil water storage and soil water contents with RZWQM-corn for the wettest irrigation level (level 4) in the 1985 growing season under the gradient line-source irrigation system.

LAI and aboveground biomass, with a slight overprediction of biomass at later growth stages. Plant height was not simulated in CERES. Leaf number was overpredicted, although the phylochron interval (PHINT) was increased from 38.9 to 50. Growth stages were adequately simulated (Figure 7.5). Simulated corn yield was 9882 kg/ha compared to observed yield of 9854 kg/ha.

Evaluation of RZWQM and CERES-Maize

After calibration, both models were used to predict corn production for other field experiments, including the irrigation levels in 1985 that differed from the dataset used for calibration, irrigation studies in 1984 and 1986 under line source irrigation system, and drip irrigation in 1985. Because RZWQM does not simulate leaf number and CERES-Maize does not simulate plant height, both models are only used to evaluate field results in terms of common simulated variables, such as yield, biomass, LAI, phenology, and ET. CERES-Maize correctly predicted corn growth stage in 1984 except for a slight delay of stages 3, 5, and 6 (Figure 7.5). RZWQM predicted a 15% plant population entering the reproductive growth stage at the observed silking date of August 6, 1984. No phenology data were available for 1986. ET data were only available for 1985. Both models showed the capability to accurately predict ET (Figure 7.6), although ET for CERES-Maize did

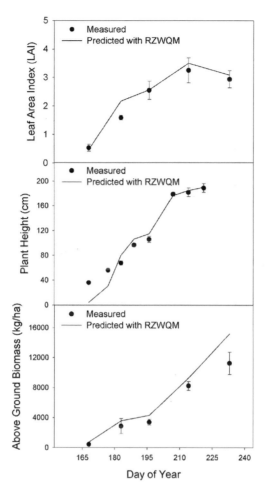

Figure 7.2 Simulated leaf area index (LAI), plant height, and biomass with RZWQM-corn for the wettest irrigation level (level 4) in the 1985 growing season under the gradient line-source irrigation system.

not respond well to the irrigation water levels. For the 1985 gradient line-source irrigation system where soil water measurements were available, RZWQM simulations of soil water content were more accurate (Figure 7.7) than CERES-Maize simulations, based on RMSE.

RZWQM predicted the yield responses to irrigation water better than CERES-Maize, especially for the year of 1985 (Figure 7.8). Although RZWQM underestimated corn yield for the 1985 drip irrigation and 1986 line source irrigation experiments, the model correctly predicted relative increases in yield with irrigation water. CERES-Maize overpredicted corn yield in all years and did not respond to irrigation water treatments. It is interesting to note that when we calibrate CERES-Maize model using the lowest irrigation level in 1985 (level 1), the plotted data points will shift to the right with almost the same scattering pattern as shown in Figure 7.8, the 1:1 line will go through the middle of the scattered data points, and no obvious bias will be observed. RMSE of simulated yields was 1381 kg/ha for RZWQM and 3609 kg/ha for CERES-Maize (Figure 7.8).

Both models provided good predictions of LAI for 1985 with comparable RMSE of 0.3 for RZWQM and 0.32 for CERES-Maize (Figure 7.9); however, both models overpredicted LAI for the 1984 growing season, with slightly better prediction from CERES-Maize model. Similarly, aboveground biomass was better predicted for the various irrigation levels in 1985 by both models than in 1984 (Figure 7.10). RMSE of simulated biomass for RZWQM was lower than that of CERES-Maize for the 1984 growing season.

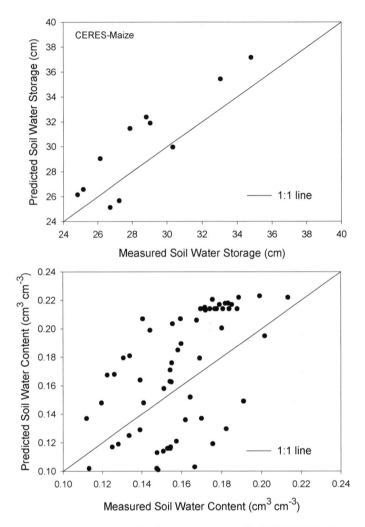

Figure 7.3   Predicted soil water storage and soil water contents with CERES-Maize for the wettest irrigation
level (level 4) in the 1985 growing season under the gradient line-source irrigation system.

Table 7.7   Cultivar-Specific Parameters Used in the CERES-Maize Model and Their Calibrated Values

Symbol	Description	Calibrated Values
P1	Thermal time from seedling emergence to the end of the juvenile phase during which the plant is not responsive to changes in photoperiod (thermal days above 8°C)	245
P2	Extent to which development is delayed for each hour increase in photoperiod above the longest photoperiod at which development proceeds at a maximum rate (days)	0.8
P5	Thermal time from silking to physiological maturity (thermal days above 8°C)	680
G2	Maximum possible number of kernels per plant	860
G3	Kernel filling rate during the linear grain filling stage and under optimum conditions (mg/day)	9.5
PHINT	Phylochron interval between successive leaf tip appearances (thermal days)	50

To evaluate how both RZWQM and CERES-Maize simulated soil water stress, the measured CWSI for the drip irrigation system from canopy temperature (Nielsen and Gardner, 1987) was compared with the water stress factor calculated by the models. RZWQM predicts a water stress factor of (1-EWP), where EWP is the ratio of actual transpiration to potential transpiration (Hanson,

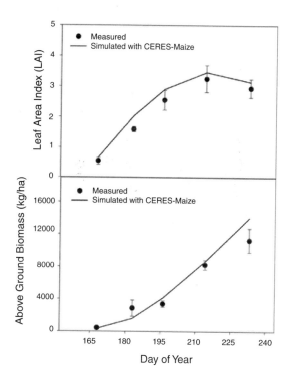

Figure 7.4 Simulated leaf area index (LAI) and biomass with CERES-Maize for the wettest irrigation level (level 4) in the 1985 growing season under the gradient line-source irrigation system.

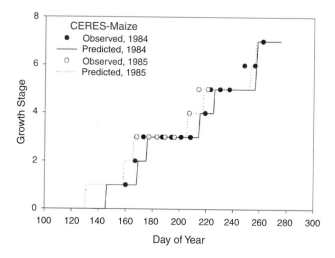

Figure 7.5 Simulated growth stage with CERES-Maize for 1984 and 1985 growing season under the gradient line-source irrigation system.

2000), and the water stress factor in CERES-Maize is calculated as (1-SWDF1), where SWDF1 is the ratio of potential uptake to potential transpiration (Ritchie, 1998). As shown in Figures 7.11 and 7.12, simulated water stress levels decreased with the amount of irrigation water applied in both models, although simulated water stress was better correlated to CWSI in RZWQM than in

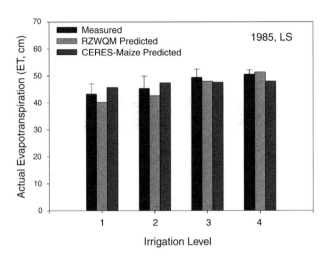

Figure 7.6   Simulated evapotranspiration (ET) (June 13 to September 25) with RZWQM-corn and CERES-Maize during 1985 growing season under the gradient line-source irrigation system (LS). See Table 7.2 for irrigation treatments.

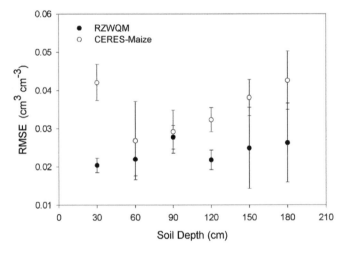

Figure 7.7   Root mean square errors (RMSE) of predicted soil water contents in various layers of the soil profile for the 1985 gradient line-source irrigation system of corn production. RMSE was averaged across treatments for each year and each irrigation system. The bars are one standard error around the means. RMSE of each soil layer is plotted at the lower soil boundary of the soil layer.

CERES-Maize. RZWQM simulated little stress from July 13 to August 11, 1985 where CWSI indicated considerable stresses. CERES-Maize did not predict water stresses from July 16 to September 12, 1985. Thus, for this application, RZWQM simulated water stresses better than CERES-Maize, which explains the better yield prediction by RZWQM.

APPLICATIONS OF RZWQM AND CROPGRO FOR SOYBEAN

Calibration of RZWQM and CROPGRO-Soybean

Calibration of RZWQM for Soybean

Researchers also used the highest irrigation levels (level 4 in Table 7.5) of the 1985 gradient line source irrigation system to calibrate the RZWQM and CROPGRO models for soybean. Soil texture

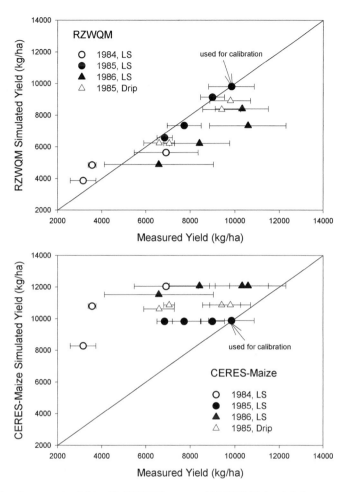

Figure 7.8   Predicted corn grain yields with RZWQM-corn and CERES-Maize. LS: line source irrigation system; Drip: drip irrigation system.

and 33 kPa soil water contents in Table 7.1 were used and soil hydraulic properties were based on soil texture class (Rawls et al. 1982). Although both corn and soybean fields were classified as Rago silt loam, different saturated soil hydraulic conductivities for corn and soybean were used because saturated hydraulic conductivity varies spatially in the field and depends on management practices (Benjamin, 1993; Lal, 1999; Rodriguez et al., 1999, van Es et al., 1999). In addition, another similar soil series named Weld (fine smectitic, mesic Aridic Argiustolls) was mixed with Rago in some fields. The Weld series has a clay loam layer at 15 to 30 cm soil depth. Therefore, both corn and soybean in RZWQM used the same soil properties except that a lower saturated hydraulic conductivity was used for soybean field, which provided better soil water content simulations.

Figure 7.13 shows simulated soil water storage and soil water contents for the calibrated data set. In general, the model simulated good soil water storage and no biased soil water content with RMSEs of 1.28 cm and 0.027 cm^3/cm^3, respectively. RZWQM adequately simulated ET from July 10 to September 9, 1985, at 37.4 cm versus 39.0 cm estimated from the soil water balance (Figure 7.14); however, better agreement between the simulated (40.1 cm) and estimated ET (40.5 cm) was obtained from July 10 to September 25, 1985. Therefore, goodness of model simulations also depends on the accuracy of estimated ET that has directly inherited errors from the measured soil water contents. RZWQM also simulated 2.0 cm of runoff and 0.7 cm of deep seepage in 1985, which were assumed to be zero when estimating ET from the soil water balance;

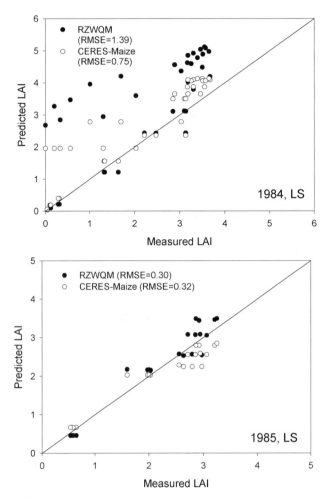

Figure 7.9 Predicted leaf area index with RZWQM-corn and CERES-Maize for 1984 and 1985 growing season under the gradient line-source irrigation system.

however, surface runoff would be considerably reduced if a higher saturated hydraulic conductivity were used (Ahuja et al. 1989). Therefore, depending on which criterion was selected for model calibration, a different saturated soil hydraulic conductivity might be used for soybean simulation in this study. In addition, it may not be valid to use a constant soil hydraulic conductivity for all the years and for all management practices because of its dependency on soil dynamics (van Es et al. 1999).

Calibrated plant parameters were based on parameters derived from RZWQM applications for soybean in the midwest of the U.S. (Table 7.6) (Hanson et al., 1999; Landa et al., 1999; Ghidey et al., 1999; Jaynes and Miller, 1999). For this experiment, the minimum leaf stomatal resistance was modified to a value of 100 s/m, based on the study of Nielsen (1990), and aboveground biomass of a mature plant was set at 13 g, based on experimental measurements. The rest of the parameters in Table 7.6 were either default or calibrated values. Although plant phenology is not the focus of RZWQM, the model also showed that 84% of plants reached maturity on September 16, 1985 at the field observed date for the R8 stage; however, the model simulated the initial reproductive growth on July 30, 1985, which was delayed according to the observed R1 stage date of July 19, 1985. The delay may have been a result of the indeterminate variety used in this study, where there was no definite flowering period. The model overpredicted LAI in the later development phases

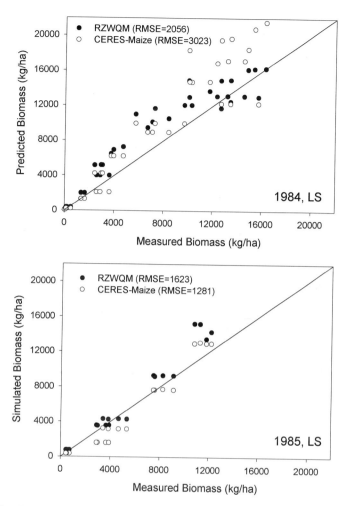

Figure 7.10 Predicted biomass with RZWQM-corn and CERES-Maize for 1984 and 1985 growing seasons under the gradient line-source irrigation system.

and plant height during the middle of the growing season (Figure 7.15). RZWQM also under-predicted plant height during the early growth phases. Aboveground biomass was adequately simulated, however. Simulated grain yield was 2686 kg/ha compared to a measured yield of 2678 kg/ha (Figure 7.16).

Calibration of CROPGRO-Soybean

The wettest treatment in 1985 line-source irrigation system was used to calibrate CROPGRO-Soybean. The soil properties shown in Table 7.1 were used without modification. Generally the model underpredicted soil water contents (Figure 7.17), which cannot be improved except by changing the upper and lower drained limits. The model provided the same results using saturated hydraulic conductivities from either method in Table 7.1. Experimentally, we observed more soil water storage in the soil profile during the soybean growing season than during the maize season (Figures 7.1, 7.3, 7.13, and 7.17). CROPGRO-Soybean simulated an ET of 31.1 cm from July 10 to September 9, 1985, which is lower than the estimated ET of 39.0 cm. The model also simulated 3.2 cm of surface runoff and 3.0 cm deep percolation. RMSEs for simulated soil water storage was 5.5 cm and for soil water contents was 0.038 cm³/cm³.

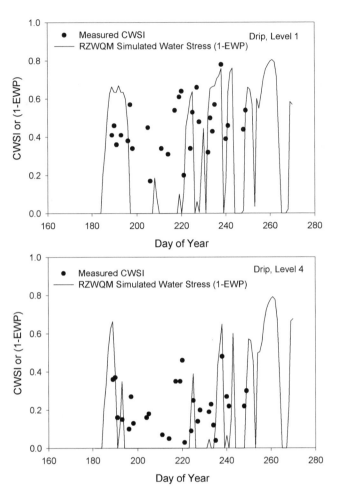

Figure 7.11   Measured crop water stress index (CWSI) and simulated water stress in RZWQM (1-EWP) for corn under the drip irrigation system in 1985. EWP is the ratio of actual transpiration to potential transpiration.

Plant growth was calibrated based on recommendations published by Boote (1999). Plant parameters were calibrated based on the default parameters for maturity group 2 as distributed with the model (Hoogenboom et al., 1994). Table 7.8 lists the calibrated values for this experiment and default values for maturity group 2. Simulated V-stage and reproductive growth stage are shown in Figure 7.18. CROPGRO-Soybean has a more accurate LAI simulation than RZWQM (Figure 7.15); however, because LAI data were not collected during leaf senescence, the model prediction of LAI during this stage is unable to be verified. Aboveground biomass simulations from both models were similar. Although CROPGRO simulated a more accurate plant height for the early growth phase, it overpredicted plant height in general (Figure 7.15). Simulated grain yield was 2647 kg/ha, which was very close to the measured yield of 2678 kg/ha (Figure 7.16).

Evaluation of RZWQM and CROPGRO-Soybean

The calibrated models were then used to predict soybean production for the other irrigation levels and the 1985 and 1986 experiments (Table 7.5). RZWQM simulated ET adequately in 1985 for the gradient line-source irrigation system after it was calibrated for the wettest treatment (Figure 7.14). Although the model generally overpredicted ET for the remaining treatments, sim-

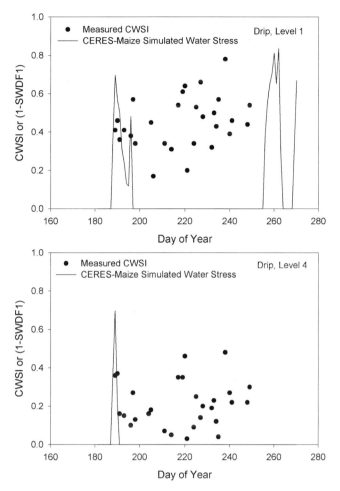

Figure 7.12   Measured crop water stress index (CWSI) and simulated water stress in CERES-Maize (1-SWDF1) for corn under the drip irrigation system in 1985. SWDF1 is the ratio of potential uptake to potential transpiration.

ulated ET responded correctly to irrigation water amounts. CROPGRO showed the most accurate simulation of ET for the drip irrigation system (Figure 7.14), although CROPGRO-simulated ET did not respond to amount of irrigation for the rain shelter irrigation system. Overall RMSEs of simulated ET among all the treatments and years were comparable, 7.3 cm for CROPGRO and 7.8 for RZWQM.

RZWQM simulated more accurate soil water contents for the 0 to 30, 30 to 60, and 60 to 90 cm soil profile for the 1985 and 1986 line-source irrigation system than CROPGRO (Figure 7.19); however, CROPGRO predicted equal or better soil water contents beyond the 90 cm soil profile except for the 120 to 150 cm layer in 1985. For the rain shelter system, CROPGRO also provided better predictions of soil water contents in soil profiles below 90 cm. RZWQM simulated more accurate soil water contents for the 30 to 60 cm soil layer. Goodness of model prediction for the 0 to 30 and 60 to 90 cm soil layers depended on the year of study. RZWQM simulated better soil water contents in 1986, whereas CROPGRO predicted better soil water contents in 1985. Overall, the differences between the two models were insignificant for the rain shelter system. For the drip system, RZWQM was better for the 0 to 30 and 30 to 60 cm soil layers, whereas CROPGRO was insignificantly better beyond 60 cm soil depth. Both models predicted soil water contents better for the rain shelter irrigation system than for the other two irrigation systems in terms of RMSE (Figure 7.19).

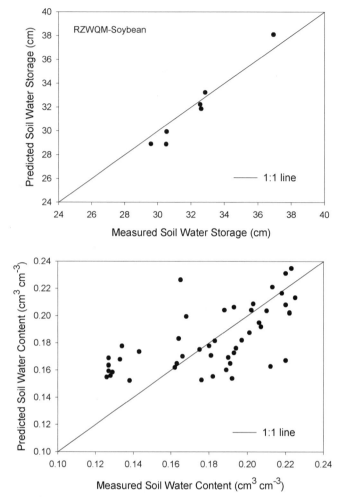

Figure 7.13   Predicted soil water storage and soil water contents with RZWQM-soybean for the wettest irrigation level (level 4) in the 1985 growing season under the gradient line-source irrigation system.

RZWQM accurately predicted good yields for all the treatments except for the 1986 drip system. CROPGRO, on the other hand, predicted yields most accurately for the 1986 drip system (Figure 7.16). RZWQM responded to irrigation water better under the rain shelter system than CROPGRO. RMSEs of simulated yields were 295 kg/ha for CROPGRO and 432 kg/ha for RZWQM. Both models adequately predicted plant canopy height for 1985, but underpredicted canopy height for 1986. Maximum canopy height was overpredicted by 30 to 100% by both models, suggesting that the models failed to account for drought effects on plant height in 1986 after model calibration in a relatively wet year of 1985.

Simulated water stresses were compared with the CWSI values as determined by Nielsen (1990). For the drip irrigation system, both RZWQM and CROPGRO simulated water stresses that responded to irrigation amount (Figures 7.20 and 7.21). RZWQM-simulated water stress matched the measured CWSI well, except for the early part of the growing season. CROPGRO simulated greater water stress than CWSI during the early growing season but less water stress in later growing season. Both models failed to simulate water stress for the rain shelter system. CROPGRO did not predict any water stress under all irrigation amounts in both 1985 and 1986, which may be responsible for the lack of response in simulated ET and yields (Figures 7.14 and 7.16). RZWQM

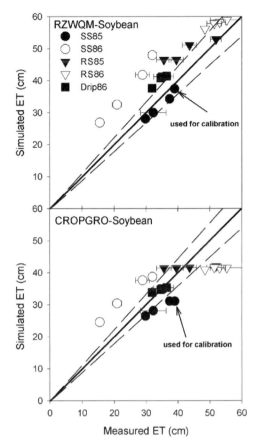

Figure 7.14 Predicted evapotranspiratioin (ET) with RZWQM-soybean and CROPGRO-Soybean (dates). LS: line-source irrigation system; RS: rain shelter irrigation system; and Drip: drip irrigation system. ET was from July 10 to September 9 for 1985 LS; from June 20 to September 10 for 1986 LS; from May 20 to September 16 for 1985 RS; from June 6 to September 12 for 1986 RO; and from June 24 to September 12 for the 1986 Drip.

simulated some water stress later during the growing season but failed to predict water stress for the early growth phase (data not shown). The simulation of stress effects in RZWQM was attributed to its better ET and yield simulations in Figures 7.14 and 7.16.

SUMMARY AND DISCUSSION

An agricultural system model is generally derived from knowledge gained in different disciplines of science and is designed to integrate the interactions among agricultural processes that have been studied individually in different scientific disciplines. Due to limited understanding and diversified theories on these processes, an agricultural system model can be developed quite differently by individual model developers. For the three models tested here, RZWQM was originally developed as a water quality model, and the generic plant growth component was used to predict biomass and yield production and interaction between plants and soils. Predictions of phenology and yield components were not the original goal of the RZWQM. On the other hand, the CERES-Maize and CROPGRO-Soybean models were developed specifically to simulate corn and soybean production, including phenology, biomass, and yield components. The soil water component was simply a medium for the plant to extract water and nutrients.

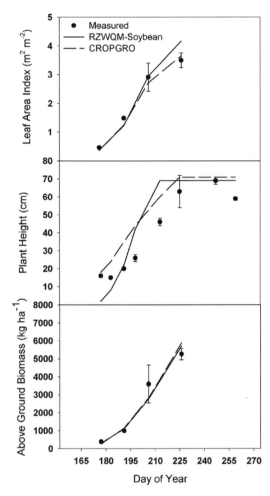

Figure 7.15 Simulated leaf area index (LAI), plant height, and biomass with RZWQM-soybean and CROPGRO for the wettest irrigation level (level 4) in the 1985 growing season under the gradient line-source irrigation system.

The CERES-Maize and CROPGRO-Soybean models were easy to use because of their simple soil water balance components and only a few cultivar-specific plant parameters required calibration (Tables 7.7 and 7.8; also see Ahuja and Ma, 2002). These few soil and plant parameters generally were able to provide good calibration of phenology, biomass, LAI, and yield. In addition, both models included a database with plant parameters categorized by cultivars (Hoogenboom et al. 1994). RZWQM, on the other hand, required more detailed soil hydraulic parameters, which may be obtained from soil texture based default values. In many cases, these default values provided reasonable soil water prediction (Ma et al., 1998; Nielsen et al., 2002). In addition, the detailed approach for soil water movement gave greater flexibility in calibrating soil water contents. Because RZWQM has a generic plant growth component, it did not have a database for each cultivar; rather it provided default plant growth parameters for tested cases in the U.S. Midwest. Therefore, users may have to calibrate additional parameters in the database besides the ones suggested by model developers (Table 7.6).

There was no objective optimization algorithm for calibrating an agricultural system model. Parameters were calibrated more or less by trial and error. CROPGRO and CERES-Maize emphasizes phenology and development, whereas RZWQM concentrates more on crop production. The soil water balance in CROPGRO and CERES-Maize was mainly calibrated through the upper and lower

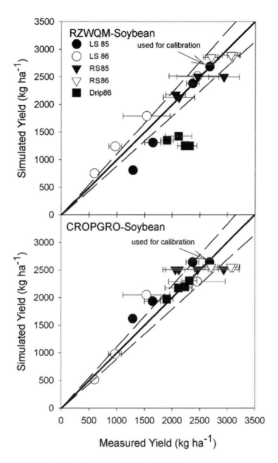

Figure 7.16 Predicted soybean yields with RZWQM-soybean and CROPGRO. LS: line source irrigation system; RS: rain shelter irrigation system; Drip: drip irrigation system.

drained limits, whereas RZWQM had a series of soil hydraulic properties that may be calibrated (Ahuja and Ma, 2002). In addition, because RZWQM has its strength in soil and water simulation, users have to be aware of the soil nitrate and chemicals in percolation and runoff waters. For the same data sets, different users may calibrate the models differently. For example, in this chapter, we used the upper and lower drained limits from Ritchie et al. (1999) for CERES and CROPGRO and interpreted as 33 kPa and 1500 kPa soil water content values to be used in RZWQM. Default 33 kPa and 1500 kPa soil water content values in RZWQM based on soil texture class would be used and interpreted as upper and lower drained limits to be applied in CERES-Maize and CROPGRO as done by Nielsen et al. (2002). Many other ways (or even better ways) can be used to calibrate the data sets. As shown in Figure 7.19, given a fixed set of soil hydraulic properties, RZWQM provided overall better soil water content prediction than CERES-Maize and CROPGRO-Soybean.

All the models could be calibrated satisfactorily, but applications depended on year and irrigation methods. For corn yield, RZWQM provided more accurate simulations and responded better to irrigation than CERES-Maize (Figure 7.8). For corn LAI, CERES-Maize predicted better than RZWQM (Figure 7.9). For corn biomass, RZWQM predictions were better in 1984 but worse in 1985 than those of CERES-Maize (Figure 7.10). Although both RZWQM and CROPGRO simulated ET equally well, RZWQM-simulated ET responded better to irrigation than CROPGRO-Soybean (Figure 7.14). For soybean yield, RZWQM provided the worst yield prediction, whereas CROPGRO provided the best yield prediction for the drip irrigation study in 1986 (Figure 7.16); however, for the rain shelter irrigation study, RZWQM predicted yield better than CROPGRO (Figure 7.16), although

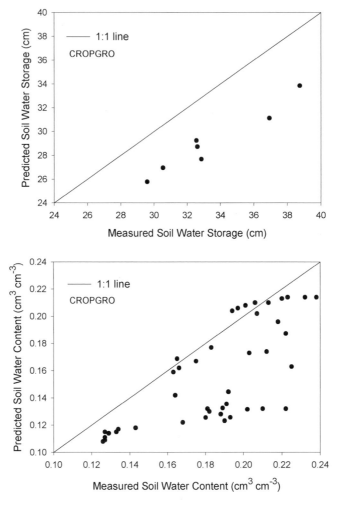

Figure 7.17 Predicted soil water storage and soil water contents with CROPGRO for the wettest irrigation level (level 4) in the 1985 growing season under the gradient line-source irrigation system.

overall RMSE was larger for RZWQM than for CROPGRO. Therefore, the benefits of developing complex and crop specific growth models, such as CROPGRO-Soybean and CERES-Maize, were minimized if other components of the agricultural systems, such as water, ET, and nutrient, were not comparable. There is a need to improve these components in CROPGRO and CERES.

All the models served the purposes for which they were designed. The crop growth component in RZWQM was designed to provide biomass and yield prediction for an agricultural water quality model. As shown in this chapter, its simulations of corn and soybean yields were adequate, once it was calibrated (Figures 7.8 and 7.16). Simulations of biomass were also reasonable (Figures 7.10 and 7.15). CERES-Maize and CROPGRO-Soybean were designed to simulate many agronomic attributes, such as phenology, leaf number, and yield components. As shown in Figures 7.5 and 7.18, both models were able to simulate the various growth and developmental stages fairly well. CROPGRO also simulated the number of leaves correctly, although CERES-Maize overestimated leaf number. CROPGRO also predicted soybean yields adequately except for the rain shelter experiments (Figure 7.16). Unfortunately, there was no experimental data to validate kernel number, kernel weight, pod number, pod weight, seed number, and seed weight as simulated by both CERES-Maize and CROPGRO-Soybean.

Table 7.8   Cultivar-Specific Parameters Used in CROPGRO-Soybean Model and Their Calibrated and Default Values for Maturity Group 2

Parameter Name	Calibrated Value	Default Value
CSDL: Critical day length for crop development (hr)	13.59	13.59
PPSEN: Sensitivity to photoperiod (1/hr)	0.249	0.249
EM-FL: Time from end of juvenile phase to first flower in photothermal days	20	17.4
FL-SH: Time from first flower to first pod greater than 0.5 cm (photothermal days)	6	6
FL-SD: Time from first flower to first seed (photothermal days)	13.5	13.5
SD-PM: Time from first seed to physiological maturity (photothermal days)	20	33
FL-LF: Time from first flower to end of leaf growth (photothermal days)	26	26
LFMAX: Maximum leaf photosynthesis rate ($CO_2/m^2/s$)	0.92	1.03
SLAVAR: Specific leaf area (SLA) (cm^2/g)	250	375
SIZLF: Maximum size of fully expanded leaf (cm^2)	180	180
XFRUIT: Maximum fraction of daily available photosynthate to seeds plus shells (dimensionless)	1.0	1.0
WTPSD: Maximum weight per seed (g)	0.19	0.19
SFDUR: Seed filling duration for a cohort of seed (photothermal days)	20	23
SDPDV: Average seed per pod	2.2	2.2
PODUR: Time for cultivar to add full pod load under optimal conditions (photothermal days)	8.0	10.0

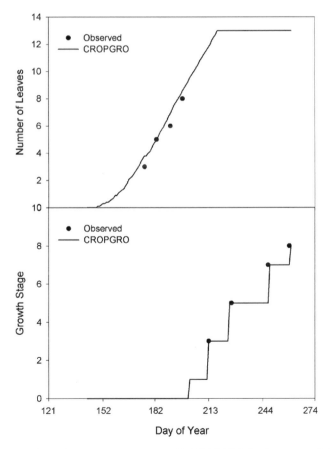

Figure 7.18   Simulated leaf number and growth stage with CROPGRO for the wettest irrigation level (level 4) in the 1985 growing season under the gradient line-source irrigation system.

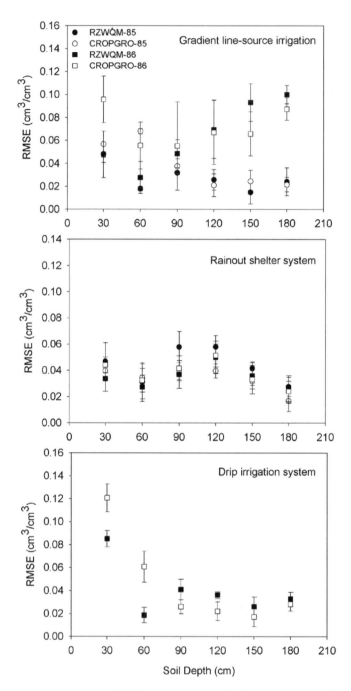

Figure 7.19 Root mean square errors (RMSE) of predicted soil water contents in various layers of the soil profile for different irrigation systems of soybean production. RMSE was averaged across treatments for each year and each irrigation system. The bars are one standard errors around the means. RMSE of each soil layer is plotted at the lower soil boundary of the soil layer.

All the models were run without nitrogen stress, which may not be true in the case of corn. The models could have responded differently if nitrogen stress had been simulated in addition to water stress (Ma et al., 2002). RZWQM included very detailed soil carbon–nitrogen components and suggested methods of initializing the soil organic carbon pools (Ahuja and Ma, 2002). RZWQM

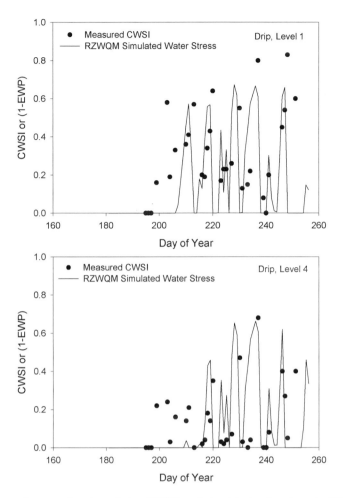

Figure 7.20   Measured crop water stress index (CWSI) and simulated water stress in RZWQM (1-EWP) for soybean under the drip irrigation system in 1986. EWP is the actual transpiration to potential transpiration.

can also be run for multiple years to initialize the pools. CERES-Maize and CROPGRO-Soybean can only run on a yearly basis, although with the DSSAT framework, the models can simulate crop sequences and rotations (Thornton et al. 1995). RZWQM also emphasizes management practices, such as tillage, crop residue management, tile drainage, manure application, and crop rotation. Thus, RZWQM has the advantage of simulating environmental impacts of agricultural systems in addition to crop production, in terms of nitrate and pesticide.

In conclusion, with simple parameterization of the models, this study compared three models for their predictions of corn and soybean production, with the same data sets, using similar initial conditions. It was also a unique study because the models were calibrated for one irrigation level in 1985 and evaluated for other irrigation levels in the same year, so that the responses of models to water stresses could be fully tested. Also, model calibration was kept to a minimum and default values were used to avoid biasing the results toward any one model. Simulation results showed that each model can be calibrated to a certain level of satisfaction, but applications of the models to other conditions, such as water amount, irrigation methods, and weather, depend on the model. Overall, RZWQM provided satisfactory yield predictions for both corn and soybean. RZWQM was better at simulating soil water contents than CROPGRO-Soybean and CERES-Maize. CERES-

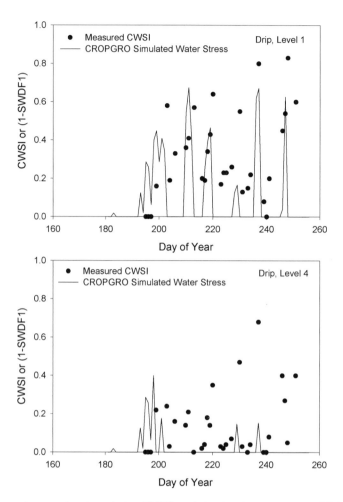

Figure 7.21   Measured crop water stress index (CWSI) and simulated water stress in CROPGRO (1-SWDF1) for soybean under the drip irrigation system in 1986. SWDF1 is the ratio of potential uptake to potential transpiration.

Maize-simulated corn yields did not respond to irrigation adequately. CROPGRO predicted soybean yields better in terms of RMSEs, but it did not respond to irrigation under the rain shelter irrigation system. RZWQM predicted water stresses better than CROPGRO-Soybean and CERES-Maize, based on CWSI.

So far, CERES-Maize and CROPGRO-Soybean have been used worldwide by international agencies and organizations. Through efforts of the IBSNAT project, the two models were adapted by and subsequently released as part of the DSSAT product. Although the IBSNAT project ended in 1993, the international collaboration has continued through the International Consortium for Agricultural Systems Applications (ICASA; www.ICASAnet.org). Through this international effort the visibility and utility of the models were considerably improved, with the distribution of DSSAT in more than 90 countries worldwide. Within the DSSAT package (a DOS-based user interface), the two models represent the grain legumes and grain cereals as part of a suite of models for more than 17 different crops. They are very easy to calibrate for crop production and generally give model users satisfaction. At present, model users are satisfied if they can only calibrate their data adequately with a few steps. Use of the models for technology transfer has limited success so far.

On the other hand, RZWQM was developed and tested mainly in collaboration with the Management Systems Evaluation Areas (MSEA) project. It has been used in only a few countries (Ma et al., 2000).

Also, its application was mainly on soil water quality. This chapter is the first study to systematically evaluate the generic crop growth components in RZWQM against the more widely used crop growth models. The development of a Windows-based user interface has promoted the use of RZWQM, and it can be downloaded free from http://arsagsoftware.ars.usda.gov. Although conclusions were drawn only from this particular study in the Central Great Plains of the U.S., more comparison studies will be needed for other experimental conditions. This study clearly showed the weaknesses and strengths of each model, and should help and encourage field scientists to use models as a tool with the analysis of their experimental results and promote technology transfer via system models.

REFERENCES

Ahuja, L.R., D.K. Cassel, R.R. Bruce, and B.B. Barnes. 1989. Evaluation of spatial distribution of hydraulic conductivity using effective porosity data, *Soil Sci.*, 148:404–411.

Ahuja, L.R., K.E. Johnsen, and K.W. Rojas. 2000. Water and chemical transport in soil matrix and macropores. In *Root Zone Water Quality Model — Modeling Management Effects on Water Quality and Crop Production. Water Resources Publications,* L.R. Ahuja, K.W. Rojas, J.D. Hanson, M.J. Shaffer, and L. Ma, Eds., LLC, Highlands Ranch, CO.

Ahuja, L.R. and L. Ma. 2002. Parameterization of agricultural system models: current approaches and future needs. In *Agricultural System Models in Field Research and Technology Transfer,* L.R. Ahuja, L. Ma, and T.A. Howell, Eds., CRC Press, Boca Raton, FL, 271–313.

Benjamin, J.G. 1993. Tillage effects on near-surface soil hydraulic properties, *Soil and Tillage Res.,* 26:277–288.

Bennett, J.M., T.R. Sinclair, R.C. Muchow, and S.R. Costello. 1987. Dependence of stomatal conductance on leaf water potential, turgor potential, and relative water content in field-grown soybean and maize, *Crop Sci.,* 27:984–990.

Boote, K.J. 1999. Concepts for calibrating crop growth models. In *1999 DSSAT version 3,* G. Hoogenboom, P. Wilkens, and G.Y. Tsuji, Eds., Vol. 4, University of Hawaii, Honolulu, HI.

Boote, K.J., J.W. Jones, G. Hoogenboom, and N.B. Pickering. 1998. The CROPGRO model for grain legumes. In *Understanding Options for Agricultural Production,* G.Y. Tsuji, G. Hoogenboom, P.K. Thornton, Eds., Kluwer Academic Publishers, Dordrecht, The Netherlands, 99–128.

Farahani, H.J., G.W. Buchleiter, L.R. Ahuja, and L.A. Sherrod. 1999. Model evaluation of dryland and irrigated cropping systems in Colorado, *Agron. J.,* 91:212–219.

Fiscus, E.L., A.N.M. Mahbub-Ul Alam, and T. Hirasawa. 1991. Fractional integrated stomatal opening to control water stress in the field, *Crop Sci.,* 31:1001–1008.

Gee, G.W. and J.W. Bauder. 1986. Particle-size analysis. In *Methods of Soil Analysis, Part 1, 2nd edition, Agron. Monogr., 9.* A. Klute, Ed., ASA and SSSA. Madison, WI, 383–411.

Ghidey, F., E.E. Alberts, and N.R. Kitchen. 1999. Evaluation of RZWQM using field measured data from the Missouri MSEA, *Agron. J.,* 91:183–192.

Godwin, D.C. and U. Singh. 1998. Nitrogen balance and crop response to nitrogen in upland and lowland cropping systems. In *Understanding Options for Agricultural Production,* G.Y. Tsuji, G. Hoogenboom, and P.K. Thornton, Eds., Kluwer Academic Publishers, Dordrecht, The Netherlands, 41–54.

Hanson, J.D. 2000. Generic crop production. In *Root Zone Water Quality Model,* L.R. Ahuja, K.W. Rojas, J.D. Hanson, M.J. Shaffer, and L. Ma, Eds., Water Resources Publications, Highland Ranch, CO, 81–118.

Hanson, J.D., K.W. Rojas, and M.J. Shaffer. 1999. Calibrating the root zone water quality model, *Agron. J.,* 91:171–177.

Hoogenboom, G., J.W. Jones, P.W. Wilkens, W.D. Batchelor, W.T. Bowen, L.A. Hunt, N.B. Pickering, U. Singh, D.C. Godwin, B. Baer, K.J. Boote, J.T. Ritchie, J.W. White. 1994. Crop Models. In *DSSAT v3, Vol. 2,* G.Y. Tsuji, G. Uehara, and S. Balas, Eds., University of Hawaii, Honolulu, HI.

Hoogenboom, G., P.W. Wilkens, and G.Y. Tsuji, Eds., 1999. *Decision Support System for Agrotechnology Transfer version 3,* Vol. 4, University of Hawaii, Honolulu, HI.

Jaynes, D.B. and J.G. Miller. 1999. Evaluation of RZWQM using field measured data from Iowa MSEA, *Agron. J.,* 91:192–200.

Kiniry, J.R., J.G. Arnold, and Y. Xie. 2002. Applications of models with different spatial scale. In *Agricultural System Models in Field Research and Technology Transfer,* L.R. Ahuja, L. Ma, and T.A. Howell, Eds., CRC Press, Boca Raton, FL, 205-225.

Lal, R. 1999. Soil compaction and tillage effects on soil physical properties of a Mollic Ochraqualf in northwest Ohio, *J. Sustainable Agric.,* 14:53–65.

Landa, F.M., N.R. Fausey, S.E. Nokes, and J.D. Hanson. 1999. Evaluation of the root zone water quality model (RZWQM3.2) at the Ohio MSEA, *Agron. J.,* 91:220–227.

Ma, L, M.J. Shaffer, J.K. Boyd, R. Waskom, L.R. Ahuja, K.W. Rojas, and C. Xu. 1998. Manure management in an irrigated silage corn field: experiment and modeling, *Soil Sci. Soc. Am. J.,* 62:1006–1017.

Ma, L., L.R. Ahuja, J.C. Ascough, II, M.J. Shaffer, K.W. Rojas, R.W. Malone, and M.R. Cameira. 2000. Integrating system modeling with field research in agriculture: applications of Root Zone Water Quality Model (RZWQM), *Adv. Agron.,* 71:233–292.

Ma, L., M.J. Shaffer, and L.R. Ahuja. 2001a. Application of RZWQM for soil nitrogen management. In *Modeling Carbon and Nitrogen Dynamics for Soil Management,* M.J. Shaffer, L. Ma, and S. Hansen, Eds., CRC Press, Boca Raton, FL, 265–301.

Ma, L., D.C. Nielsen, L.R. Ahuja, K.W. Rojas, J.D. Hanson, and J.G. Benjamin, 2002. Evaluation of the RZWQM for corn responses to water stress under various irrigation levels. Trans. ASAE. (in review).

Nielsen, D.C. 1990. Scheduling irrigations for soybeans with the crop water stress index (CWSI), *Field Crop Res.,* 23:103–116.

Nielsen, D.C. 1997. Water use and yield of canola under dryland conditions in the central Great Plains, *J. Prod. Agric.,* 10:307–313.

Nielsen, D.C. and B.R. Gardner. 1987. Scheduling irrigations for corn with the crop water stress index (CWSI), *Applied Agric. Res.,* 2:295–300.

Nielsen, D.C., L. Ma, L.R. Ahuja, and G. Hoogenboom. 2002. Simulating soybean water stress effects with RZWQM and CROPGRO models, *Agron. J.,* (in press).

Rawls, W.J., D.L. Brakensiek, and K.E. Saxton. 1982. Estimation of soil water properties, *Trans. ASAE,* 25:1316–1320, 1328.

Rawls, W.J., D. Gimenez, and R. Grossman. 1998. Soil texture, bulk density, and slopes of soil water retention curve to predict saturated hydraulic conductivity, *Trans. ASAE,* 41:983–988.

Ritchie, J.T. 1998. Soil water balance and plant water stress. In *Understanding Options for Agricultural Production,* G.Y. Tsuji, G. Hoogenboom, P.K. Thornton, Eds., Kluwer Academic Publishers, Dordrecht, The Netherlands, 41–54.

Ritchie, J.T., U. Singh, D.C. Godwin, and W.T. Bowen. 1998. Cereal growth, development and yield. In *Understanding Options for Agricultural Production,* G.Y. Tsuji, G. Hoogenboom, P.K. Thornton, Eds., Kluwer Academic Publishers, Dordrecht, The Netherlands, 79–98.

Ritchie, J.T., A. Gerakis, and A. Suleiman. 1999. Simple model to estimate field-measured soil water limits, *Trans. ASAE,* 42:1609–1614.

Ritchie, S.W., J.J. Hanway, and G.O. Benson. 1986. How a corn plant develops, *Special Report No. 80,* Iowa State University, Ames, IA.

Rodriguez, M.B., M.A. Taboada, and D. Cosentino. 1999. Influence of growing plants and nitrogen fertilizer on saturated hydraulic conductivity, *Commun. Soil Sci. Plant Anal.,* 30:1681–1689.

Rojas, K.W., L. Ma, J.D. Hanson, and L.R. Ahuja. 2000. RZWQM98 User Guide. In *Root Zone Water Quality Model,* L.R. Ahuja et al., Eds., Water Resources Publications LLC, Highlands Ranch, CO, 327–364.

Singh, U. et al. 2002. Decision support tools for improved resource management and agricultural sustainability. In *Agricultural System Models in Field Research and Technology Transfer,* L.R. Ahuja, L. Ma, and T.A. Howell, Eds., CRC Press, Boca Raton, FL, 91–118.

Thornton, P.K., G. Hoogenboom, P.W. Wilkens, and W.T. Bowen. 1995. A computer program to analyze multiple-season crop model outputs, *Agron. J.,* 87:131–136.

Tsuji, G.Y., A. du Toit, A. Jintrawet, J.W. Jones, R.M. Ogoshi, and G. Uehara. 2001. Benefits of models in research and decision support with examples of applications and case studies. The IBSNAT Experience. In *Agricultural System Models in Field Research and Technology Transfer,* L.R. Ahuja, L. Ma, and T.A. Howell, Eds., CRC Press, Boca Raton, FL, 71–89.

Tsuji, G.Y., G. Hoogenboom, and P.K. Thornton, Eds., 1998. *Understanding Options for Agricultural Production,* Kluwer Academic Publishers, Dordrecht, The Netherlands, 400.

Tsuji, G.Y., G. Uehara, and S. Balas, Eds., 1994. *DSSAT version 3,* University of Hawaii, Honolulu, HI.

Van Es, H.M., C.B. Ogden, R.L. Hill, R.R. Schindelbeck, and T. Tsegaye. 1999. Integrated assessment of space, time, and management-related variability of soil hydraulic properties, *Soil Sci. Soc. Am. J.,* 63:1599–1608.

CHAPTER 8

The Co-Evolution of the Agricultural Production Systems Simulator (APSIM) and Its Use in Australian Dryland Cropping Research and Farm Management Intervention

Robert L. McCown, Brian A. Keating, Peter S. Carberry, Zvi Hochman, and Dean Hargreaves

CONTENTS

INTRODUCTION

Farmers, their advisors, and agricultural researchers face the challenge of understanding farms and the broader systems in which they reside well enough to make innovations appropriate to maintaining or improving system performance. In a complex, dynamic environment, "appropriate innovation" is a moving target as more efficient technology becomes available, prices shift, new knowledge about the state of the system emerges, market and environmental regulatory standards and procedures evolve, etc. Although the ability to cope with these challenges is much of what has long constituted *expertise* in farm management, such complexities and uncertainties of modern farming have constituted much of the case for computer models and decision support systems (e.g., Wagner and Kuhlmann 1991; Ikerd 1991; Parker, 1999)

For nearly 20 years, champions of decision support systems have tended to explain low farmer adoption and use by the low ownership of computers in the farming community. But recent studies show that farmers are no longer lagging behind the general community in computer ownership. In 1999, computers were in use on approximately half of Australian farms and one in five were using the Internet — comparable usage to people in metropolitan areas (Australian Bureau of Statistics, 1999). Personal computers have become integral to farm business financial accounting and record keeping, and when connected to the Internet, by facilitating information gathering and diverse types of transactions. But this has not been accompanied by similar growth in the use of decision support systems (Hoag et al. 1999; Parker, 1999).

Because many agricultural decision support systems rely on an underlying simulation model of a production system, this has implications for model research and development. Since their origin in the late 1940s, these models of physical and biological processes in and around agricultural production have been integral to enormous progress in understanding of agricultural production processes and environment. Models have played significant roles in research problem setting and, occasionally, in applying scientific understanding to public policy. But expectation of eventual benefits to farmers of the decision support system has been a significant basis of financial support for modeling during the 1980s and into the 1990s. Although evidence is diffuse, our perception is that this support has faded since the mid-1990s with the failure of any significant market to emerge among farmers or their advisors.

More conspicuous has been the experience of ICASA (International Consortium for Agricultural Systems Applications), formed in 1995 to aggregate and position elite modeling groups to market a service to international agricultural research and development organizations (www.icasanet.org). Most observers would agree that there has been a disappointing level of financial support for ICASA among agencies expected to benefit from further development of models and applications in systems research.

In Australia, where demand for decision support systems has been no greater than elsewhere, significant support for modeling has been part of a major institutional shift away from research stations to research with a farming systems perspective and whose conduct involves farmers (Carberry, 2001). In addition to this driver of research reform, i.e., relevance to efficient farm management, there is the increasingly important driver of pressing environmental problems. On both fronts, the adequacy of traditional agricultural research approaches is being questioned, and there seems to be a new recognition by research stakeholders that simulation modeling, imbedded in the appropriate methodology, may be an important element of needed RD&E innovation.

This chapter summarizes the collective and ongoing experience of a group of researchers concerned with exploring how simulation may benefit the management of dryland farming systems in Australia. Over 15 years much change has taken place both in the models and in the ways in which researchers are using them. The primary aim here is to relate and reflect upon two intertwining threads of learning:

1. What capabilities in a simulator are important and feasible to develop and maintain?
2. In using simulation to aid farmers, and other system managers who influence or are influenced by farm production system performance, what are the keys to genuine usefulness that creates an ongoing demand for simulation?

Although major activities and learnings have not always been strictly sequential, it is convenient in telling the story here, to organize experience as phases. In the initial phase the team used what they judged at the time to be the most appropriate crop model for research aimed at contributing to better farming practices in semiarid dryland farming systems. This was a period in which both the power of dynamic simulation models in enhancing management research and the limits of the existing models for dealing with important phenomena in such systems became apparent. This phase was followed by one of rethinking and re-engineering models to overcome significant model limits. Overlapping with this phase was one that focused on using models for aiding farm decision making by working closely with farmers and their advisors on farms. What was learned in this activity has led, in a current phase, to diversification of the way simulation models are used to assist client managers and to new markets for simulation-aided services. In this ongoing phase, simulator development continues in response to expanding diversity of crops and practices and demand by users for features that aid effective and efficient data management, simulation, presentation, and Internet transmission. Demands from important "system analysis and design" activities now compete with demands from "management discussion facilitating" and "farm management consulting" activities as well.

FIRST PHASE: USING CROP MODELS TO DESIGN BETTER FARMING PRACTICES AND DISCOVERING MODEL LIMITATIONS

Research on Semiarid Dryland Farming Systems

In the decade of the 1980s, simulation modeling became important to research on dryland farming systems in semiarid, tropical regions of Australia and Kenya. Although very different socioeconomic systems, they were similar in two important ways:

1. Economic feasibility of investment and intensification for greater production was made problematic by climatic risk.
2. Testing feasibility, either in practice or in an experimental research program, was problematic due to high rainfall variability.

Feasibility of Cropping in Semiarid Northern Australia

Until recent years, the agricultural development potential of northern Australia has been uncertain, and this uncertainty allowed rapid recovery of optimism following failed development ventures (Chapman et al., 1996). A sufficient lapse of time, favorable economic conditions, and a conspicuous run of good seasons seem to have been sufficient to conclude that earlier failures were due mainly to factors other than an unsuitable climate. During the period 1978 to 1992, a significant research effort was made to reevaluate the potential for cropping in the Australian semiarid tropics. Previous failed initiatives had highlighted climatic risks, but there was disagreement about the importance of low rainfall versus the affordability of adequate soil conservation measures and the importance of these in relation to infrastructure constraints. Annual crops in this climate are at high risk especially during establishment and anthesis due to highly variable rainfall and high radiation load resulting in high evaporative demand and high soil and air temperatures (Abrecht and Bristow, 1996). The potentially arable soils are of low fertility, low water holding capacity and of poor structural stability, and, under conventional tillage, highly vulnerable to serious water erosion (Dilshad et al., 1996).

In the Northern Territory (NT), this research initially concentrated on soil management systems, but later expanded to include climatic variability and production risk. In 1985, we undertook simulation modeling to complement field research on the climatic and soil constraints to dryland cropping and to develop and evaluate cropping practices that reduced risks and costs. This research was in three key areas:

1. Quantifying the yield and economic returns from conventional dryland maize and sorghum enterprises (Muchow et al., 1991)
2. Testing the feasibility of a new dryland cropping system which centered on the use of no-tillage technology and the integration of livestock into the cropping system with the inclusion of pasture *legumes as leys and intercrops (McCown et al., 1985)
3. Exploring the potential for a pulp and paper industry based on a totally new crop, kenaf (*Hibiscus cannabinus* L.) (Carberry et al., 1993a)

In addition, models developed in the NT were used to explore dryland cropping prospects on the dryer margins of established cropping regions in north Queensland (Carberry et al., 1991a).

Based on the criteria of being conceptually appropriate to the research issue, of having affordable input requirements, and capable of realistic simulation of performance, CERES-Maize (Jones and Kiniry, 1986) was initially selected in 1985 and tested for northern Australia (Carberry et al., 1989). After testing in this severe environment, CERES-Maize was modified to improve simulation of the effects of soil water deficit and extreme temperatures on crop establishment, phenology, leaf area development, grain set, and plant mortality (Carberry and Abrecht, 1991). The nitrogen supply routines were also modified with a number of improvements, principally to permit the simulation of surface residue dynamics (Dimes et al., 1996).

As part of this early phase of model evaluation and development, new models for sorghum (Birch et al., 1990; Carberry and Abrecht, 1991), kenaf (Carberry and Muchow, 1992), and *Stylosanthes hamata* (cv. Verano) (Carberry et al., 1992) were developed and validated for use in northern Australia. Together with the modified CERES-Maize, these enabled a systems analysis approach to evaluation of prospects for cropping in northern Australia. For maize and sorghum, these studies include the simulation of yields and assessment of risks to cropping at different locations, for different genotypes, for a range of planting times, and for different tillage strategies (Cogle et al., 1990; Carberry and Abrecht, 1991; Muchow and Carberry, 1991; Carberry et al., 1991b; Muchow et al., 1991). These evaluations provided a sobering picture of low expected economic returns. The prospects for intensive dryland cropping in semiarid northern Australia appeared bleak unless the cost–price situation for coarse grain crops changed dramatically.

But the hypothetical integrated crop-beef grazing system under evaluation recognized the viability of the existing local beef production system based on extensive natural pastures as well as the success of crop-sheep grazing systems based on legume pasture leys in Australia's Mediterranean climatic regions. This hypothetical system for the Australian semiarid tropics (SAT) featured legume pastures grown in rotation with crops of maize and sorghum, which were used to fatten cattle (McCown et al. 1985). Crops were sown directly into chemically killed pastures that provided both protection from high soil temperatures and a source of mineralizable nitrogen. An understory of volunteer legume was permitted to establish from hard seed to form an intercrop with the grain crop. Cattle grazed native grass pastures on surrounding land during the wet season and were brought back to the cropland to graze crop residues and the volunteer legume pasture during the dry season. Although most aspects of the proposed integrated system were tested in agronomic experimentation over several years (McCown, 1996; Carberry et al., 1996a), the long-term potential under this system was also tested in simulation analyses (Carberry et al., 1993b, 1996b; Jones et al., 1996). Such analyses indicated that a combination of legume pasture and sorghum grain production was superior in terms of both gross margin returns and long-term soil fertility status compared with conventional coarse grain production.

Another regional development proposal has been for a pulp and paper industry based on the fiber crop, kenaf, but there has been uncertainty about effects of weather on long-term continuity of adequate supply of fiber to mills. A major feasibility study, funded by the NT government, was undertaken to assess the climatic risks to dryland kenaf production, using a simulation model of kenaf developed and validated for this task (Carberry et al., 1993a). The kenaf model was run, using long-term historical weather data, to determine optimal sowing strategies and expected yields at four representative sites in the NT. A conflict existed between sowing early, with resulting long duration and high yield potential but high probability of plant mortality, and sowing later, with more reliable plant population but shorter duration and lower yields.

Since the late 1980s, *ex ante* analyses of the climatic risk to innovative production systems using locally specified and validated simulation models have contributed to development of policy both by government and private investors for dryland cropping industries in the semiarid tropics of northern Australia. The cautionary results of the analyses have complemented other cautionary information, and no such industries of national significance have yet been established (Chapman et al., 1996). The analyses have provided an enhanced understanding of the climatic realities, largely based on simulation with well-tested models (validated, in part, by failed commercial ventures). The models are a resource to assist future evaluations of new, inevitably risky, development propositions in the future.

Feasibility of Purchased Fertilizer in Smallholder Crop Production in Kenya

In 1985, in conjunction with the Australian Centre for International Agricultural Research (ACIAR), we began working with researchers from KARI (Kenya Agricultural Research Institute) on prospects for improvement in food security in semiarid eastern Kenya (McCown et al., 1992). The focus was on maize production strategies on smallholder farms where traditional soil fertility management systems had broken down due to population growth and farm fragmentation. Was there a place for commercial fertilizer inputs in these areas where the climatic risk was generally seen to preclude this option for soil fertility maintenance in a predominantly subsistence system? Over the next 6 years, we sought to answer this central question as well as explore a range of other management issues relevant to productivity in the maize-based farming systems of the region.

As we experienced the dominance of rainfall variability in this 500 to 700 mm bimodal rainfall environment (where two maize crops a year are attempted) on experimental results and the overall performance of different technologies, we realized that a maize model would be a valuable first step. The reasoning that had led to the selection of CERES-Maize (Jones and Kiniry 1986) for analysis of system performance in northern Australia appeared equally valid in Kenya. This model addressed a range of factors that were generally important in maize production (plant population, genotype characteristics, time of sowing, and water and nitrogen constraints), but a model of this type had not been "stretched" to mimic conditions in a smallholder agriculture setting in such a harsh environment before. In common with the work proceeding in parallel in northern Australia, the abiotic environment placed more severe constraints on crop growth and yield than was the case for high-input systems in temperate and humid tropical environments, which provided much of the basis for CERES-Maize initial development.

A major on-station experimental program over the 1985 to 1989 period resulted in many changes to CERES-Maize that improved its predictive performance in these semiarid, low-input environments (detailed in Keating et al., 1992a). Important changes included:

- Development of plant mortality routines to capture severely limiting water and nitrogen effects during vegetative growth
- Making the phenology model more sensitive to extreme water and nitrogen stress
- Making the grain number model better reflect yield response to plant population, and in particular the low plant populations that characterized resource-poor systems
- Redesign of the leaf area determination routines (Keating and Wafula, 1991) to improve performance over a diverse range of maize maturity classes
- Development of genotype parameter coefficients and related algorithms that were effective for the germplasm under consideration, which were generally outside the range then covered by CERES-Maize

The model that emerged from this testing and revision (referred to as CM-KEN to distinguish it from its parent CERES-Maize) provided a good overall predictive capability. Of special importance for our main objective of evaluating the economics of purchased fertilizer in this farming system was its competence in simulating the interactions between the management factors of plant population, water supply, and nitrogen supply. Over the period 1988 to 1992, this model proved to be a core tool in exploring implications of innovative management options, complicated by interactions between N fertilizer inputs and water climate. The first use of a simulation model to explore plant density effects under variable climates was reported by Keating et al. (1988), and the first application of this generation of simulation model to N fertilizer use studies in smallholder agriculture was reported by Keating et al. (1992b, 1991). This work showed that good long-term returns appeared to be possible, but the fact that N fertilizers were moderately high-risk in the season of fertilizer application was inescapable; and much depended on in-season management control in relation to an uncontrollable water supply.

A major part of the work in Kenya evaluated a tactical planning and fertilizing strategy called "response farming" as a means of reducing risk associated with crop production. This strategy involved adjusting plant populations and nitrogen fertilizer inputs in "response" to the timing and amount of early season rainfall. The concept had been developed earlier in this district (Stewart and Faught, 1984), and while intuitively appealing, there was no way of evaluating its efficacy in reducing risk without an adequate simulation capability. This required a maize model capable of processing complex management rules concerning events and actions in advance of crop planting, and subsequently imposing management actions such as thinning or fertilizer side dressing during early crop development. All these management actions needed to be conditional on the timing, pattern, and level of early season rainfall. CM-KEN and a visual and interactive version of CERES-Maize, CERES V/I (Hargreaves and McCown, 1988) were elaborated to enable them to be used to explore the benefits and risks associated with response farming. Analysis of individual seasons showed important benefits from tactical adjustments in response to goodness of season. But if the time scale of analysis was extended to *sequences* of seasons, the most important benefit was from routine use of *some* N fertilizer (Wafula et al., 1992; McCown et al., 1991). The most important tactical benefit of response farming was the saving of costs by the withholding of fertilizer application in those seasons where indicated prospects were poor (Keating et al., 1992b).

Despite what appeared to be a desirable and feasible production strategy, researchers were increasingly conscious of the social, economic, institutional, and cultural factors that were likely to be inhibiting fertilizer usage in these systems (McCown et al., 1992). These realizations were important in shaping later work to include a much greater emphasis on the human side of the farming system, and to be more measured in our expectations for biophysical simulation modeling and decision support systems to underpin useful change in real-world practice (McCown et al., 1993). We return to this shift of emphasis after discussion of how this experience influenced simulation software development.

Lessons for Model Development

By 1990, a comprehensive crop modeling capability for the semiarid tropics had been developed. It was done cost effectively by appropriate modification of an established product, i.e., CERES Maize; however, coverage of the major crops was limited (maize, sorghum, kenaf, and the forage legume, *Stylosanthes hamata* cv.Verano), and we could not address key "systems issues" such as tillage, erosion, nutrients other than nitrogen, crop rotations, and competition between intercrops or crops and weeds.

These systems issues revealed inadequacies of the scope and architecture of the model of a crop. By the late 1980s, researchers became increasingly cognizant of the limitations of their approach of continuously elaborating the code of a model of a crop to address systems issues. They were interested in interactions between crops sequentially and spatially and wanted to properly account for the effect cropping was having on the soil. In addition, they were increasingly ambitious in the conditional management strategies and tactics that they wanted to represent within the models. Their software was evolving in ways that made maintenance and further development awkward and costly. Adding each extra item of functionality caused ripples to go through the entire model code, necessitating an infeasible level of testing and repairs to ensure prior functionality was maintained. It was hard to see how continuation could deal with new needs, and by 1990, they opted for fundamental reengineering.

PHASE OF SOFTWARE REDESIGN AND REDEVELOPMENT

A major lesson from the efforts in Phase 1 was the discovery of the limitations of good crop simulation models when taken so far from the environmental and research domains in which they were developed. Research in Kenya and northern Australia provided grounds for key modifications. Modifications to CERES-Maize were sufficiently numerous and substantial to justify recognizing the derivative CM-Ken and CM-Sorghum (SAT) as distinct products (Keating et al., 1991; Carberry and Abrecht, 1991). A particularly important bridge to the subsequent phase of developing a cropping systems *simulator*, the early term used by Baker and Curry (1976), was the reengineering of CERES-Maize to enable interactive simulation. In Visual-Interactive (V/I) CERES-Maize, events and trends of state variable are dynamically displayed and runs can be interrupted at any time in order to interrogate output files, change settings, etc. (Hargreaves and McCown, 1988). The experience using V/I CERES-Maize to deal with complex management rules helped shape approaches in later manager modules.

Reengineering efforts progressed in the late 1980s to design a cropping systems simulator that overcame limits of existing available software. The assessment was that existing models fell into two distinct classes (McCown and Williams, 1989). One class was crop-oriented and aimed at accurate simulation of crop yields over a wide range of environmental conditions and genetic attributes, e.g., the CERES family of models.

The focus of the second class of software was on simulation of soil processes and management, e.g., EPIC (Williams, 1983) and NTRM (Shaffer et al., 1982). Between them, EPIC and NTRM simulated soil erosion and sedimentation, organic matter changes, nutrient dynamics, and they were actively expanding capabilities for dealing with other soil phenomena. The absence of such soil management orientation in our adapted CERES models was a severe limitation. A further attractive feature of both EPIC and NTRM was provision of a wide range of crops that could simulate diverse crop sequences. But the crucial deficiency was that this comprehensiveness and flexibility was achieved by use of simplified crop routines that did not have the degree of sensitivity to environmental extremes required for risk analysis in our climates (Williams et al., 1989; Steiner et al., 1987).

A peculiar need arose from the fact that, although analysis of risk in dryland cropping required high crop model sensitivity to environmental extremes, management of the risk to a crop generally involves actions or events that take place well prior to the planting of the crop, and indeed may be associated with the previous crop. (DSSAT later provided a framework for efficiently and flexibly accessing a suite of environmentally sensitive crop models as well as parameter and input data [Uehara and Tsuji, 1991], but failed to achieve the required 'systems' functionality of the soil-oriented simulators.) A simulator that combined the strengths of both classes of models was needed, and to get this, it appeared that the team had to make it.

McCown and Williams (1989) reported plans and progress concerning a design that featured three attributes:

1. Crop models with sufficient sensitivity to extremes of environmental inputs to predict realistic yield variation for risk analyses
2. Models to simulate trends in soil productivity as influenced by management, including crop sequences, intercropping, and crop residue management
3. Modular software that enables efficient evolution of the simulator by research teams

The core need was for an architecture that overcame the conflict between comprehensive system representation (agronomically important soil phenomena and a wide range of crops, as in EPIC and NTRM) and comprehensive treatment of crop physiology that conferred sensitive prediction of crop yield as in CERES. Because any given run configuration would require only a fraction of the total code, in the interest of minimizing run time the structure needed to be such that only required code would be processed. This was achieved by the concept of "plug-in, pull-out" modularity. The basic system being simulated is the soil profile as influenced by weather, crops, and management. Even when primary interest is in crop production, this architecture is advantageous because of the simplicity with which bare fallows, crop sequences, or crop mixtures can be configured by controlling only *what* crop modules are plugged in *when*. Multiple crop modules plugged in together compete for light, water, and nutrients using only the code for each crop behaving singularly and with access to resources regulated by a simple "arbitrator."

The formation of the Agricultural Production Systems Research Unit (APSRU) in 1991 brought together a CSIRO cropping systems team with a team at the Queensland Department of Primary Industries that had developed PERFECT (Littleboy et al., 1992). PERFECT combined a soil-oriented system with crop models sufficiently elaborate to have the desired sensitivity for risk analysis but, similar to CSIRO software, suffered a lack of good design and engineering process. The resultant and ongoing joint venture has produced APSIM (Agricultural Production System sIMulator) (McCown et al., 1996).

APSIM v.1 was designed around the plug-in, pull-out modular concept, shown as a "hub–spoke" construct in Figure 8.1. All modules communicate with each other only by messages passed via the "engine" at the hub. Crops appear in the system as a consequence of management decisions, find the soil in some state, expire as a matter of course or are terminated by the manager, leave residues, and leave the soil in a different state. Using a standard interface protocol, this design enables easy removal, replacement, or exchange of modules without disruption to the overall operation of the system. It surpasses its predecessors in ease of representation of complex crop sequences or mixtures and dynamic simulation of the temporal and spatial interactions. From 1991 to 2001, and as a result of substantial investment, APSIM has developed in multiple directions. Space does not allow a historical account of these developments, but in a later section, an overview of APSIM as it stands in 2001 is provided.

PHASE OF REENGINEERING MODEL-BASED DECISION SUPPORT FOR FARMERS

In 1991, interest in decision support systems was at its zenith, and it was expected that a major focus of this new APSRU modeling team would be to produce appropriate decision support software for farmers. The starting point for this activity was a workshop with a group of elite farmers and extension experts. A surprising degree of skepticism and criticism expressed by farmers about the relevance of models to their management made this event a profoundly sobering experience. It appeared that the gap between farmers' management techniques and scientists' visions of the potential for simulation in management might be too wide to be bridged by mere talking. The stakes were too high to stop short of testing the hypothesis about the utility of simulation for farm management in real management situations.

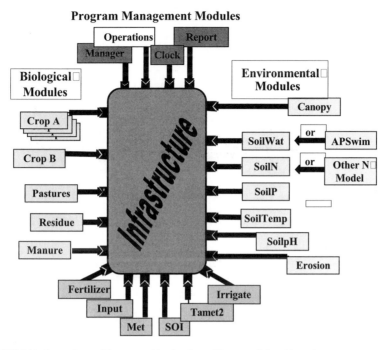

Figure 8.1   APSIM hub–spoke architecture with plug-in, pull-out modules. More than one crop module plugged in invokes aboveground competition via "Canopy" and belowground competition for water and N. APSwim is based on the Richards equation.

The focus for this phase of research had shifted to the subtropical Darling Downs region of northeast Australia, characterized by intensive commercial dryland farming systems on productive self-mulching vertisols and in a highly variable rainfall regime. This trial was structured as a program of action research with the aim to find a way to use simulation in risky farm management — a methodical trial and error development of a methodology (in the manner of Checkland, 1981). The essential features of the situation were:

- Management dominated by uncertain rainfall
- An information technology with potential to alleviate this problem
- Weak farmer enthusiasm for the existing methodology for using this technology (the decision support system)
- Researcher conviction that a successful methodology was most likely if simulation modeling was researched in the context of farm management practice

Significantly, the activities did not initially feature simulation, but centered on reducing management uncertainty by enhanced monitoring of soil water and nitrogen (Dalgliesh and Foale, 1998). Measurements were often in the context of simple management experiments using treatment strips in commercial crops. This focus, besides capturing the farmers' attention regarding matters of perceived importance and managerial deficiencies, provided researchers with data needed for subsequent simulation of specific farms and fields. The soil data often provided satisfying explanations of differences in crop performance. But the inevitable question of "What if we had done this last year?" highlighted the limitation of this approach and opened the door for answering the "What if?" question using simulation. The subsequent evolution of simulation-aided group discussions about farm management is the core element of a methodology that evolved over a period of several years, i.e., the FARMSCAPE approach (Farmers, Advisors, Researchers, Monitoring, Simulation, Communication, And Performance Evaluation) (Hochman et al., 2000).

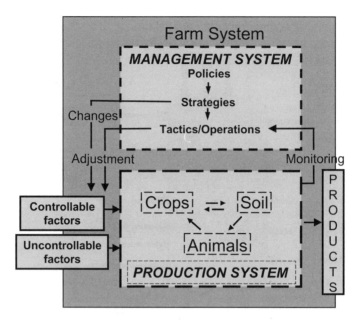

Figure 8.2   A simplified model of the farm system, depicting management as normative, instrumental, and cybernetic. (After Sorensen, J.T. and Kristensen, E.S., *Global Appraisal of Livestock Farming Systems and Study on Their Organizational Levels: Concept, Methodology and Results,* Commission of European Communities,Toulouse.)

The FARMSCAPE approach departs radically from the traditional concept of scientific decision support for farmer practice. Of operational significance, the simulator is "run" by an intermediary — a facilitator or service provider — not by the farmer. Departures of a conceptual nature can be seen with the aid of the representation of a farm system in Figure 8.2. Of the several subsystems that could be depicted as comprising a farm, only two are shown here. Agricultural science and simulation models are about the production system. Decision support is about what managers should do with regard to the production system, informed by scientific understanding. But in Figure 8.2, the operational emphasis is on the *cybernetic* relationships between the management system and the production system, characterized by actions to control production, feedback from the production system gained by monitoring in various ways, and further adjusting actions. The monitoring and site-specific simulations in the FARMSCAPE approach reflect the local and responsive nature of management and the need for simulation to capture this if it is to be seen by farmers as relevant to their management situations. The most significant learning by researchers from the FARMSCAPE experience concerns the importance to farmers of simulations being situated in their practice to be meaningful. An important aspect of this is specification of the simulator using local soil and climate data. Equally important is the origin of the simulation as an inquiry by a farmer seeking understanding or foresight. "Policies" in Figure 8.2 indicate the reality of a context of high order personal–household–cultural guides and constraints for management of production systems. Although well outside the boundaries of simulators of production systems and decision support systems, their influence in real decision making is implicit in the dynamics of the participatory "What if…? Analysis and Discussion" (WifAD) sessions.

For a simulation to be taken seriously by a farmer, it must be more than notionally relevant; it must be seen as significant for changed management action. But for this to happen, the action–outcome inferences must be credible. Every farmer with whom the scientists worked had to establish the simulator's credibility before enthusiasm and strong demand for WifADs developed. In the main, this was achieved by demonstration of successful simulation either using special collaborative projects to collect the necessary data from relevant commercial crops (often overlaid by simple treatment comparisons) or by simulations that matched with either farm records or memory of past crop yields (Carberry and Bange, 1998); however, with time, there appears to be an increasing readiness of farmers to accept as meaningful the validating experiences of respected farmer colleagues in their district.

The reluctance and skepticism of farmers to take simulations seriously and participate in their practice highlights deficiencies of the typical decision support system. For reasons of economy in the overall process, these are generally not designed to fit real situations, but only deal with the logic of situations (Checkland, 1981). Even if the scientific and economic logic of such products is sound, if farmers do not find treatments of issues locally meaningful, the products will not be used.

Finally, the FARMSCAPE approach represents a paradigm shift in interfaces between scientific knowledge and practical knowledge. The main process in a decision support system is intervention in farmer practice with a science-based recommendation for best practice. The main process in a WifAD is very different. If a production system simulator can be conditionally accepted as a substitute for a real production system in Figure 8.2, then management possibilities can be tested virtually very quickly and cheaply. Figure 8.2 can be seen as a learning cycle: starting on the left and moving counter-clockwise, actions as adjustments, or, alternatively, deeper structural changes, are taken. Production System consequences of the action are simulated, outcomes observed, overall implications deliberated in the management system, and a new priority for action constructed. This is similarly tested in the next cycle.

This use of a simulator in action learning has been described by Bakken et al. (1994), p. 246:

> The goal of a learning laboratory is to provide an environment that will help enrich managers' mental models using tools such as the management simulators. Learning laboratories help managers leverage their domain-rich knowledge by allowing them to play through simulated years, reflect on their actions, modify their mental models, then repeat the process. By compressing time and space, flight simulators can accelerate learning by enabling them to conduct many cycles of action and reflection.

Formal evaluation of participants' experiences in the FARMSCAPE research program has indicated farmers appreciate that this is the nature of WifADs (Coutts et al., 1998). This learning from the behavior of the production system (Figure 8.1) places great importance on simulated behavior being realistic, and prior investment in establishing this for the situation in question is indispensable. Researchers also found that farmers often benefit from analyses and discussions that make the functioning of the production system more intelligible. Just as stated by Bakken et al. (1994, p. 250):

> The simulator also demands structural explanations of the "action → result" link that will force participants to search for a better understanding of the underlying forces that produce a given set of outcomes.

WifADs are free-flowing discussions that follow the directions of farmers' interests within the domains of the system represented in the simulator. Four functional types have been distinguished:

1. Yield benchmarking
2. Production decision support
3. Marketing decision support
4. Analysis of consequences of possible management change (Hochman et al., 2000)

These simulator applications address variously the two types of complexities (Senge, 1990). The contribution to yield benchmarking is reduction of detail complexity. This serves to aid insight (the reduction of detail that masks structure) that helps explain why the crop did what it did and what effects altered management actions could have made. In the last three types of WifADs, the function is reduction of dynamic complexity (Senge, 1990, p. 71) — complexity that includes variable outcomes from the same action taken repeatedly in the same circumstances. Instrumental to the contribution of simulation in reducing this complexity has been an emergent method in climate forecasting based on the Southern Oscillation Index (Stone et al., 1996). Additional atmospheric pressure information, available for all years of weather records, provides a basis for identifying a set of "analog years" in which simulated outcomes for a specified action are more homogenous than for the entire population of years (Hammer et al., 2000).

Shifting from the notion of producing decision support systems for farmers to use of situated simulation in a FARMSCAPE approach is in line with a major paradigm shift in systems thinking from "hard" to "soft" (Checkland, 1981). It is also in line with a shift in cognitive science from using computers mainly to compensate for managers' cognitive deficiencies in their decision making to mainly using computers to aid in constructing new understandings and new possibilities for future management (Clancey, 1997; Winograd and Flores, 1986). Such a shift to the FARMSCAPE approach has resulted in significant and demonstrable achievements in bringing benefits to farmers, broader agriindustries and the research community in northern Australia (Carberry, 2001).

Another general lesson that emerged in this phase of using and adapting existing models was the importance to the value of simulation in real farm management research that data for parameterization, initial conditions, and weather inputs are from the site. In contrast with much modeling in crop physiology research in which environmental settings can be abstracted to "scenarios," simulations were used primarily to virtually enlarge the sample size of years in which field monitoring and experimentation were conducted. There was recognition that efficient systems of weather monitoring, soil characterization, and soil monitoring needed to evolve together with simulation capability.

These developments in the use of the simulator to enable farmer education and management have placed a number of calls upon the model development effort. The demand for a comprehensive simulator, addressing the major crops and constraints in the farming system in realistic ways, is something that is needed in both management support and systems analysis and design applications. The reengineered approach to decision support that FARMSCAPE represents has placed additional demands on user interfaces, graphics tools, and database resources. The APSIM suite of tools contains a user-friendly interface, a flexible graphics tool, and a database tool for storing, manipulating and sharing soil properties data. These interface tools have much in common with the interface trappings of some decision support systems; however, because they are designed for a trained intermediary in the FARMSCAPE application instead of casual use by an untrained user, the interfaces can generally be more flexible and powerful.

THE CURRENT PHASE: MULTIPLE THEMES FOR ENHANCING SIMULATION OF PRODUCTION SYSTEMS

Since 1995, four themes have characterized CSIRO's research and development:

1. Continued development of APSIM as software in response to needs arising from different applications, from opportunities and investment aimed at improving its science base and from growing experience in software engineering
2. Continued use of APSIM in R&D for functional design of potentially superior agricultural systems by identifying feasible or optimal strategies from a larger set of possibilities
3. Development of a delivery system for farm-situated simulations — FARMSCAPE training and accreditation for agribusiness consultants
4. FARMSCAPE Online — Development of Internet video-conference interactions between researchers and farmer groups centered on "What if?" analysis and discussion for specific situations of farmer participants

APSIM Development: 1995 to 2000

By 2000, APSIM had reached version 1.6. Vegetation modules had been developed for barley, canola, chickpea, cowpea, fababean, mungbean, navy bean, hemp, wheat, lucerne, maize, peanut, millet and pigeonpea (in association with ICRISAT), sorghum, sunflower, sugarcane, and cotton (the OZCOT model in association with CSIRO Plant Industry). A FOREST module has been used for *Eucalyptus*, *Pinus*, and other woody vegetation. A MICROMET module is available for treatment of energy and water fluxes in mixed canopy situations. Soil and related modules include models for soil water, nitrogen and phosphorus balance, soil surface residue decomposition, soil erosion and soil acidification. In some cases, alternative model representations are available as options. For instance, SOILWAT is a layered tipping bucket soil water balance model (Probert et al., 1997), and SWIM (in association with CSIRO Land and Water) is an implementation of Richard's Equation for water movement and the Convection–Dispersion Equation for solute movement (Verburg, 1996). APSIM vegetation modules generally include water and nitrogen as limiting factors, with phosphorus limitation currently under development and, at present, only operational for maize. Powerful and flexible control of management in the simulation remains a feature in APSIM, with a pedigree that traces back to the Response Farming rules in CM-KEN and V/I CERES-Maize. The MANAGER module in APSIM utilizes a custom-built script language and compiler that enables users to comprehensively specify complex and conditional management rules for all aspects of a simulation.

Investment in software engineering process for the APSIM effort was stepped up in the mid-1990s. This was in recognition of the complexity of the task of managing a large software project that involved simultaneous development efforts by different programmer and modeling teams. A version control and regression testing system is central to the software engineering process. This system enables any past version of APSIM to be recreated and ensures code changes take place in an ordered way. Regression tests are automatically run every evening and reports provided on any changes in system performance. When unexpected or undesirable changes in performance are detected, action is taken immediately to investigate the causes. Other key aspects of APSIM software engineering process include formal documentation and peer review procedures, code analysis tools, code auto-documentation tools, as well as Web-based defect reporting and change request logging procedures. An APSIM support web site (www.APSIM-Help.tag.csiro.au) provides assess to documentation and other support materials for users as well as restricted access to software engineering support materials for developers.

APSIM has been made available to individuals and groups *via* a license system that ensures and orderly development and support effort. As of February 2001, over 250 such licenses have been issued and the model has been used in all Australian States, and with national and international agencies in Africa and Asia.

Version 2 of APSIM was released in February 2001. The key new capability is support for the development of multipoint simulations, as the underlying software infrastructure can create multiple instances of any module. This new capability is starting to be applied to the simulation of multi-paddock crop-livestock systems and the simulation of agroforestry systems. APSIM v2 infrastructure is written in C++ and modules are contained within dynamic linked libraries (DLLs). These developments mean that modules written in different programming languages can generally be linked into the software system. The latter infrastructure developments are part of a joint effort with the GRAZPLAN/GrassGro developers in CSIRO Plant Industry (Donnelly et al., 1997) to develop a common modeling protocol to facilitate linkages between different modeling software entities. This protocol will be available on www.APSIM-Help.tag.csiro.au from June 2001.

APSIM in Farming Systems Analysis and Design: 1995 to 2000

As APSIM has developed from a limited single crop simulator to a comprehensive simulator of farming systems, the scope of the issues to which it has been applied has widened. All applications are characterized by an intent of system analysis to identify some feasible or optimal strategies or designs from a larger set of possibilities. This is policy research (Figure 8.3) and is quite different from the action research that characterizes the FARMSCAPE approach. Some key applications of APSIM in formal diagnosis and design are summarized in Table 8.1.

Knowing how to manage a situation

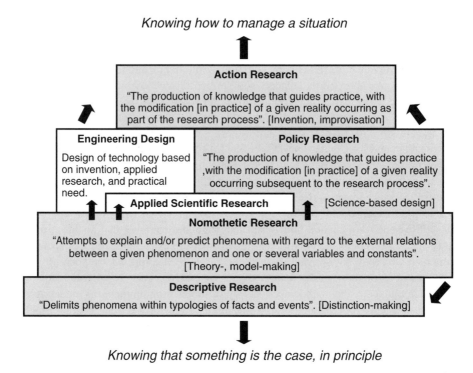

Figure 8.3   Different ways to conduct research to achieve different types of knowledge. (The four-category typology is from Oquist, P., *Acta Sociologica*, 1978.)

This portfolio of applications continues to grow as new needs arise. The important finding is that, unlike our experience in the late 1980s, the integrity of the simulator software is maintained as additional capability is added. Software maintenance costs have not blown out as the model's scope has broadened. If anything, these costs have been reduced in recent years as the benefits of investment made in the mid-1990s in improved software engineering process have manifested throughout CSIRO's software development and maintenance activity.

FARMSCAPE Training and Accreditation

By 1998, over 230 farmers had engaged in FARMSCAPE research activities, the research had created a market demand by farmers for such interactions, and the systems research priority had shifted to development of a sustainable system for delivery of a customized service. It became necessary to scale up a FARMSCAPE service fast enough to prevent farmer disillusionment due to unmet high expectations, but at the expense of the quality of system simulation and human interactions necessary to retain the level of interest and confidence that created the demand.

Implicit in this mode of using simulation is a professional who is skilled in using the simulator and interpreting simulations of real farming situations, and who, generally, (but not necessarily) leads the "What-if?" discussions with farmer groups. Evolution in this direction has taken place in a farm service environment characterized by decline in publicly funded extension and increase in various forms of commercial consulting. Farmers who were enthusiastic about FARMSCAPE were prepared to pay for a service. In order to pilot the feasibility of provision of such a service, an in-business action research activity, analogous to earlier on-farm research with farmers was initiated. Researchers worked for a period within a commercial advisory firm that had an interest in gearing up to provide a FARMSCAPE service to farmers. The aim was for researchers and consultants to learn together in order to invent together feasible approaches for commercial service provision as well as training and support by researchers for service providers.

Table 8.1   Summary of Major Applications of APSIM to the Design or Enhanced Performance of Agricultural Systems

Application	Reference
Assessment of long term consequences of crop-pasture rotations, including changes in soil C/N status	Carberry et al., 1996b
Assessment of long-term impacts of tillage and fertilizer management on crop productivity and soil fertility	Probert et al., 1995
Assessment of management and climate on soil acidification	Hochman et al., 1998b
Assessment of long-term impacts of residue retention on sugarcane productivity and soil organic matter dynamics	Thorburn et al., 1999
Identification of optimal crop management practice to minimize nitrate leaching to groundwater from fertilizer in irrigated sugarcane production systems	Keating et al., 1997
Assessing water balance of cereal-lucerne systems	Dunin et al., 1999
Design of strategies to address soil structure degradation in cereal–pasture rotations	Connolly and Freebairn, 1996
Maize-weed interactions in relation to fertilizer effectiveness in smallhoder systems in Zimbabwe	Shamudzaria et al., 1999; Keating et al., 1999
Optimizing irrigation inputs in sugarcane production systems	Muchow and Keating, 1998
Design of on-farm water storage in sugarcane production systems	Lisson et al., 2000
Assessment of nitrogen fertilizer strategies in sugarcane systems	Verburg et al., 1996
Design of effluent irrigation practices for sugarcane, pastures and tree crop systems	Snow et al., 1999
Assessment of severity of drought for government policy implementation on assistance under exceptional circumstances	Keating and Meinke, 1998
Evaluation of seasonal climate forecast systems	Meinke et al., 1996
Evaluation of cropping potential in new regions	Cogle et al., 1990
Assessing the benefits associated with intercropping systems	Carberry et al., 1996c
Design of farming systems to minimize deep drainage below root zones to restrict development of dryland salinity	Paydar et al., 1999; Ringrose-Voase et al., 1999; Asseng et al., 1998
Assessing the erosion and productivity impacts of maize/shrub legume hedge-row cropping in the Philippines	Nelson et al., 1998a,b

The outcome of deliberations on training and support was a plan to:

1. Establish a FARMSCAPE training, accreditation, and support program to provide both the scientific and technical training and on-going support for advisors to provide a service enhancement based on competent APSIM simulations.
2. In collaboration with accredited advisors, continue to learn and develop more cost-effective and sustainable mechanisms for delivery of the FARMSCAPE approach in assisting farmers' learning, planning, and decision making.
3. Progress FARMSCAPE tools and methods to suit a range of industry users.
4. Evaluate the success of and learn from the FARMSCAPE approach to RD&E delivery and its impacts on management decisions.

Two aspects of commercial consultant situations strongly influenced the choice of methodology:

1. Trainees are already accomplished professionals in their businesses, and their inputs, drawing on their expertise and experience, were an important determinant of the adapted service product
2. Trainees are busy practitioners in a highly competitive environment.

The research approach that nicely maps onto these realities is that of action research (Figure 8.3), which features learning-in-action rather than being trained by instruction (Schon, 1983). Learning projects are collaboratively designed and conducted within the consultant's practice, with the high degree of mentoring and support required initially declining over time. The research team treats each individual project as a case study, which is documented, evaluated, discussed, and compared with others as a way of learning how FARMSCAPE suits consulting, what adaptations consultants make to the approach, and what changes might be needed in APSIM or in the training program.

After publicly advertising, four companies were selected from eight applicants to participate in the initial training and accreditation program now in progress, each with two participants. The training program and associated on-going support was designed to provide:

- A high level of expertise in the use of APSIM and a sound appreciation of the underlying science
- Internet access to weather data for regions of interest in the Australian cropping belt that is both long-term and updated regularly enough for yield forecasting
- Internet access to a GIS of measured soil properties needed for APSIM and the means to add significant locations to this database
- The ability to measure initial soil conditions cost-effectively

The training program consists of the following modules/competency areas.

1. Soil monitoring and data management — principles, techniques, and quality assurance
2. Weather monitoring and data management — principles, techniques, and quality assurance
3. APSIM — the program and the science
4. Simulation applications in farm management
5. Analysis of simulation results and quality assurance
6. Flexible representation of results and communication with decision makers

The core of the program consists of on-farm projects negotiated with trainees and built around trainees' services to selected farmer clients. The initial project was the systematic monitoring of a crop or crops that enabled project participants to track soil water and nitrogen supply and crop growth through the season, providing the data needed to simulate the crop using APSIM and to test model performance. Procedural manuals and interactive coaching from APSRU staff, often through Internet video conferences, enabled trainees distributed over a large geographic area to carry out their projects with a high degree of success. This provided a practical, experience-producing framework within which the formal technical and theoretical training modules were flexibly presented to maximize practical relevance and significance. For the researchers pioneering a novel activity, it provided a valuable suite of case studies whose analysis will guide the program redesign for the next intake of trainees. Case data include logs of all soil monitoring advisory activity and APSIM simulations applied to clients' problems. In addition, regular review of trainee logs aid the trainers in insuring that trainees get the required experience with FARMSCAPE tools and techniques, assist in assessing the progress of trainees, and to learn how they adapt the FARMSCAPE approach to their business situations.

At the time of this writing, the first trainees were halfway through the program. Key research findings included:

- Some individuals were experiencing periodic difficulties meeting the demands of training in competition with urgent demands from their clients.
- Trainees were generally demonstrating high levels of competence in assessing tasks and involving clients in them.
- Job mobility for trainees and changing company ownership has proved to be a challenge for retaining the same trainees through a prolonged training program.
- Both trainees and their employers reported high levels of satisfaction with the program.

A critical, but subtle, lesson was that, in order for training activities to compete for attention in a high pressure commercial environment, an ongoing need exists for nurturing and reinforcement of the early expectations of trainees and their managers regarding valuable new service to farmer clients.

FARMSCAPE Online

In parallel with training and accreditation of intermediaries, means of conducting FARMSCAPE interactions with farmers using Internet conferencing are being developed (Hargreaves et al., 2001). Importantly, the latter is not in competition with the training and accreditation of intermediaries, but, instead, exploration of a medium that may prove to be valuable to the trained intermediaries, as well as valuable for interactions by researchers with farmers on matters outside technical and business consulting. The key research question concerns the degree of loss in value of online WifADs relative to face-to-face meetings. The approach is one of action research in which researchers play the role of a commercial advisors.

As described previously, although the FARMSCAPE approach features simulation of production system behavior using abstract models, experience has shown that effective interactions begin with concrete management issues and actions whose treatment is later enhanced in interactions using simulation. The most common starting point involved practical means of reducing uncertainty about soil water and nitrogen supply. One of the challenges of the online project has been to provide this through a combination of a soil-monitoring workshop followed by support for soil and weather monitoring for a season.

Team involvement typically begins with researchers conducting a face-to-face soil workshop for farmers in a district. The workshops are a mixture of instruction about local landscape geo-history and hands-on activities in practical soil sampling, processing soil cores, and calculations and data processing. A central aim is to provide farmers with the opportunity to appreciate their soil resource in new ways. The cores allow participants to see soil properties at depths not often accessed and the utility of measurements beyond those made normally in their practice. Simply breaking cores sequentially down the profile to trace rooting depth and feeling the relative wetness or dryness throughout the profile is a start to new appreciations for many. Researchers report back on data in an online meeting at a later date and implications are discussed. Measurements are a combination of what is valuable in its own right for decision making and what enables specification of APSIM for the situation (Dalgliesh and Foale, 1998).

The aim is to engage farmers in a process of active learning. Evaluation shows that most farmers who attend these workshops have mental models of soil water congruent with the prevailing technique for measurement — the pushing of a pointed steel rod about a meter long into the soil as far as the farmer can push it. This measures the depth of wet, i.e., soft soil, following a rainfall event. The activity undertaken during the soil workshop provides the opportunity for farmers to evaluate an alternative to this representation of the soil water environment — one that features a "water budget" concept with water stored in the empty volume of the soil "sponge." This concept leads logically to plant available water capacity (PAWC), which links with the way APSIM simulates soil water change. We have found that many farmers find the storage concept and the metaphor of the soil as a bucket, with a capacity and a content that varies from empty to full, is more useful than the depth of wet soil. "How big is the bucket?" is an increasingly common question in farmers' discussions of their soils. (Probes, nevertheless, continue to be valuable for quick checks after a rainfall event.)

Evaluations undertaken after the half-day soil workshops reveal that many participants are motivated to increase monitoring activity. The team's research program has followed this energy by providing support of a limited number of enthusiastic farmer groups for a cropping season in monitoring soil and weather, including access to a hydraulic soil-coring rig, an automatic weather station, and an electronic balance. Monitoring programs have been developed jointly with these farmers and are centered on issues of significance nominated by them. Issues suggested by farmers include evaluating the variables of row spacing, planting rate, nitrogen application, and planting date, as they relate to yield and gross margins, particularly via effects on the often scarce resource of soil water.

An online meeting using Microsoft NetMeeting™ takes place between a group of farmers assembled around a host farmer's computer and a researcher at his or her office. This technology supports audio and video communication as well as computer screen sharing. Bandwidth limitations in rural Australia generally require the use of separate audio via a simultaneous telephone connection with hands-free speakers. The researcher shares graphs of field data using specifically designed Excel spreadsheets. Farmers discuss these, often in relation to their personal and practical experiences with a particular crop, soil state, and weather.

How severe is the loss in quality of experience in an online meeting relative to a face-to-face meeting? This has been posed in evaluation interviews as "If an online meeting was free, how much extra would you pay for a face-to-face meeting?" The characteristic answer has been "Why would I pay more when this is as good as?" This response has been somewhat surprising to the researchers, because some exercise of tolerance with technical problems and effort in repairing human communications is often required. But aside from early problems attributable to inexperience, all concerned expect that most problems will be solved by increased telecommunication bandwidth for rural areas in Australia in the future.

DISCUSSION

Most researchers would agree with the proposition that agricultural systems research is a distinctive form of agricultural science. But there would probably be less agreement on what makes systems knowledge distinctive. The simple dichotomous structure of human knowledge proposed by Ryle (1949) appeals to us as the basis for a significant distinction. Ryle distinguished between "*knowing that* something is the case in the world" and "*knowing how* to bring about, or maintain, a desired state in the world."

The importance of both types of knowledge in a farm system can be discussed in terms of Figure 8.2. The analytical knowledge of agricultural science concerning the nature of the production system exemplifies "knowing that." The focus of systems research is on "knowing how" to achieve and maintain the desired state of the production system. Although this distinction is useful, it is not absolute. There is interaction and overlap between the two in both research and farming practice. These days, the ways that high-performing farmers see and think about their production systems is strongly influenced by knowing that production systems are structured as science claims them to be. This theoretical knowledge augments the primary structure of management that is based on feedback linkages (monitoring and action in Figure 8.2) underpinned by know-how, based on the historical patterns of system behavior (Senge, 1990).

In their 15-year systems research experiment, the authors found simulation modeling can provide a unique bridge between the knowledge of agricultural science and the know-how of farming practice. Effective bridging depends on good science, embodied in good process models. But in this role, good systems practice requires good compromise of both comprehensiveness of process treatment and the practicalities of model specification and testing. This results in models that are more often *functional* than scientifically mechanistic (Simon, 1996). Practicalities also drive the need for adding value to models by imbedding them in simulators (Baker and Curry, 1976; Banks et al., 1991), which make data management, model reconfiguration, and simulation output reporting efficient and effective.

Simulators convert modeled relationships to meaningful, albeit virtual, histories of system behavior. Such artificial histories have proved of particular value in providing a sort of *artificial* experience to both professionals and farmers. Whereas a farmer has some cumulative *actual* experience of the nature and variability of his or her environment, creation of a simulated history can provide advisors and researchers seeking to service the farmer with a helpful substitute. In highly variable climates, patterns are hard to perceive and a sense of "how often" or rules of thumb for action concerning weather tend to be weakly developed [a possible example of the Outcome-Irrelevant Learning Structure (OILS) of Einhorn (1982)]. Using a simulator and local climatic records, patterns in local histories are arguably more readily perceived than in real farming life, which is protracted and largely undocumented. Team members found that farmers value simulated histories, appropriately analyzed and graphed, as a means of evaluating the prospects of a contemplated management change (Hochman et al., 1998a).

In an entirely different arena, that of public policy, descriptions of problems and their backgrounds have been created that enable policy analysts to view certain simulated histories as relevant and, possibly, significant to actual futures, and expansion of this role for simulation is of high priority in our research on sustainable ecosystems (Table 8.1).

In real life, there is no substitute for experience gained through encounters with system behavior. Simulation can sometimes enhance this experience, through artificial experience, as described previously, but it can also enhance it by providing insight to deep structure — theory valuable in explaining behavior and in anticipating future behavior. When system simulation is providing a (partial) substitute for actual (risky) experience, a good simulator makes it easy to look inside to see how it works, and such understanding is often highly valued by farmers and advisors.

This involvement in human learning and planning practice is a long way from our starting points in crop and soil process research. How does simulation based on abstract models play these various roles in the representation of "knowing that," and in the construction of both of new knowledge and enhanced know-how? This is a systems question, but it is at a level of abstraction such that answers must be expected to be largely philosophical. Although an answer at this level has not proved to be essential prior to conduct of the research, it takes on greater significance as we try to make holistic sense of this new model-based research and justify it in a modeling community that is expecting something quite different. It involves both deeper understanding of the natures of both simulation and management or an actual family farm.

Although an instantiation of the production system in Figure 8.2 would refer ultimately to a set of soils, crops, etc., somewhere in the world as a framework for discussing models and simulation, its reference is more directly an abstract set of such objects and their relationships to one another (Kleene, 1952 quoted by Kliemt, 1996, p. 15):

> By a system S of objects we mean a set D of objects among which are established certain relationships. The system is abstract if the objects of the system are known only through the relationships of the system. [ ] …what is established in this case is the structure of the system, and what the objects are, in any respects other than how they fit into the structure, is left unspecified. Moreover, any further specification of what the objects are gives a representation (or model) of the abstract system, i.e., a system of objects which satisfy the relationships of the abstract system and have some further status as well.

This construction of the theoretical structure of the system (creating subsets of objects identified only by parsimonious mathematical relationships) is the essence of nomothetic research (Oquist, 1978). This activity creates smaller, intelligible worlds, in part by abstracting away detail complexity (Senge 1990) to provide general structure, and, in part, by selecting only relevant objects and aspects to study (Schutz, 1963, quoted by Blaikie, 1993, p. 42).

> It is up to the natural sciences to determine which sector of the universe of nature, which facts and events therein, and which aspects of such facts and events are… relevant to their specific purpose… Relevance is not inherent in nature as such; it is the result of the selective and interpretive activity of man [sic] within nature or observing nature….

Agricultural scientists seek to understand "what is the case" in production systems, focusing on "what is relevant to the activity of agriculture," in principle. Models of the production system represent elements whose structural relationships, provided with Kleene's "further status" by appropriate specification, are capable of mimicking relevant facts and events. This capability then provides a double-sided capability — to explain and to predict the behavior of the abstract system S, which represents, in principle, real production systems. The fact that a structural–functional model can be specified for local conditions provides a systems research tool for partially bridging the gap between scientific knowledge and the know-how in farming practice. In the typology of research of Oquist (1978), depicted as the gray portions of Figure 8.3, nomothetic research, which creates relevant abstract systems and models based on "knowing that," contribute to policy research, applied scientific research, and engineering design. The latter two are aimed at contributing to the "knowing how" of practice. Whereas agricultural science is concerned with descriptive, nomothetic and applied research, systems research extends beyond these in the direction of management know-how.

According to Checkland (1981), "hard" systems research embraces engineering design and policy research aimed at optimizing the logic of human activity, or practice. As hard systems activities, operations research and decision support systems provide guides to practice that are structured by the underlying abstract models — logic for action based on theory about the nature of the external environment as constructed by science. Taylor and Evans (1985) termed this "potentially knowing how." Recommendations for optimal action, explicit or implicit, so derived, are scientifically normative. They represent the best efforts of a scientific design approach to real world practice from outside the problem context, using the logic of the situation. But actual know-how has passed the filter of meaningful practical experience in the management situation. The profound nature of the gap between such external, potential know-how and internal, actual know-how is only beginning to be appreciated by hard systems researchers (McCown, 2001). Good science studies the local out of necessity, but uses clever techniques to move beyond the local to make general statements. But, perversely, this very abstraction contributes to the gap that is often lamented. The biologist and philosopher, Gregory Bateson, saw this as one of the most profound aspects of the distillation process of history as well as science: "...there is a deep gulf between statements about an identified individual and statements about a class. Such statements are of different logical type, and prediction from one to the other are always unsure" (Bateson, 1980).

The FARMSCAPE approach uses nomothetic research and the hard systems tool of production system simulation to deal realistically with the structure of the production system. But instead of using science to design optimal practice, the simulator is used to enable meaningful and adaptive experience of managers — a soft systems approach to construction of managers' "knowing how" (but often with a by-product of "knowing that" insights by participants). The paradigm shift is epitomized by the fact that soft systems knowledge is structured, not by abstract biophysical relationships, but by the intentionality of the manager (Caws, 1988) — implicit in the policies portion of the management system of Figure 8.1. One of the main arguments for the superiority of the decision support system over previous operations research optimization for managers was that the former recognized that many management problems were less than fully structured and so could not be adequately modeled. This lack of structure referred to the degree to which managers' policies included deviations from the striving for maximum profits — deviations that are important in farming practice (Frost, 2000).

Decision support systems in the hands of farmers notionally allowed other preferences to be exercised, guided by knowledge of structure as revealed in simulated events and patterns of events. Although this rationale was a step in situating models within the management system (Figure 8.2), our work with farmers in FARMSCAPE indicates that a central reason for the low use of decision support systems was the failure of simulations to be adequately situated in local practice physically or a lack of means for the user to test if this was the case. There is a substantial cost to specifying a model to simulate a real situation, and this cost has been considered by many to be prohibitive (e.g., Boote et al., 1996). Such costs include not only the demands of creating data of adequate quality but, generally, operating the model in this more open, flexible mode with acceptable risks of operator error. FARMSCAPE addresses this by having a flexible simulator operated by a professional intermediary trained in its operation, pragmatic data requirements, and efficient field measurement systems. This investment in physical situatedness is complemented by socially situating simulations in WifADs. One aspect of this social situating is analysis and discussion of farmers' perceived problems; another is the satisfaction of farmers' needs for evidence that the simulated system performance conforms to their own experienced performance or performance records.

It is important that justifiable criticism of the logic of applying models to design of decision support systems for individuals not detract from the justified logic of using of models in nomothetic and policy research when the aim is general knowledge and public policy. This was the original logic of hard systems that grew out of a World War II need for more rational plans for responding to enemy air attacks. This was followed in peacetime by enormous contributions by operations research and systems engineering to design of better policies and physical systems in industry and government. This intellectual and technological movement diffused into agriculture in the late 1940s. (In an interview with the late Professor C.T. DeWit shortly before his death, the pioneer of modeling in agriculture said he was strongly influenced as a graduate student by a new professor who had come from Shell Oil — an expert in simulating petroleum distillation.) Enormous progress has been made in the past 50 years in the ability to simulate agricultural production systems, building on the experience from industrial systems. Competent models imbedded in efficient simulators will undoubtedly be even more important in the future to enable learning from virtual mistakes instead of costly real ones. But systems researchers face a public application context that has become increasingly complex and problematic.

This trend can be tracked by two successive books with the same title: *Operational Research and the Social Sciences*, which are proceedings of conferences that brought together operations researchers and social scientists — first, in 1964 (Lawrence, 1966) and again in 1989 (Jackson et al., 1989). At the time of the first conference, hard, scientific modeling efforts concerning important systems problems were attracting criticism for work that ignored or underestimated the social nature of most problems of importance. By the time of the second conference, it was noted by the editors that (Jackson et al., 1989, p. v):

> Few, these days, regard OR [operational research] as being simply applied mathematics. The recognition that OR is a process of intervention in organizations and human affairs is now wider and more explicit. There has been a penetration and diffusion of ideas from the social sciences into OR, reflected most strongly in the body of writing about soft OR methods and soft systems thinking. Social and political skills are now recognized as critical to the success of OR practice and this is particularly so as operational researchers have sought to extend their client base outside that conventionally served. The rise of computer technology, embraced by OR, has required some thought to be given to its powerful impact on and consequences for organizations, people, and processes of decision. As a consequence, perhaps, of broadening its methodological base in attempting to extend its impact, OR has encountered the problem of competing "paradigms" — a condition long experienced by the social sciences.

Systems activities in agriculture have historically often lagged behind the main systems movement by a decade or two. During the past 10 years, CSIRO's own hard systems team has become very cognizant of the soft paradigm, and its thinking about systems and research and intervention practices have changed significantly. Although the team experienced paradigm competition (Ridge and Cox, 2000; Woods et al., 1997), members are impressed with what might be called the paradigm cooperation of Mingers and Gill (1997), which offers, in the metaphor of Sellars (1963), "stereoscopic vision." The aim is to use the best hard tools and methods science can provide in a "soft" philosophical and communication matrix that will enable appropriate response to the human setting of the problem and our own inescapable inclusion in this setting.

The authors find that scientists commonly perceive their program of soft and hard systems approaches as an unconventional mix of research and extension. They assert it is more helpful to see it as a way of doing research differently in an age when scientific knowledge is increasingly expected be directly linked to, and justified by, relevant and significant new know-how in practice. The exciting new realization for us is that a simulator based on scientific knowledge can be instrumental in enabling and facilitating situated cooperative learning by practitioners and researchers.

ACKNOWLEDGMENTS

We are indebted to the farmers, extensionists, agribusiness consultants, and research colleagues, with whom we have worked and without whom we would have had an impoverished experience. We acknowledge the especially significant contributions of Neal Dalgliesh, Merv Probert, Mike Foale, Mike Robertson, Perry Poulton, John Dimes, and Roger Jones.

REFERENCES

Abrecht, D.G. and K.L. Bristow. 1996. Coping with soil and climatic hazards during crop establishment in the semi-arid tropics, *Aust. J. Exp. Agric.*, 36: 971–983.

Asseng, S., I.R.P. Fillery, G.C. Anderson, P.J. Dolling, F.X. Dunin, and B.A. Keating. 1998. Use of the APSIM wheat model to predict yield, drainage and NO3 leaching in a deep sand. *Aust. J. Agric. Res.* 49:363–377.

Australian Bureau of Statistics (ABS). 1999. Use of technology on farms (1998–99). *ABS Report,* 8150.

Baker, C.H. and R.B. Curry. 1976. Structure of agricultural simulators: a philosophical view, *Agric. Syst.*, 1:201–218.

Bakken, B., J. Gould, and D. Kim. 1994. Experimentation in learning organizations: a management flight simulator approach. In *Modeling for Learning Organization*, Morecroft, J.D.W. and J.D. Sterman, Eds., Productivity Press, Portland, Oregon, 243–266.

Banks, J., E. Aviles, J.R. McLaughlin, and R.C. Yuan. 1991. The simulator: new member of the simulation family, *Interfaces,* 21:76–86.

Bateson, G. 1980. *Mind and Nature – A Necessary Unity,* Bantam, New York.

Birch, C.J., P.S. Carberry, R.C. Muchow, R.L. McCown and J.N.G. Hargreaves. 1990. Development and evaluation of a sorghum model based on CERES-Maize in a semi-arid tropical environment, *Field Crops Res.,* 24:87–104.

Blaikie, N. 1993. *Approaches to social enquiry,* Polity Press, Cambridge.

Boote, K.J., J.W. Jones, and N.B. Pickering. 1996. Potential uses and limitations of crop models, *Agron. J.,* 88:704.

Carberry, P.S., R.C. Muchow, and R.L. McCown. 1989. Testing the CERES-Maize simulation model in a semi-arid tropical environment, *Fields Crop Res.,* 20:297–315.

Carberry, P.S., R.L. McCown, J.P. Dimes, B.H. Wall, D.G. Abrecht, J.N.G. Hargreaves, and S. Nguluu. 1992. Model development in northern Australia and relevance to Kenya. In *A Search for Strategies for Sustainable Dryland Cropping in Semi-Arid Eastern Kenya, Proc. of a Symposium,* Probert, M.E., Ed., Nairobi, Kenya, December 10–11, 1990. Australian Centre for International Agricultural Research, Canberra, Australia, 34–41.

Carberry, P.S., A.L. Cogle, and R.L. McCown. 1991a. An assessment of cropping potential for land marginal to the Atherton Tablelands of north Queensland. A final report prepared for the Rural Industries R&D Corporation.

Carberry, P.S., R.C. Muchow, and R.L. McCown. 1991b. Cultivar selection for rain-fed maize in northern Australia. *Proc. of 1st Australian Maize Conference.* Maize Association of Australia, 54–57.

Carberry, P.S. and D.G. Abrecht. 1991. Tailoring crop models to the semiarid tropics. In *Crop Production: Models and Management in the Semiarid Tropics and Subtropics.* R.C. Muchow and J.A. Bellamy, Eds., Climatic Risk in CAB International, Wallingford, U.K. 157–182.

Carberry, P.S. and R.C. Muchow. 1992. A simulation model of kenaf for assisting fibre industry planning in northern Australia: 3. Model description and validation, *Aust. J. Agric. Res.,* 43:1527–1545.

Carberry, P.S., B.A. Keating, and R.L. McCown. 1992. Competition between a crop and undersown pasture — simulations in a water-limited environment. *Proc. 6th Australian Agronomy Conference* Sept. 19–24, 1993, Australian Society of Agronomy, Parkville, Australia, 206–209.

Carberry, P.S., R.C. Muchow, and R.L. McCown. 1993a. A simulation model of kenaf for assisting fiber industry planning in northern Australia: 4. Analysis of climatic risk, *Aust. J. Agric. Res.,* 44:713–730.

Carberry, P.S., R.L. McCown, and N. De Groot. 1993b. Can profitable cropping systems be developed in the semi-arid tropics of northern Australia? *Proc. 7th Australian Agronomy Conference,* Australian Society of Agronomy, Parkville, Australia, 236–239.

Carberry, P.S., A.L. Chapman, C.A. Anderson, and L.L. Muir, Eds., 1996a. Conservation tillage and ley farming systems in the Australian semi-arid tropics, *Aust. J. Exp. Agric.,* 36:915–1089.

Carberry, P.S. et al. 1996b. Simulation of a legume ley farming system in northern Australia using the Agricultural Production Systems Simulator, *Aust. J. Exp. Agric.,* 36:1037–1048.

Carberry P.S., S.G.K. Adiku, R.L. McCown, and B.A. Keating. 1996c. Application of the APSIM cropping systems model to intercropping systems. In *Roots and Nitrogen in Cropping Systems of the Semi-Arid Tropics, Japan,* Ito, O., C. Johansen, J.J. Adu-Gyamfi, K. Katayama, J.V.D.K. Kumar Rao, and T.J. Rego, Eds., Agricultural Research Centre for Agricultural Sciences, Ibaraki, Japan, 637–648.

Carberry, P.S. and Bange, M.P. 1998. Using systems models in farm management. In *Proc. 9th Australian Cotton Conference,* August 10–14, Gold Coast Australia, The Australian Cotton Growers Research Organisation, 153–160.

Carberry, P.S. 2001. Are science rigour and industry relevance both achievable in participatory action research? In *Proc. of the 10th Australian Agronomy Conference,* Australian Society of Agronomy, Hobart, Australia. Jan. 28–Feb. 1, 2001.

Caws, P. 1988. *Structuralism: A philosophy for the human sciences.* Humanities Press International,Atlantic Highlands, New Jersey.

Chapman, A.L., J.D. Sturtz, A.L. Cogle, W.S. Mollah, and R.J. Bateman. 1996. Farming systems in the Australian semi-arid tropics — a recent history, *Aust. J. Exp. Agric.,* 36:915–928.

Checkland, P.B. 1981. *Systems Thinking, Systems Practice,* John Wiley & Sons, Chichester, U.K..

Checkland, P.B. 1991. From optimising to learning. In *Critical Systems Thinking: Directed Readings,* Flood, R.L. and M.C. Jackson, Eds., John Wiley & Sons, Chichester, U.K.

Clancey, W.J. 1997. *Situated Cognition: On Human Knowledge and Computer Representations,* Cambridge University Press, Cambridge.

Cogle, A.L., P.S. Carberry, and R.L. McCown. 1990. Cropping potential assessment of land marginal to the Atherton Tablelands, North Queensland, Australia. In *Climatic Risk in Crop Production: Models and Management in the Semi-Arid Tropics and Subtropics. Poster Papers from the International Symposium,* R.C. Muchow and J.A. Bellamy, Eds., CSIRO Division of Tropical Crops and Pastures, St. Lucia, 6–7.

Connolly, R.D. and D.M. Freebairn. 1996. Managing soil structure, particularly with respect to infiltration, for long-term cropping system productivity. In *Multiple Objective Decision Making for Land, Water, and Environmental Management. Proc. of the 1st International Conference on Multiple Objective Decision Support Systems for Land, Water, and Environmental Management: Concepts, Approaches, and Applications,* El-Swaify, S.A. and D.S. Yakowitz, Eds., St. Lewis Press, New York, 379–384.

Coutts, J.A., Z. Hochman, M.A. Foale, R.L. McCown, P.S. Carberry. 1998. Evaluation of participative approaches to RD&E: A case study of FARMSCAPE. *Proc. of the 9th Australian Society of Agronomy Conference,* Australian Society of Agronomy, Wagga Wagga, Australia, July 20–23, 1998. 681–682.

Dalgliesh, N.P. and M.A. Foale. 1998. *Soil Matters: Monitoring Soil Water and Nutrients in Dryland Farming Systems,* CSIRO, Agricultural Production Systems Research Unit, Toowoomba, Queensland, Australia.

Dilshad, M., J.A. Motha, and L.J. Peel. 1996. Surface runoff, soil and nutrient losses from farming systems in the Australian semi-arid tropics, *Aust. J. Exp. Agric.,* 36:1003–1012.

Dimes, J.P., R.L. McCown, and P.G. Saffigna. 1996. Nitrogen supply to no-tillage crops as influenced by mulch type, soil type and season, following pasture leys in the semi-arid tropics, *Aust. J. Exp. Agric.,* 36:937–946.

Donnelly, J.R., A.D. Moore, and D.M. Freer. 1997. GRAZPLAN: Decision Support Systems for Australian Grazing Enterprises I. Overview of the GRAZPLAN Project and a Description of the MetAccess and LambAlive DSS. *Agric. Syst.,* 54:57–76.

Dunin, F.X., J. Williams, K. Verburg, and B.A. Keating. 1999. Can agricultural management emulate natural ecosystems in recharge control in south eastern Australia? *Agroforestry Syst.*. 45:343–364.

Einhorn, H.J. 1982. Learning from experience and suboptimal rules in decision making. In *Judgement Under Uncertainty: Heuristics and Biases,* Kahneman, D., P. Slovic, A. Tversky, Eds., Cambridge University Press, Cambridge, U.K. 268–283.

Frost, F.M. 2000. Value orientations: impact and implications in the extension of complex farming systems, *Aust. J. Exp. Agric.,* 40:511–518.

Hammer, G.L., N. Nicholls, and C. Mitchell, Eds. 2000. *Application of Seasonal Climate Forecasting in Agricultural and Natural Ecosystems — The Australian Experience*, Kluwer Academic Publishers, Dordrecht, The Netherlands.

Hargreaves, J.N.G and R.L. McCown. 1988. V/I CERES-Maize: A Visual-Interactive Version of CERES-Maize, Tropical Agronomy Technical Memorandum, CSIRO Division of Tropical Crops and Pastures, Brisbane, Australia.

Hargreaves, D.M.G., Z. Hochman, N. Dalgliesh and P. Poulton. 2001. FARMSCAPE online- developing a method for interactive Internet support for farmers situated learning and planning, In *Proc. 10th Australian Agronomy Conference,* Hobart, Australia, (www.regional.org.au/au/asa/2001/5/a/hargreaves.htm). Jan. 28–Feb. 1, 2001.

Hoag, D.L, J.C. Ascough, and W.M. Frasier. 1999. Farm computer adoption in the Great Plains, *J. Agric. Appl. Econ.,* 31:57–67.

Hochman, Z., R. Skerman, G. Cripps, P.L. Poulton, and N.P. Dalgliesh. 1998a. A new sorghum planting strategy resulting from the synergy of farmer systemic knowledge and systematic simulation of sorghum production systems. In *Proc. of the 9th Australian Agronomy Conference,* Wagga Wagga, Australia, 411–414.

Hochman, Z., S. Braithwaite, M.E. Probert, K. Verburg, and K.R. Helyar. 1998b. SOILpH — a new APSIM module for management of soil acidification. In *Proc. 9th Australian Agronomy Conference,* Wagga Wagga, Australia, 709–712. July 20–23, 1998.

Hochman, Z., J.A. Coutts, P.S. Carberry, R.L. McCown, RL. 2000. The FARMSCAPE experience: simulations aid participative learning in risky farming systems in Australia. In *Cow up a Tree: Knowing and Learning for Change in Agriculture: Case Studies from Industrialised Countries,* Cerf, M., D. Gibbon, B. Hubert, R. Ison, J. Jiggins, M.S. Paine, J. Proost and N. Roling, Eds., INRA Editions, Versailles Cedex, France, 175–188.

Ikerd, J.E. 1991. A decision support system for sustainable farming, *Northeastern J. Agric. and Resour. Econ.,* 20:109–113.

Jackson, M.C., P. Keys, and S.A. Cropper. 1989. *Operational Research and the Social Sciences,* Plenum Press, London.

Jones, C.A. and J.R. Kiniry. 1986. *CERES-Maize: A Simulation Model of Maize Growth and Development,* University Press, College Station, Texas.

Jones, R.K., M.E. Probert, N.P. Dalgliesh, and R.L. McCown. 1996. Nitrogen inputs from a pasture legume in rotations with cereals in the semi-arid tropics of northern Australia: experimentation and modelling on a clay loam soil, *Aust. J. Exp. Agric.,* 36:985–994.

Keating, B.A., B.M. Wafula, and R.L. McCown. 1988. Simulation of plant density effects on maize yield as influenced by water and nitrogen limitations. In *Proc. of the Int. Congr. of Plant Physiol.,* February 15–20, 1988, New Delhi, India.

Keating, B.A. and B.M. Wafula. 1991. Modelling the fully expanded area of maize leaves, *Field Crops Res.,* 29:163–176.

Keating, B.A., D.C. Godwin, and J.M. Watiki. 1991. Optimization of nitrogen inputs under climatic risk. In Climatic Risk in Crop Production — Models and Management for the Semi-arid Tropics and Sub-Tropics, Muchow, R.C. and J.A. Bellamy, Eds., CAB International, Wallingford, U.K., 329–357.

Keating, B.A., B.M. Wafula, and J.M. Watiki. 1992a. Exploring strategies for increased productivity — the case for maize in semi-arid Eastern Kenya. *In A Search for Strategies for Sustainable Dryland Cropping in Semi-arid Eastern Kenya, Proc. of a Symposium,* Nairobi, Kenya, December 10–11, 1990. Probert, M.E., Ed., Australian Centre for International Agricultural Research, Canberra, Australia, 90–99.

Keating, B.A., R.L. McCown, and B.M. Wafula. 1992b. Adjustment of nitrogen inputs in response to a seasonal forecast in a region of high climatic risk. In *Systems Approaches for Agricultural Development, International Symposium on Systems Approaches to Agricultural Development,* Penning de Vries, F.W.T. et al., Eds., Bangkok, December 1991, 233–252.

Keating, B.A., K. Verburg, N.I. Huth, M.J. Robertson. 1997. Nitrogen management in intensive agriculture: sugarcane in Australia. In *Intensive Sugarcane Production: Meeting the Challenges beyond 2000.* Keating, B.A. and J.R. Wilson, Eds., CAB International, Wallingford, UK., 221–242.

Keating, B.A. and H. Meinke. 1998. Assessing exceptional drought with a cropping systems simulator: a case study for grain production in north-east Australia, *Agric. Syst.,* 57:315–332.

Keating, B.A., P.S. Carberry, and M.J. Robertson. 1999. Simulating N fertiliser response in low-input farming systems: 2. Effects of weed competition, In *Proc. Int. Symp. Modeling Cropping Syst.,* Lleida, Spain, 21–23, June 1999.Donatelli, M. C. Stockle, F. Villalobus and J.M. Villar Mir, Eds. Universitat de Lleida, 205–206.

Keene, S.C. 1952/71. *Introduction to Metamathematics.* Bibliotheca Mathematica I, North-Holland Publishing Co., Amsterdam, The Netherlands, 8.

Kliemt, H. 1996. Simulation and rational practice. Modelling and Simulation in the Social Sciences from the Philosophy Point of View. In Hegselmann, R.,U. Mueller, K.G. Troitzsch, Eds., Klewer, Dortrecht, 13–27.

Lawrence, J.R., Ed., 1966. *Operational Research and the Social Sciences,* Tavistock, London.

Lisson, S., L.E. Brennan, K.L. Bristow, M. Schuurs, T. Linedale, M. Smith, D. Hughes, B.A. Keating. 2000. DAMEA$Y — A framework for assessing the costs and benefits of on-farm storage based sugarcane production systems, *Aust. Soc. Sugar Cane Technol.,* 22:186–193.

Littleboy, M., D.M. Silburn, D.M. Freebairn, D.R. Woodruff, G.L. Hammer, J.K. Leslie. 1992. Impact of soil erosion on production in cropping systems: I. Development and validation of a simulation model. *Aust. J. Soil Res.,* 30:757–74.

McCown, R.L., R.K. Jones, and D.C.I. Peake. 1985. Evaluation of a no-till, tropical legume ley-farming strategy. In *Agro-Research for the Semi-Arid Tropics: North-West Australia,* R.C. Muchow, Ed., University of Queensland Press, St. Lucia, 450–469.

McCown, R.L. and J. Williams. 1989. AUSIM: A cropping systems model for operational research. *SSA ImACS 1989 Biennial Conference on Modelling and Simulation,* Australian National University, September 25–27, 1989, Canberra, Australia.

McCown R.L., B.M. Wafula, L. Mohammed, J.G. Ryan J.N.G. Hargreaves. 1991. Assessing the value of a seasonal rainfall predictor to agronomic decisions: the case of response farming in Kenya. In *Climatic Risk in Crop Production — Models and Management for the Semiarid Tropics and Subtropics,* Muchow, R.C. and J.A. Bellamy, Eds., CAB International, Wallingford, U.K., 383–409.

McCown, R.L, Keating, BA, Probert, ME, and Jones, RK. 1992. Strategies for Sustainable Crop Production in Semi-Arid Africa, *Outlook on Agric.,* 21:21–31.

McCown, R.L., P.G. Cox, B.A. Keating, G.L. Hammer, P.S. Carberry, M.E. Probert, D.M. Freebairn. 1993. The development of strategies for improved agricultural systems and land use management. In *Opportunities, Use, and Transfer of Systems Research Methods in Agriculture to Developing Countries,* Goldsworthy, P. and F. Penning de Vries, Eds., Kluwer, Dordrecht, The Netherlands, 81–96.

McCown, R.L. 1996. Being realistic about no-tillage, legume ley farming for the Australian semi-arid tropics, *Australian Journal of Experimental Agriculture,* 36:1069–1080.

McCown, R.L, G.L. Hammer, J.N.G. Hargreaves, D.L. Holzworth, D.M. Freebairn. 1996. APSIM: a novel software system for model development, model testing, and simulation in agricultural systems research, *Agric. Syst.,* 50:255–271.

McCown, R.L. 2001. Learning to bridge the gap between scientific decision support and the practice of farming: evolution in paradigms of model-based research and intervention from design to dialogue, *Aust. J. Agric. Res.,* 52:549–571.

Meinke, H., R.C. Stone, and G.L. Hammer. 1996. SOI phases and climatic risk to peanut production: a case study for northern Australia, *Aust. J. Agric. Res.,* 48:789–793.

Mingers, J. and A. Gill. 1997. *Multimethodology: The Theory and Practice of Combining Management Science Methodologies,* John Wiley & Sons, West Sussex, U.K.

Muchow, R.C. and P.S. Carberry. 1991. Climatic constraints to maize productivity in Australia. In *Maize in Australia — Food, Forage and Grain. Proc. of 1st Australian Maize Conference,* April 15–17. 1991. J. Moran, Ed., Maize Association of Australia, Brisbane, Australia, 8–12.

Muchow, R.C., G.L. Hammer, and P.S. Carberry. 1991. Optimizing crop and cultivar selection in response to climatic risk. In *Climatic Risk in Crop Production: Models and Management in the Semi-Arid Tropics and Subtropics,* R.C. Muchow and J.A. Bellamy, Eds., CAB International, Wallingford, U.K., 235–262.

Muchow, R.C. and B.A. Keating. 1998. Assessing irrigation requirements in the Ord sugar industry using a simulation modelling approach, *Aust. J. Exp. Agric.,* 38:345–354.

Nelson, R.A., Dimes, J.P., Paningbatan, E.P. and Silburn, D.M. 1998a. Erosion/productivity modelling of maize farming in the Philippine uplands. Part I: parameterising the Agricultural Production System Simulator, *Agric. Syst.,* 58:129–146.

Nelson, R.A., Dimes, J.P., Silburn, D.M., Paningbatan, E.P. and Cramb, R.A. 1998b. Erosion/productivity modelling of maize farming in the Philippine uplands. Part II: Simulation of alternative farming methods, *Agric. Syst.,* 58:147–163.

Oquist, P. 1978. The epistemology of action research, *Acta Sociologica,* 21:143–163.

Parker, C. 1999. Decision support systems: lessons from past failures, *Farm Management,* 10:273–289.

Paydar, Z., A.J. Ringrose-Voase, N.I. Huth, R.R. Young, A.L. Bernardi, B.A. Keating, H.P. Cresswell, J.F. Holland, I. Daniels. 1999. Modelling deep drainage under different land use systems: 1. Verification and systems comparison. In *Proc. of MODSIM'99,* December 6–9, 1999, Hamilton, New Zealand, 37–42.

Probert, M.E., B.A. Keating, J.P. Thompson, W.J. Parton. 1995. Modelling water, nitrogen and crop yield for a long-term fallow management experiment, *Aust. J. Exp. Agric.,* 35:941–950.

Probert, M.P., J.P. Dimes, B.A. Keating, R.C. Dahal, W.M. Strong. 1997. APSIM's water and nitrogen modules and simulation of the dynamics of water and nitrogen in fallow systems, *Agric. Syst.,* 56:1–26.

Ridge, P.E. and P.G. Cox. 2000. Market research for decision support for dryland crop production. In *Case Studies in Increasing the Adoption of Sustainable Resource Management Practices,* Shulman, A.D. and R.J. Price, Eds., Land and Water Resources Research and Development Corp. Occasional Paper 04/99. Land and Water Resources Research Development Corporation, Canberra, Australia, 125–182.

Ringrose-Voase, A.J., Z. Paydar, N.I. Huth, R.G. Banks, H.P. Cresswell, B.A. Keating, R.R. Young, A.L. Bernardi, J.F. Holland, J.F. I. Daniels. 1999. Modelling deep drainage of different land use systems. 2. Catchment wide application. In *Proc. of MODSIM'99,* December 6–9, 1999, Hamilton, New Zealand, 43–48.

Ryle, G. 1949. *The Concept of Mind,* Penguin Books, Harmondsworth, Middlesex, U.K.

Schon, D.A. 1983. *The Reflective Practitioner: How Professionals Think in Action,* Basic Books.

Schultz, A. 1963. Common Sense and Scientific Interpretation of Human Action, In *Philosophy of the Social Sciences,* Natanson, M.A., Ed., Random House, New York, 302–346.

Sellars, W. 1963. *Philosophy and the Scientific Image of Man: Science, Perception, and Reality,* Routledge & Kegan Paul, London.

Senge, P.M. 1990. *The Fifth Discipline: The Art and Practice of the Learning Organization,* Doubleday Currency, New York.

Shaffer, M.J. 1988. Estimating confidence bands for soil-crop simulation models, *Soil Sci. Soc. Am. J.,* 52:1782–1789.

Shaffer, M.J., S.C. Gupta, D.R. Linden, J.A.E. Molina, C.E. Clapp, W. Larson. 1982. Simulation of nitrogen, tillage, and residue management effects on soil fertility. In *Analysis of Ecological Systems: State-of-the-Art in Ecological Modeling,* Lauenroth W.K., G.V. Skogerboe, M. Flug, Eds. *Proc. 3rd. Int, Conf. On State-of-the-Art in Ecological Modeling,* Colorado State University, Boulder, CO, May 24–28, 525–543.

Shamudzarira, Z., M.J. Robertson, P.T. Mushayi, B.A. Keating, S. Waddington, C. Chiduza, P. Grace. 1999. Simulating N fertiliser response in low-input farming systems: 1. Fertiliser recovery and crop response, *ESA Symposium on Modelling Cropping Systems,* June 1999, Spain.

Simon, H.A. 1979. From substantive to procedural rationality. In *Philosophy and Economic Theory,* Hahn, F. and M. Hollis, Eds., Oxford Univ Press. Oxford, U.K., 65–86.

Simon, H.A. 1996. *The Sciences of the Artificial,* 3rd ed., MIT Press, Cambridge, Massachusetts.

Snow, V.O. et al. 1999. Nitrogen dynamics in an eucalypt plantation irrigated with sewage effluent or bore water, *Aust. J. Soil Res.,* 527–544.

Sorensen, J.T. and Kristensen, E.S. 1992. Systemic modelling: a research methodology in livestock farming. In *Global Appraisal of Livestock Farming Systems and Study on their Organizational Levels: Concept, Methodology and Results, Proc. Symp organized by INRA-SAD and the DIRAD-IEMVT*, Toulouse, France, July 7–10, 1990, Gibon, A. and B. Matheron, Eds., Commission of European Communities, 1992, 45–57.

Steiner, J.L., J.R. Williams, and O.R Jones. 1987. Evaluation of the EPIC simulation model using a dryland wheat-sorghum-fallow crop rotation, *Agron. J.,* 79:732–738.

Stewart J.I. and W.A. Faught. 1984. Response farming of maize and beans at Katumani, Machakos District, Kenya: recommendations, yield expectations, and economic benefits, *E. Afr. Agric. For. J.,* 44:29–51.

Stone, R.C., G.L. Hammer, and T. Marcussen. 1996. Prediction of global rainfall probabilities using phases of the Southern Oscillation Index, *Nature,* 384:252–255.

Taylor, J.C. and G. Evans. 1985. The architecture of human information processing: empirical evidence, *Instructional Science,* 13:347–359.

Thorburn P.T., M.E. Probert, S.N. Lisson, A.W. Wood, B.A. Keating. 1999. Impacts of trash retention on soil nitrogen and water: an example from the Australian sugar industry. In *Proc. of South African Society of Sugarcane Technologists,* 73:75–79.

Uehara, G. and G.Y. Tsuji. 1991. Progress in crop modeling in the IBSNAT Project. In Climatic risk in crop production: Models and management in the semi-arid tropics and subtropics. R.C. Muchow and J.A. Bellamy, Eds., CAB International, Wallingford, U.K., 143–156.

Verburg, K. 1996. Methodology in soil water and solute balance modelling: an evaluation of the APSIM-SoilWat and SWIMv2 models. *Divisional Report No. 131,* CSIRO Division of Soils, CSIRO, Adelaide, Australia.

Verburg, K., B.A. Keating, K.L. Bristow, N.I. Huth, P.J. Ross, and V.R. Catchpoole. 1996. Evaluation of nitrogen fertiliser management strategies in sugarcane using APSIM-SWIM. In *Sugarcane: Research towards Efficient and Sustainable Production,* Wilson, J.R., D.M. Hogarth, J.A. Campbell, and A.L. Garside, Eds., CSIRO Division of Tropical Crops and Pastures, Brisbane, Australia, 200–202.

Wafula, B.M., R.L. McCown, and B.A. Keating. 1992. Prospects for improving maize productivity through response farming. In A search for strategies for sustainable dryland cropping in semi-arid eastern kenya. *Proc. of a Symposium,* Probert, M.E., Ed., Nairobi, Kenya, December 10–11, 1990. Australian Centre for International Agricultural Research, Canberra, Australia, 101–107.

Wagner, P. and F. Kuhlmann. 1991. Concept and implementation of an integrated decision support system (IDSS) for capital-intensive farming, *Agric. Econ.,* 5:287–310.

Williams, J.R. 1983. The physical components of the EPIC model. In *Soil Erosion and Conservation,* El-Swaify, S.A. and W.C. Moldenhauer, Eds., Soil Conservation Society of America, _____, 272–284.

Williams, J.R., C.A. Jones, J.R. Kiniry, and D.A. Spanel. 1989. The EPIC crop growth model, *Trans. ASAE,* 32:497–511.

Winograd, T. and F. Flores. 1986. *Understanding Computers and Cognition: A New Foundation for Design,* Addison-Wesley, Reading, Massachusetts.

Woods, E.J., P.G. Cox, and S. Norish. 1997. Doing things differently: the RD&E revolution. In *Intensive Sugarcane Production: Meeting the Challenges beyond 2000,* Keating, B.A., J.R. Wilson, Eds., CAB International, Wallingford, U.K., 469–490.

CHAPTER 9

Applications of Crop Growth Models in the Semiarid Regions

Mannava V.K. Sivakumar and Ariella F. Glinni

CONTENTS

INTRODUCTION

The World Atlas of Desertification (UNEP, 1992) defines semiarid regions as the areas where the ratio of mean annual rainfall (R) to mean annual potential evapotranspiration (PET) varies between 0.2 to 0.5. In an assessment of population levels in the world's drylands, the Office to Combat Desertification and Drought (UNSO) of the United Nations Development Programme (UNDP) showed that the semiarid regions account for approximately 18% of the total area in the world and are inhabited by 874 million people or approximately 16% of the world's population. The world's semiarid regions are home to about 18% of the total population in Africa, 14% in the Americas and the Caribbean, 18% in Asia, 5% in Australia and Oceania, and 5% in Europe (UNSO, 1997). Population growth rates are high in the semiarid regions, and hence, the need to develop effective strategies to improve and sustain agricultural productivity in these regions is of paramount importance.

The low, variable, and undependable rainfall and the relatively poor soils with their inherently low soil nutrients needed by plants that characterize the semiarid regions call for efficient soil and water management strategies to achieve good yields at the farm level. Although drought occurs during periods of insufficient rainfall, water logging can occur during periods of excessive rainfall. Crop drought stress occurs in the crops grown in these semiarid regions due to many factors including high reference ET (ET_o), low extractable soil water in the root zone, poor root distribution, restricted canopy size, and other plant and environmental factors. A range of pests and diseases attack these crops in the semiarid regions, and effective and timely crop protection strategies are crucial to reduce the yield losses that occur frequently. Hence, the tasks of finding an ideal crop/cropping system, and suitable soil and water management strategies that can increase and sustain high levels of crop productivity in the semiarid regions, should take into account a number of soil, climate and crop factors that interact in different ways.

Scientists have traditionally approached the above task by designing field trials that examine the impact of one or more factors on crop productivity; however, given the complexity of factors involved and the possible combinations that need to be studied in the semiarid agriculture, such traditional experimental approaches can indeed be quite expensive and time-consuming. As it is not feasible to study the full range of crop–soil–weather–management interactions in experiments, applications of models based on concepts of systems analysis (De Wit, 1982) has been a promising alternative (Penning de Vries, 1994). A model is defined as a simplified representation of a system, and a system is a distinct part of presumed reality that contains interrelated elements (De Wit, 1982). Sinclair and Seligman (1996) defined crop modeling as the dynamic simulation of crop growth by numerical integration of constituent processes with the aid of computers. It is described as a technology used to construct a relatively transparent surrogate (or substitute) for a real crop, one that can be analysed or manipulated with far greater ease than the complex and cumbersome real crop. Time and expense are the prime considerations (Roberts, 1976) in addition to the introduction of systemic errors in experiments.

Crop modeling evolved in the late 1960s as a means of integrating knowledge about plant physiological processes in order to explain the functioning of crops as a whole (Bouman et al., 1996). Insights into various processes were expressed using mathematical equations and integrated in so-called simulation models. In the early years of crop simulation models, much euphoria surrounded the initial excitement and the perceived potential of crop models and claims of the crop models were made that crop models could present effective solutions to various strategic and tactical questions concerning on-farm crop management. By 1996, a more realistic appraisal of the potential of crop-growth models was made, and it was concluded that there is much we do not know about the mechanistic structure of the workings of plants and their interactions with their environment (Passioura, 1996). In a thought-provoking article on "The Quest for Balance in Crop Modelling," Monteith (1996) pointed out that model building can draw attention to gaps in understanding and thereby stimulate new experimental or theoretical work. He also explained that models can provide a framework for interpreting the output from field experiments in different environments, and that they can be used to explore, with due caution, ways of improving management or minimizing risk. Sinclair and Seligman (1996) went a bit further and concluded that user-friendly models can fulfill a vital heuristic function in teaching, research, and management planning.

Following development of object-oriented programming and the recent advances in internet applications, Pan et al. (1997) developed an object-oriented and Internet-based generic plant growth simulator for research and educational purposes on the World Wide Web. With a Java-embedded Web browser, the user can link to the run-time model from the Web site of the authors.

Despite the rapid advances described previously, providing a comprehensive description of the applications of crop growth models in the semiarid regions is not an easy task. In this chapter, we have attempted to describe some of the salient features of the climate, soils and crops in the semiarid regions of the world. This is followed by a brief description of the potential for use of crop growth models for practical applications in the semiarid regions. Needless to say, in this short chapter it is rather difficult to do justice to applications of crop growth models in the vast area of semiarid regions. Nonetheless, we have attempted to describe the current applications of crop growth models in the semiarid regions with suitable examples, especially from the semiarid tropics.

AGROCLIMATIC CONDITIONS IN SEMIARID REGIONS

Climate

The semiarid climates usually constitute a transition between desert and sub-humid to humid climates and are characterized by a short period of rainfall during the year. It is this brief period of seasonal rain, which causes them to be semiarid, instead of arid (Trewartha,1968). The semiarid regions are characterized by a low and highly variable rainfall and a high demand for water imposed by the consistently high temperatures and solar radiation. The annual potential evapotranspiration demand (PET) at the land–water surface exceeds the annual rainfall amount; the actual ET is, of course, less than rainfall.

The semiarid regions of northern Australia, west Africa, southwestern Africa and northwestern India-Pakistan receive their maximum rainfall during summer. In contrast, the semiarid regions of North Africa, southern Australia, southern Iran, western Asia, northwestern Mexico, and adjacent parts of the southwestern U.S. are dominated upon by westerlies with their cyclones and fronts in winter. Because of the cool season, evaporation is less, and consequently the modest amount rain that falls is relatively more effective for plant growth. Semiarid regions in the middle latitudes are found in the deep interiors of the continents with relatively severe seasonal temperatures and thus large annual ranges.

Rainfall in the semiarid regions has to be characterized not only by average behaviour but by interannual and interseasonal variability. In low-rainfall years, there may be droughts; in high-rainfall years or even for short periods in low-rainfall years there may be floods or excessive rainfall. Deficient rainfall years may be followed by similar years or by years with excess rainfall.

The scale of rainfall variability in the semiarid regions determines to a large extent the kind of crops/cropping systems that can be grown and the magnitude of their vulnerability to the rainfall vagaries. Temporal or time-dependent variations in rainfall are common, and can be represented by three time scales: annual, monthly, and daily. The coefficient of variation of annual rainfall ranges between 15 and 30%. Variability in monthly rainfall is larger, because the rainfall is usually limited to 3 to 5 months. Rainfall variability reaches its maximum at the level of daily rainfall. Rainfall in the semiarid regions is also characterized by high spatial variability. Few intensive studies on this aspect in the semiarid regions are available. Sivakumar and Hatfield (1990) reported that over a 500-ha research farm in Niger (West Africa) deviations of the order of 30 to 40% in individual raingages from the central observatory were observed in isolated cases, deviations up to 80% were noted.

In certain semiarid regions, persistence in the rainfall deviations is significant. For example, in the semiarid West Africa rainfall shortages persisting over one to two decades were observed (Nicholson, 1982). Rainfall fluctuations are also associated with a geographic pattern. Sivakumar (1989) showed that the reduction in mean annual rainfall in both Niger and Burkina Faso after 1969 was characteristic of the entire region. After 1969, the rainfall isohyets were displaced further south showing that the rainfall changes affect large areas.

Because the semiarid climates are found in such a wide variety of latitudes and continental locations, few valid general comments can be made concerning their temperature regimes. In general, the clear skies and dry atmosphere tend to make them severe for their latitude, with relatively extreme seasonal temperatures and thus large annual ranges (Swindale, 1982).

Soils

The effectiveness of the low and variable rainfall in the semiarid regions is further determined by the soil type and its physical and chemical properties. There are many kinds of soils in the semiarid regions. As Swindale (1982) described, not all of them have ustic soil moisture regimes and not all have isohyperthermic or hyperthermic soil temperatures. Many are unsuited for agriculture. The brief description of the soils of the semiarid regions presented below is taken mainly from the publications on soils of the semiarid tropics.

Arenosols or coarse textured soils containing more than 65% sand and less than 18% clay comprise 11% of the arable soils in the semiarid tropics. Luvisols, soils with base-rich argillic B horizons, occupy 15% of the semiarid tropics, occurring in wetter climates than the arenosols. Vertisols occupy 7% of the arable soils in the semiarid tropics. Ferralsols are the most weathered and extensive soils of the semiarid tropics, occupying 33% of the region. Acrisols occur mainly in regions where seasonal rainfall is high, or in humid areas occurring within the semiarid regions.

Arenosols or the sandy soils are characterized by low fertility, low water-holding capacity and, hence, lack of water, and poor physical conditions. These soils are generally also low in organic matter, nitrogen and phosphorus (Jones and Wild, 1975). Luvisols or alfisols are low in nitrogen and phosphorus and have problems with surface crusting which reduces infiltration, affects seedling emergence, and reduces plant stand. The poor structural stability at the surface also makes them quite susceptible to erosion by water when tilled for cropping. Vertisols have a high available water-holding capacity, but are very susceptible to erosion on tillage.

Analysis of 31 soil samples from the major millet growing regions of the semiarid regions of West Africa showed that the total sand content varied from 71 to 99% with a mean of 87% (Bationo et al., 1993). These soils have production constraints imposed by physical and climatic processes, e.g., crusting, drought, erosion by wind and water, and high soil temperatures. One striking feature of these soils is the inherently low soil fertility, which is expressed through their low levels of organic matter, total nitrogen, and effective cation exchange capacity.

Much of the recent research in the Sahel indicates that rainfall per se is not necessarily the crucial limiting factor to agricultural production. Rather it is the proportion of rainfall which enters the soil water reservoir and its subsequent utilization by plants. Arenosols have a low water-holding capacity, which imposes a severe drought risk if extended dry periods occur during the crop growing season. On the other hand, luvisols present a different management problem because of their tendency to form a hard crust on the surface resulting in reduced infiltration and increased runoff. Charreau (1974) showed that as much as 32% of the mean annual rainfall could be lost as runoff on a well-tilled, cropped luvisol while on a bare soil the losses could be as high as 60%.

It is common knowledge that similar amounts of rainfall, on dissimilar soil types, could lead to different levels of available water for crop growth. At Bengou in southern Niger, measured soil water profiles in two soil types, a loam and a sandy loam, at the time of planting and harvesting of millet during the 1986 growing season (Sivakumar and Wallace, 1991) showed important differences with the same seasonal rainfall of 784 mm. Hence, strategies for sustainable agriculture may have to be site-specific.

Crops

Sorghum (*Sorghum bicolor* L. Moench*)*, pearl millet (*Pennisetum glaucum* L. R. Br.) and maize (*Zea mays* L.) are the major rainfed cereals grown in the semiarid regions. Sorghum is an important and widely adapted small grain cereal grown between 40°N and 40°S of the equator. It is mainly a rainfed crop in the lowland, semiarid areas of the tropics and subtropics, and a post-rainy season crop is grown principally on residual soil water, particularly in India (Craufurd et al., 1999), with limited rainfall. Pearl millet is widely grown as a food crop in subsistence agriculture in Africa and on the Indian subcontinent. Pearl millet has a number of advantages that have made it the traditional staple cereal crop in subsistence or low resource agriculture in hot semiarid regions like the West African Sahel and Rajasthan in northwestern India. These advantages include tolerance to drought, heat, and leached acid sandy soils with very low clay and organic matter content (Andrews and Kumar, 1992). Generally, it is grown on sandy soils in association with cereals such as sorghum and/or with legumes such as cowpea (*Vigna unguiculata*).

Rice (*Oryza sativa* L.) and sugarcane (*Saccharum officinarum* L.) are grown under irrigation and in the river deltas and wheat (*Triticum aestivum* L.) is grown under irrigation in the winter season mainly at higher latitudes. Wheat is also grown as a rainfed crop in the semiarid zone (Anon., 1981). Under these conditions, drought accompanied by high temperatures occurs frequently later in growth and the post-anthesis water shortages drastically reduce the grain yield.

The major grain legumes grown in the semiarid regions are pigeonpea (*Cajanus cajan* L. Millsp.), chickpea (*Cicer arietinum* L.), cowpea (*Vigna unguiculata* L. Walp.) and mung bean (*Vigna radiata* L. R. Wilcz.). Pigeonpea is an important tropical grain legume commonly intercropped with cereals. For example, in the semiarid regions of India, it is predominantly intercropped with sorghum, but in the lower rainfall areas with lighter and shallower soils it is grown with pearl millet. In the uplands, it is grown in combination with rice. Chickpea is grown over a wide range of agroclimatic environments in the arid and semiarid regions and is traditionally grown in the northern hemisphere mostly between 20°N and 40°N latitudes. It is best adapted to the cool winter temperatures of the semiarid tropics and the spring to early summer seasons of the Mediterranean region and has considerable importance as food, feed, and fodder.

Cowpea is an important seed and fodder crop commonly grown in agriculturally difficult conditions in the diverse cropping systems of both the semiarid and the humid tropics (Steele et al., 1985). Most farmers in West Africa grow medium- to long-duration, photoperiod-sensitive cultivars, usually in association with cereals such as pearl millet or sorghum. Cowpea is also an important grain legume grown in South and Southeast Asia, to meet the protein requirements of small farmers, forage requirements of ruminants and to improve the soil fertility of rice lands. Comparitive studies of species performance (Muchow, 1985a, b, c) showed that cowpea is a drought-avoiding species, whereas soybean (*Glycine max* L. Merr.) has high yield potential but has only a limited capacity to avoid drought in African conditions.

Groundnut or peanut (*Arachis hypogaea* L.), soybean, safflower (*Carthamus tinctorius* L.), sesame (*Sesamum indicum* L.), and mustard (*Brassica juncea*) are the main oilseed crops grown in the semiarid region, while cotton (*Gossypium hirsutum* L.) is the main fiber crop. Groundnut is grown in diverse agroclimatic environments in the semiarid regions. Soybean is a major crop with an increasing worldwide value as animal feed and human food source.

POTENTIAL FOR USE OF CROP GROWTH MODELS FOR PRACTICAL APPLICATIONS IN THE SEMIARID REGIONS

The summary of climate, soil, and crop characteristics described earlier shows that the farmers in the semiarid regions have the rather difficult task of managing their crops on generally poor soils in harsh and risky climates. Scientists and research managers striving to find solutions to the complex problem of soil and crop management in the unreliable semiarid climates also need tools that can assist them in taking an integrated approach to finding solutions. Policy makers and administrators need simple tools that can assist them in policy management. In this regard, a systems based effort can assist in understanding, predicting, and manipulating the outcomes from agricultural systems for ecological, agricultural, and economic gains (Singh et al., 1999).

Crop growth models can serve as effective tools in problem solving for the different user groups listed above: the farmers, scientists and policy makers. The applications that they seek vary. According to Boote et al. (1996), there are three primary uses or reasons for crop modeling — research knowledge synthesis, crop system decision management, and policy analysis. They believed that only recently have crop models been used as grower decision support tools.

Hoogenboom (2000) defined the management applications of crop simulation models as strategic applications, tactical applications and forecasting applications. Strategic applications were explained as those where crop models are run prior to planting of a crop to evaluate alternative management strategies in contrast to tactical applications, which are run before planting or during the crop-growing season. In forecasting applications, the crop models are used to predict yields either prior to planting or during the growing season. To the different applications listed by Hoogenboom, one can add research management applications. Depending on their needs, the different user groups could seek either one or more of the different applications, i.e., strategic, tactical, forecasting, and research.

Model Applications for On-Farm Decision Making

A number of decisions have to be made at the farm level such as which crop/cropping system to chose, which variety, when to cultivate the soil, when to sow, when to weed, when to fertilize, when to apply pesticides and insecticides, when to schedule irrigations, what depth of irrigations, when to harvest, etc. Crop growth models can be effectively used to derive simple decision rules for farmers in a decision support system framework to provide answers to these questions. As Tsuji et al. (1998) explained, models allow evaluation of one or more options that are available with respect to one or more management decisions.

Crop growth models can be used to predict crop performance in regions where the crop has not been grown before or not grown under optimal conditions. As shown by van Keulen and Wolf (1986), such applications are of value for regional development and agricultural planning in developing countries. Modeling is efficient since time frames of many years can be simulated quickly and inexpensively for many locations and management strategies (ARS, 1993). In addition, an unlimited number of management strategies can be considered.

A crop model can calculate probabilities of grain yield levels for a given soil type based on fall and winter rain and probabilities of these climatic conditions for the upcoming season, before investing in fertilizer (Kiniry and Bockholt, 1998). Assessments of the consequences of different timings and dosages of fertilizer applications can also be conducted. This kind of application is very important for improving the efficiency of fertilizers and biocides for specific cases, and for reducing environmental pollution. Model based management strategies, such as these, are already in use on large scales for optimising fungicide application in crops (Zadoks et al., 1984; Rosa et al., 1992). In addition, the biological outputs and management inputs can be combined with economic factors to determine the risk associated with the various management practices that are being evaluated (Lansigan et al., 1997; Thornton and Wilkens, 1998). Applications can also include investment decisions, such as those related to the purchase of irrigation systems (Boggess and Amerling, 1983).

The main goal of most modeling applications is to predict final production in the form of either grain yield, fruit yield, root or tuber yield, biomass yield for fodder, or any other harvestable product (Hoogenboom, 2000). Simulation models were used to predict crop yield, to extrapolate and to interpolate crop performance over large regions and to create links with other sciences. Maas (1988) showed that updating the initialization of a simple crop model with accumulated remote sensing data provided improved estimates of final yield than updating based on crop measurements.

Model Applications for Research

Although farm level applications are clearly the most important ones for crop-growth models, they are also widely used in research. Given the need for increased and sustained agricultural productivity in the semiarid regions, especially in the developing countries, efforts are being made to increase the research investment. In most developing countries, however, financial and human resources limit the number and the quality of experiments, and soils and weather change over short distances. Hence, Timsina et al. (1993) stated that conclusions from such experiments have only limited applicability, no matter how carefully they have been conducted. In this regard, crop models are particularly valuable for synthesizing research understanding and for integrating up from a reductionist research process (Whisler et al., 1986).

As Penning de Vries (1977) explained, the use of a simulation model of a system contributes to our understanding of the real system because it helps to integrate the relevant processes of the system studied and to bridge areas and levels of knowledge. The task of building models demands more complete acquisition and assimilation of knowledge. Conversion of conceptual models into mathematical simulation models reveals gaps in our knowledge of agricultural processes. Hence the interdisciplinary nature of simulation modeling efforts leads to increased research efficiency and improved research direction through direct feedback.

In the last two decades, simulation models have been principally used to determine the potential growth and establish the biological limits of agricultural production. A good example of how modeling helped research comes from the Wageningen Agricultural University in The Netherlands. In the 1960s, the first attempt to model photosynthetic rates of crop canopies was made (De Wit, 1965). Results obtained from this model were used among others, to estimate potential food production for some areas of the world and to provide indications for crop management and breeding (De Wit, 1967; Linneman et al., 1979). This was followed by the construction of an ELementary CROp growth Simulator (ELCROS) by De Wit et al. (1970). This model included the static photosynthesis model and crop respiration was taken as a fixed fraction per day of the biomass, plus an amount proportional to the growth rate. In addition, a functional equilibrium between root and shoot growth was added (Penning de Vries et al., 1974). The introduction of micrometeorology in the models (Goudriaan, 1977) and the quantification of canopy resistance to gas exchanges allowed the models to improve the simulation of transpiration and evolve into the BAsic CROp growth Simulator (BACROS) (De Wit and Goudriaan, 1978). BACROS model was subsequently used as a reference model for developing other models and as a basis for developing summary models such as SUCROS (Simple and Universal CROp growth Simulator) (van Keulen et al., 1982).

One of the first application-oriented research challenges for modeling was the Dutch/Israeli project titled "Actual and Potential Production of Semiarid Grasslands," which was initiated by De Wit in 1970 (Alberda et al., 1992). In this project, crop modeling was used to quantify and formalize, as far as possible, the relevant processes involved in water-limited production, and to extrapolate and apply the resultant knowledge to agricultural production systems (van Keulen et al., 1982). ARIDCROP (van Keulen, 1975), which was based on the concepts elaborated in ELCROS and BACROS, was developed to simulate the growth and water use of fertilized natural pastures in the Mediterranean region. This model was successfully incorporated into an integrated model of grazing system comprising separate management and biological sections, which was used to examine the consequences of contrasting management strategies in intensive agropastoral systems in a semiarid region (Ungar, 1990).

For prediction of annual pasture production in semiarid conditions in which growth is limited by rainfall and nitrogen, a preliminary model PAPRAN (Production of Arid Pastures limited by RAinfall and Nitrogen) was developed (Seligman and van Keulen, 1981; van Keulen, 1982). This model development was based on the combination of a relatively simple set of supply and demand functions giving the description of N uptake and redistribution in plant tissue with those of ARIDCROP.

Another very useful research application of crop growth models is in the domain of crop breeding strategies. Models can be used to examine the sensitivity of crop response to changes in plant characteristics so as to better define breeding strategies and goals. Accordingly breeders can survey the impact that breeding may have for specific characteristics (Landivar, 1979; Ng and Loomis, 1984). O'Toole and Stockle (1987) described the potential of simulation models in assessing trait benefits using the case of winter cereals which have a wide genetic base and physiological features that enable the crop to survive and reproduce in stress-prone environments. Crop simulation models provide the breeders with an analytical tool for integrating and quantifying the knowledge embodied in conceptual models. Crop growth models have been used in plant breeding to simulate the effects of changes in the morphological and physiological characteristics of crops and thus to aid in the identification of ideotypes (Donald, 1968) for different environments (Dingkuhn et al., 1993; Hunt, 1993; Kropff et al., 1995). Crop growth models that have been parameterized for new cultivars in field experiments can be used to simulate the long-term yield stability of these cultivars at a location under the expected range of climatic conditions (Hunt, 1993; Palanisamy et al., 1993).

The combination of crop growth models with pest, disease, and weed models can be used to investigate interactions between both systems (Rabbinge and Rijsdijk, 1983). Strong interactions between crop growth and disease or pest development make it potentially interesting for crop management, for example in the appropriate choice of spraying insecticides and fungicides and in avoiding unnecessary treatments. Kiniry et al. (1991) described the ALMANAC (Agricultural Land Management Alternatives with Numerical Assessment Criteria) model which contains a general crop growth model in which genotype-specific coefficients describe differences in the growth of different crops and crop cultivars. These coefficients control the simulation of development and senescence of leaf area, conversion of intercepted photosynthetically active radiation to biomass, growth of the root system, nutrient composition of the tissue, development of economic yield, and sensitivity of the crop to temperature, water, and nutrient stresses. ALMANAC simulates the water balance, nutrient balance, and plant growth, and additional detail for light competition, population density effects, and vapor pressure deficit effects which enable it to simulate the growth and seed yield of two competing plant species in a wide range of environments. Kiniry et al. (1991) showed that for maize, both simulated and measured mean yields with weeds are 86% of the weed-free yields.

Coop et al. (1991) developed a decision support system linking simulation models, databases and a user interface for benefit cost analysis of chemical treatment of the Senegalese grasshopper, *Oedaleus senegalensis* Krauss, to assist in the training, analysis, and management of grasshopper treatment programs. The analysis indicated that optimal timing was 5 to 10 days earlier than the actual treatments. Crop yield reports from treated and nontreated areas, a crop loss assessment conducted in Batha, Chad, in October 1987, and a breakeven analysis provided further evidence that the campaign was successful and cost effective at most sites, as indicated by the model results.

Model Applications for Policy Management

One very useful application of crop simulation models is for policy management. The issues for policy could range from global issues such as climate change impacts to field-level issues such as the effect of crop rotation strategies on the long-term changes in soil quality. Typical applications include agroecological zonation, regional yield forecasting and scenario studies for exploring the effects of environmental or socioeconomic changes on agriculture (Bouman et al., 1996). World food production studies (Buringh et al., 1979; Penning de Vries et al., 1995), agroecological zonation (Aggarwal, 1993; van Keulen and Stol, 1991) and explorations of the effects of climate change on crop production (Wolf, 1993; Matthews et al., 1995) employed crop growth models. Boote et al. (1996) concluded that a number of policy uses have been made in the areas of climate change, water use, erosion, soil nutrients, and pesticide use. Thornton et al. (1997) showed that in Burkina Faso, crop simulation modeling using satellite and ground-based data could be used to estimate millet production for famine early warning which can allow policy makers the time they need to take appropriate steps to ameliorate the effects of regional food shortages on vulnerable urban and rural populations. Some of the important policy management issues in agriculture include resource allocation, land use planning and environmental protection. The Decision Support System for Agrotechnology Transfer (DSSAT) developed by the International Benchmark Sites Network for Agrotechnology Transfer (IBSNAT) helps in seeking solutions to such specific issues (IBSNAT, 1988a). The DSSAT itself (IBSNAT, 1988b; Jones, 1993; Tsuji et al., 1994) is a shell that allows the users to organize and manipulate crop, soils, and weather data and to run crop models in various ways and analyze their outputs. Among the semiarid crops included in the DSSAT are the CERES cereal model for maize, sorghum, pearl millet, rice, and wheat as well as the CROPGRO model for groundnut, soybean, and peas (Tsuji et al., 1994).

In the semiarid regions, climatically induced production uncertainties also cause concern after the produce leaves the farm gate. Processing and marketing bodies require information that enables them to plan strategically for the season ahead (Meinke and Hammer, 1997). Using groundnut as an example, Meinke and Hammer (1997) presented a generic methodology to forward-estimate regional crop production and associated climatic risks based on phases of the Southern Oscillation Index (SOI) in Australia. Combining knowledge of SOI phases in November and December with output from a dynamic simulation model allows the derivation of yield probability distributions based on historic rainfall data. This information is available shortly after planting a crop and at least 3 to 5 months prior to harvest. Meinke and Hammer (1997) showed that in years when the November to December SOI phase is positive, there is an 80% chance of exceeding average district yields. Conversely, in years when the November to December SOI phase is either negative or rapidly falling, there is only a 5% chance of exceeding average district yields, but a 95% chance of below average yields. This information allows the industry to adjust strategically for the expected volume of production.

The repercussions of global changes for agriculture and natural ecosystems are potentially serious and simulation models are appropriate tools to explore these effects. Models used in these studies range from descriptive models that couple the information from general circulation models (GCMs) with the current knowledge regarding the environmental constraints that limit the area of cultivation of crops (Bindi et al., 1992; Parry et al., 1990), to explanatory models that predict the more detailed effects of warming and of increasing CO_2 on crop development and yield (Adams et al., 1990; Miglietta and Porter, 1992). Simulation models can be used to explore the effects of the increase in temperature and CO_2 concentrations on crop development, growth and yield, harvest index and water use, and can help breeders to anticipate future requirements (Goudriaan et al., 1984). In evaluating long-term sustainability issues, models such as CropSyst (Donatelli et al., 1997) and Erosion Prediction Impact Calculator (EPIC) of Jones et al. (1991) were found useful.

CURRENT APPLICATIONS OF CROP GROWTH MODELS
IN THE SEMIARID REGIONS

Maize

Most of the applications of simulation modeling thus far for maize were centered around the CERES-Maize model, while models such as EPIC, ALMANAC, CropSys, ADEL-Maize, WOFOST, and PUTU-Maize have also been used to simulate the growth and yield of maize. A brief description of these applications is presented below.

The model CERES-Maize (Crop Environment REsource Synthesis), first published by Jones and Kiniry (1986), was included in DSSAT. It has been also integrated into the higher level modeling environment CropSyst (Crop Systems) as described by Gommes (1999). Hodges et al. (1987) tested the ability of the CERES-Maize model to estimate annual fluctuations in maize production for 51 weather stations in the 14 states of the Corn Belt in the U.S. for the years 1982 to 1985. Their results indicated that the model might be used for large area yield and production estimation in the U.S. with minimal regional calibration.

The CERES-Maize model was used by Jagtap et al. (1998) for testing six different crop densities, at Ibadan in southwestern Nigeria, because of its relatively limited requirements of computer resources and data while presenting enough sensitivity for environmental factors. The analysis of planting densities ranging from 2.96 to 13.3 plants per m^2 showed that the optimum density for highest for LAI was 9.4 plants m^2, with a linear decrease of the absolute growth rate with higher densities. The use of the model in sub-saharan Africa, however, could have constraints related to the need to feed daily weather data in the model and access to soil data and IT equipment.

Another application of the CERES-Maize model was made in southern Africa. The model was tested for examining the vulnerability of maize yields to climate change in different farming sectors in Zimbabwe (Muchena and Iglesias, 1995). Maize crop production was simulated under different climate scenarios generated by General Circulation Models (GCMs) and the sensitivity was tested by 2° and 4°C increase in the daily temperature, including the impact of higher levels of CO_2 in the climate change scenarios, as the model includes an option to simulate the physiological effects of CO_2 on photosynthesis and water use efficiency. This study concluded that maize yields decrease significantly under all scenarios of warming tested (GISS, GFDL, and UKMO), even when the direct beneficial effects of increased CO_2 and water use were included.

Climate data from four different agroecological zones in Zimbabwe were analyzed with respect to ENSO phases and used to drive the CERES-Maize model, parameterized for soil conditions typical of the zones, using two nitrogen fertilizer treatments and three planting dates (Phillips et al., 1998). Their study showed that while average simulated maize yields were generally lowest in El Niño years, variability in rainfall pattern and standard deviation of yields at the site level was high within each ENSO phase, indicating that more precise seasonal climate predictions would be necessary for forecasts to be valuable in crop management decisions in Zimbabwe; however, simulation results pointed toward the relative importance of predicting favorable cropping seasons as opposed to poor ones with respect to better nitrogen management and yield improvement for the more marginal sites.

An innovative use of CERES-Maize was made for predicting crop response to salinity stress in southern Italy by Castrignanò et al. (1998) by integrating a new saline stress index related to the predawn leaf water potential. After calibration of the crop-specific model to adapt to the peculiar environmental conditions of the Mediterranean region (salinity and water stress, high evaporation demand), the model calculated the final grain yield and the seasonal evaporation correctly, but was underestimating aboveground biomass and maximum LAI.

Furthermore, the CERES-Maize model was evaluated for its predicting capacity of the effect of nitrogen deficiency on crop growth duration and yield by Singh et al. (1999). The authors proposed to enhance the sensitivity of the N stress indices with a modified version of the model, which simulates the effect of N deficiency on the phyllochron (leaf appearance rate) and phenological stages. The results were compared with field trials in tropical locations and showed the effect of N management on yield and risk prevention.

A modified version of CERES-Maize, which included the effects of limited soil aeration on crop growth and development by a root distribution weighting factor, was tested to improve the simulation of site-specific crop development and yield by Fraisse et al. (1999) in seven environmentally diverse sites in the U.S.

Paz et al. (1999) employed the CERES-Maize model-based technique to determine variable rate nitrogen for maize. The model used to characterise spatial maize yield variability was reported to give good predictions after a calibration process of three years in the U.S.

As CERES-Maize was designed mainly for simulation of hybrid maize, it has been further developed into the CERES-IM model in the U.S. for specific simulations of seed-producing inbred maize which takes into account the inbred specific field operations and traits (Rasse et al., 2000).

The model EPIC-phase is the result of the modification of the EPIC model for the simulation of effects of water and nitrogen stress on biomass and yield (Cabelguenne et al., 1999). It takes into account the sensitivity of the crops to water and N stress during the course of their development cycle. A 9-year validation process, using experimental data from a long term cropping systems experiment, was carried out at three levels of cropping intensity. Although the same input data was used, the results showed that EPIC overestimates crop production, in particular in conditions of severe water stress. Crop parameters were introduced in the model related to the water extraction capacities peculiar to each crop, the growth period was divided into four phases and the conversion efficiency of intercepted radiation was made a function of biomass, and the sensitivity of the harvest index to water and N stress was introduced for each phase of growth. The simulations were then close to the measured values. These trials were carried out in southwest France. In another study, Cabelguenne et al. (1996) described the use for tactical irrigation management by combining real-time, EPIC-phase with weather forecasts.

A study was undertaken by Dhakhwa et al. (1997) to assess the effects of global warming and CO_2 fertilization on maize growth with crop models. CERES-Maize and EPIC were used for simulating the yield and biomass of maize under projected future climate change scenarios derived from two GCMs, the GFDL, and UKMO. Both models were modified to account for the beneficial physiological effects of increased CO_2 concentration on crop growth and transpiration. The CERES, unlike EPIC, simulates different plant components such as stem-biomass, leaf-biomass, and ear-biomass at various phases of the growth stages. When only the direct effects of CO_2 were considered, the CERES predicted yield increases of about 14 to 18%. Simulated yield and biomass decreased under both GCM scenarios, mainly due to the effects of the higher temperature. Both models could be modified to reflect changes in biomass, yield, root, and WUE (water use efficiency) response of direct and indirect effects of climate change. The positive effects of the hypothesized differential day–night warming were more pronounced when EPIC was used than when CERES was used to simulate various plant components.

EPIC-phase was tested together with CROPWAT for their ability to simulate maize grain reduction caused by water stress under semiarid conditions by Cavero et al. (2000). The simulation of evapotranspiration, harvest index, leaf area index and final biomass was evaluated. The results showed that EPIC-phase overestimated the biomass in the more water stressed treatments, due to overestimation of LAI; however, following improvements made especially with regard to the effect of water stress on LAI growth, the model became more consistent for calculating yield reduction due to water stress and hence for semiarid regions.

The CROPSYST (cropping systems simulation model) simulates the growth of a variety of crops on a daily base, which can be linked to GIS software and a weather simulator (Stockle et al., 1994). It calculates the soil water budget, soil–plant–nitrogen budget, crop phenology, crop canopy, root growth, biomass production, crop yield, residue cycle, soil erosion by water, and pesticide fate. In the trials Stockle carried out, the crop water use was calculated properly, while the predicted nitrogen did not exactly match measured values from leaching experiments. Nevertheless, the simulated biomass and yield of corn indicated the potential of CropSyst, provided it is validated, as a promising tool to analyse management practices for water and nitrogen.

Fournier and Andrieu (1999) proposed a new simulation model based on L-system. The L-system formalism is a language to perform visual simulations of plant growth; it codes plant development as the parallel functioning of plant subunits (the modules) and enables production of dynamic three-dimensional (3-D) outputs of plants. ADEL-Maize combines a 3-D model of maize development with physical models computing the microclimate on the 3-D structure. The 3-D architectural and process-based ADEL (architectural model of development based on L-systems) simulates the development of maize as function of the temperature of organs. The time scale varies from a few hours to one day, enabling the mechanistic description of both the physiological process and the changes in the plant environment. The previous version of ADEL, which simulated development as a function of temperature, was complemented with a module for the regulation of growth by dry matter availability. The dry matter production is based on the concept of light use efficiency and allocation and is a function of the sink strength of each organ. It is therefore calculated according to organ size and temperature. In ADEL-Maize the canopy is considered as a set of individual plants. Fournier and Andrieu (1999) found the simulation of the effects of plant density with this model of dry matter management promising. The model requires only three parameters obtained from direct measurements. The modular approach and the 3-D representation allow an accurate calculation of the light interception by single plants.

Choudhury (2001) carried out research on the simulation model RUE (radiation use efficiency) calculating gross photosynthesis and net carbon accumulation by wheat before anthesis. He assessed the applicability of the RUE model to study the radiation–and–carbon–use efficiencies of maize, sorghum and rice, with regard to their growth in different environmental conditions and also to C3 crops such as rice (while maize and sorghum are C4 crops). Although most calculated RUE appeared to be consistent with the observations, for sorghum the RUE was about 20% lower than for maize. Therefore, the model needs to be further improved.

A methodology for assessing risks associated with crop production and fertilizer use in the tropics using statistical methods combined with crop modeling was developed by Rötter et al. (1997). It was aimed at quantifying yields and financial risks and opportunities for crops cultivated in different agroecological zones. The different production goals for non-irrigated maize, a ranking of the severity of damages and farmers' attitudes toward risk were taken into account in the risk assessment approach. Subsequently, Rötter and van Keulen (1997) studied the maize yield response to fertilizer application in arable land in Kenya in different agroecological zones and under different management practices, specially regarding the nitrogen and phosphorus inputs. The assessment of potential risks by using crop growth models was aimed at providing advice to smallholders for their farming management choices, in particular concerning the fertilizer application and the rentability of custom applications.

The model WOFOST (World Food Studies) calibrated to Kenyan maize cultivars was used on eight study sites, with data from the Fertiliser Use Recommendation Project (FURP) database and with climatic data from the Kenyan Meteorological Department and soil data from the Kenya Soil Survey (Rötter and van Keulen, 1997). The authors concluded that techniques to minimize farming risks such as the recommended "response farming" (adjusting cultivation practices to rainfall conditions early in the season), was beneficial in reducing losses. The use of the El Niño Southern Oscillation (ENSO) index was found more promising for predicting water- and nutrient-limited maize yields. The experimental period of 4 years was considered too short for giving indications with respect to the effects of fertilizer applications on the sustainability of the production system. Moreover, the farmer time scale of experiences with these risks are about 30 to 50 years, in which time other factors intervene, such as the increasing population pressure on the arable land. In order to obtain complete information on the economic benefit derived from the fertilizer application, the experiments should include trials of maize intercropped with legumes and rotations with other crop. Rötter and van Keulen (1997) recommended also taking into account the uncertainty associated with the unforeseeable price variations, in particular under semiarid conditions.

A trial using the WOFOST model to quantify agricultural resources was carried out by Shisanya and Thuneman (1993) in Machakos District of southern Kenya. The study was aimed at addressing the problem of increasing food production on marginal land with small holder farms and with low external inputs. WOFOST calculated the crop production for different zones, all characterized by a maize–bean intercropping pattern, by testing different planting dates for maize varieties adapted to dry conditions in order to provide technical advise to farmers. This model was used for a dynamic crop production simulation and as a explanatory tool for the higher crop failure and lower yields observed during long rainy season (March to May) compared to the short rainy season (October to December). Within this bimodal rain pattern, the short rainy seasons appeared to be more reliable and the results confirmed the advantage of early planting.

An attempt to build a framework for forecasting the extent and severity of drought in maize was made in a semiarid region in South Africa by de Jager et al. (1998). This system, already in use since 1994 in the Free State Province of South Africa, is based upon the phase of the SOI and has been applied to quantify and map drought hazard in maize by running maize crop growth models in a GIS framework. The data were grouped into 9800 homogenous natural resource zones. For each zone, the computed maize grain yield forecasts were compared to long-term cumulative probability functions of yield to determine their probabilities of nonexceedance, which were then used to identify drought severity areas (de Jager et al., 1998). To ensure climate, soil and crops specificity, the two crop growth models CERES-Maize and PUTU-Maize (previously developed by de Jager) and weather data were used. For a user-friendly access to information on drought conditions for farmers and decision makers, a specific GIS was developed for drought monitoring. This system is widely accepted and well received by the users; however, no test of accuracy had yet been carried out, but was planned for the near future.

Sorghum

Modeling applications with sorghum were initially made using SORGF model and subsequently with the SORKAM model. Limited applications were also made with other models such as ALMANAC and SorModel. A brief description of these applications is given below.

The first attempt at modeling the growth and yield of sorghum was made by Maas and Arkin (1978) from Temple, Texas, in their description of SORGF, a dynamic grain sorghum growth model. The initial testing phase of SORGF in Texas was followed by multi-locational trials in the semiarid regions of India conducted by the International Crops Research Institute for the Semiarid Tropics (Huda et al., 1980, 1982). It was shown that there is a good agreement between simulated and measured sorghum yields.

In tests of SORGF model to simulate growth of sweet sorghum under Australian conditions, Ferraris and Vanderlip (1986) concluded that SORGF underpredicted yield under good growing conditions and overpredicted yield when water stress limited plant growth. They concluded that tillering was predicted poorly by SORGF and that overestimation of leaf senescence resulted in erroneous prediction of leaf area index, particularly late in the growing season or under conditions of water deficit.

Subsequently, the Temple, Texas group developed an improved grain sorghum growth model SORKAM (Rosenthal et al., 1989). Differences in management practices, cultivars and locations were accounted for by SORKAM through the use of a number of input parameters. Management inputs included planting date, plant population, seeding depth, row spacing, and irrigation. Cultivar input parameters in the model were number of leaves, maturity class, photoperiod sensitivity, and tillering and seed number coefficients. Climatic data requirements included solar radiation, maximum and minimum temperatures, and precipitation. Soil data inputs included maximum and plant available water by soil layer, Stage 1 and Stage 2 soil water evaporation coefficients, soil albedo, maximum rooting depth, runoff curve number, field slope, and potential evapotranspiration correction factor.

Heiniger et al. (1997a) made improvements in the tillering (nonlinear temperature function) and grain production (linear function of plant growth) routines in SORKAM. In tests of the ability of SORKAM model to simulate forage sorghum yields for a wide range of environmental conditions, Fritz et al. (1997) found that the model generally overpredicted grain and total drymatter yields. They concluded that the inability to simulate phenological development accurately under water-deficit conditions and to partition dry matter into grain and stover are additional weaknesses in the model. In studies of the use of SORKAM to develop guidelines for replanting grain sorghum, Heiniger et al. (1997b) found that SORKAM could capture only 27 to 79% of grain yield variability. Yield predictions from different plant populations within a planting date were particularly inaccurate. They concluded that poor yield predictions were the result of improper computation of tiller number and faulty partitioning of biomass to caryopsis weight and to use SORKAM to generate replant guidelines, the authors recommended that improvements must be made in modeling the relationships among yield components and the source-sink relationship that determines caryopsis weight. To address this problem, Heiniger et al. (1997c) developed a new grain growth equation that relates grain filling rate to the rate of change of plant dry matter per caryopsis during the effective grain filling period. The revised model accounted for 48 to 72% of the observed variability as opposed to 15 to 57% for the original model.

Kiniry and Bockholt (1998) evaluated the ability of the ALMANAC model to simulate plot grain yields of sorghum at eight locations under diverse weather conditions and soils in Texas. Model inputs included parameters for the soil type, planting dates, planting rates, and locally measured weather data. Mean simulated grain yield for each site was within 10% of the mean measured grain yield in all cases. The narrow range of measured yields was given as the reason why the models did not account for a significant amount of the year-to-year variability in measured grain yield.

Baez-Gonzalez and Jones (1995) developed SorModel, a dynamic and deterministic growth model for sorghum based on plant growth and water flow relationships. The growth and development process of the crop was mainly based on the SUCROS developed by van Keulen et al. (1982). Because SUCROS calculated potential crop production when water or plant nutrients were not limiting factors, Baez-Gonzalez and Jones (1995) interlinked it with soil water flow processes to make it possible to calculate dry matter production under conditions that are not always optimum for the growth and development of the plants. The SorModel was used to predict forage dry matter production in semiarid Mexico and the results showed reasonable agreement between the observed and simulated dry matter production.

Apart from the full growth simulation models, such as SORGF and SORKAM, simple models have also been developed for addressing specific tasks of sorghum crop management.

Hodges et al. (1979) modeled dry matter accumulation of a grain sorghum crop with photosynthesis and respiration equations requiring only daily meteorological variables, leaf area index and stage of development. Reduction of rate of photosynthesis due to high temperature and water stresses was included in the model. They showed that dry matter predicted by the model was within 10 to 15% of the dry matter measured in Kansas, Texas, and Nebraska. Grain yield, which was assumed to be 80% of the panicle weight, was related to measured grain yield ($R^2 = 0.58$).

Sinclair et al. (1997) developed a simple, mechanistic model to interpret measurements of the growth and yield of sorghum at different levels of nitrogen and water supply. The nitrogen model was developed considering experimental results obtained in the tropical climate of Katherine, Australia. Comparison of the model results with those obtained in an irrigation-nitrogen application experiment conducted in Hyderabad, India, led to two interesting hypotheses. First, the irrigation level of the well-watered treatment appeared to be inadequate to avoid drought stress at the end of the growing season. Second, about 4 g N m^{-2} of soil N was unavailable to the crop in each of the irrigation treatments. This hypothesis was based on the observation that at high applications of N, the model predicted yield well, but at low applications predictions substantially exceeded measurements and the uptake of N was also overestimated. Muchow et al. (1994) assessed climatic risks relative to planting date decisions for sorghum in a subtropical rainfed region.

Pearl Millet

The most common growth model in use for simulating the growth and yield of pearl millet is the CERES-Pearl Millet model. Limited applications have also been reported using the CROPSYST and PmModels.

Ritchie and Alagarswamy (1989a, 1989b, 1989c) described the CERES-Pearl Millet model, in particular the genetic coefficients and simulation of phenology and growth and development. Ritchie et al. (1998) elaborated the procedures used in CERES crop models, including pearl millet, to estimate crop growth, development, and yield.

Thornton et al. (1997) developed a prototype pearl millet yield estimation system for 30 provinces of Burkina Faso using the CERES-Pearl Millet model and remotely sensed estimates of rainfall in real-time, embedded in a geographic information system. They showed that early warning of impending poor harvests in this manner were useful for policy makers to take appropriate action to ameliorate the effects of regional food shortages on vulnerable rural and urban populations.

Ram Niwas et al. (1996) tested the CERES-Pearl Millet model using three pearl millet cultivars at New Delhi, India. The predicted days for anthesis showed a deviation from 1 to 4 days, with a mean deviation of 2.3 days for all the varieties and 2 seasons. The number of days required for maturity varied from those observed by 5 to 8 days. The predicted biomass and grain yields agreed well with observed data. The authors concluded that CERES-Pearl Millet model can be used to study the suitability of genotypes in a particular region.

Badini et al. (1997) used the crop growth simulation model, CropSyst, to simulate the soil water budget components and millet production potential in Burkina Faso, both spatially and temporally, by coupling the model with databases of soil type, long-term weather, and crop management using a geographic information system (GIS). The model consists of several integrated components and different management options (Stockle and Nelson, 1993). From the cropping model outputs, Badini et al. (1997) quantified and mapped two agroclimatic indices (Aridity Index and Crop Water Stress Index) that show the water-limited growth environment of the millet crop throughout Burkina Faso.

Baez-Gonzalez and Jones (1995) developed PmModel, a dynamic and deterministic growth model for pearl millet based on plant growth and water flow relationships. As in the SorModel, the growth and development process of the crop was mainly based on the SUCROS model developed by van Keulen et al. (1982). The PmModel was used to predict forage dry matter production in semiarid Mexico and the results showed reasonable agreement between the observed and simulated dry matter production.

Overman and Robinson (1995) used a logistic model by coupling dry matter and plant nitrogen accumulation of pearl millet through a common response coefficient c. The model was shown to describe accurately the response of millet dry matter, plant N removal, and plant N concentration to applied N. Furthermore, the model was shown to closely describe the relationship between yield and plant N removal.

Cotton

The two most common growth models used in applications for cotton are the GOSSYM and COTONS models. One application involved the use of AMAPpara model. These applications are briefly described below.

The GOSSYM (GOSSYpium siMulator) model was developed initially as a physiologically based simulation model (Baker et al., 1983), integrating all aspects related to growth, development, physiology, and agronomy in order to be used as an experimental tool. Coupled with COMAX (CrOp MAnagement eXpert), an expert system, it provides management recommendations for irrigation, fertilizer, and harvest aid applications. It has been further adapted to become a crop management tool. Reddy et al. (1997) provide a detailed description of applications of GOSSYM.

Among the recent developments, researchers of CIRAD (Centre de Coopération Internationale en Recherche Agronomique pour le Développement) and USDA-ARS (Agricultural Research Service) have jointly developed a model, COTONS, by combining the mechanistic-physiological approach of GOSSYM with an architectural model (Jallas et al., 1999). It combines the functions and concepts of a cultivated plant type model and a 3-D plant architecture visualization tool. A large number of potential applications are seen by the authors, specifically with regard to the 3-D aspects that allow "virtual" cultivation, like for plant protection, plant mapping, and applications of multifactor combinations including conversion-efficiency architecture and growth regulators.

The COTONS model enables the tactical choices that the farmer has to make during the cultivation period in order to address the gap between the decisions adopted by the farmers within the constraints of his farm management and the technical advice of experts. However the advice to strategic and tactical farming decisions should be adapted to the diversity of environmental and socioeconomic situations. In this regard, field trials have been made in western Africa by integrating these advises in the agricultural extension services. Currently, the COTONS model is being built in a GIS in Burkina Faso (Jallas et al., 1999), to provide predictions on a more regional level, which could then be coupled with databases for soil, climate, etc.

Another hydraulic architecture-based growth model for cotton plants developed by a CIRAD team is the AMAPpara model (de Reffye et al., 1999). It describes long-term plant growth as the cumulative output of the cyclic interactions between plant physiological functioning and architectural development. It is based on the classical relationship between transpiration and biomass production, under the assumption of constant water use efficiency. Allometric rules are used to derive the geometry of the organs as a function of their volume or biomass. The model considers simultaneously the topological and geometrical structures of the plant and relates them to the environmental conditions. The number of new organs is predicted from the cumulative temperature according to the architectural model, and the volume and geometry of each organ are computed according to biomass production and allocation, using sink-source and allometric concepts. The feedback between plant growth and architecture is modeled through a recurrence equation, which links successive growth cycles to each other, although this remains based only on a plant morphogenetic model that is not sensitive to the environment (de Reffye et al., 1999). The parameters of the AMAP are estimated from the observation of plant architecture and morphology at the end of their growth. Real time could be introduced into such models thanks to the strong links that exist between the temperature and growth rate. According to the authors, the calibration of the model for cotton as described, gave satisfactory results quantitatively and qualitatively.

Groundnut

The only groundnut simulation model that has been used in applications so far is the PNUTGRO model which was developed and tested at the University of Florida in Gainesville (Boote et al., 1987). It is a process-level model to simulate growth and yield of groundnut and includes vegetative and reproductive development, carbon balance, nitrogen balance, and water balance models as the major components. The basic structure of the model and the underlying differential equations have been explained in detail by Wilkerson et al. (1983) and Boote et al. (1987). To simulate groundnut response to row spacing and plant population, Boote et al. (1988, 1989, 1992) revised the light interception and canopy assimilation subroutines to include the hedgerow approach developed by Gijzen and Goudriaan (1989), which was simplified for inclusion in PNUTGRO. This revised model was referred to as the PNUTGRO hedgerow version. This approach predicts canopy light interception, projected shadow cast by the canopy, and the fractions of the leaf area that are sunlit and shaded to estimate carbon assimilation by the crop.

Singh et al. (1994a) evaluated the performance of PNUTGRO under different levels of water availability in various seasons and sowing dates at four locations in semiarid tropical India. The model predicted the occurrence of flowering and podding within ±5 days of observed values at locations where growth stages were recorded most frequently. Predictions of growth stages beyond podding were less accurate because of difficulties, associated with the indeterminate nature of the crop, to record growth stages after pod growth has started in the soil. Changes in vegetative growth stages, total dry matter accumulation, growth of pods and seeds, and soil-water were predicted accurately by the model. Predicted pod yields were significantly correlated with observed yields ($R^2 = 0.90$). It is interesting to note that the authors stress that PNUTGRO can be used to predict groundnut yields in different environments under biotic stress-free conditions; however, groundnut grown in the semiarid regions, in the developing countries in particular, is attacked by a number of pests and diseases that are not incorporated in the model.

The hedgerow version of PNUTGRO was evaluated (Singh et al.,1994b) for predicting phenological development, light interception, canopy growth, dry matter production, pod and seed yields of groundnut as influenced by row spacing and plant population. The model predicted the occurrence of vegetative and reproductive stages, canopy development, total dry matter production, and the accurate partitioning to pod and seed. Correlation between simulated and observed pod yield was significant ($R^2 = 0.61$).

Chickpea

The chickpea growth models that had been developed so far include the CHIKPGRO model and another mechanistic model, which are briefly described in this subsection. Singh and Virmani (1996) used the hedgerow version of groundnut model PNUTGRO, earlier described by Boote et al. (1988, 1989), to develop the chickpea model, CHIKPGRO. Major soil and plant processes included in the model were soil water balance, root growth and extension, vegetative and reproductive development, water stress effects on reproductive development, photoperiod response to flowering, canopy growth and expansion of leaves, photosynthesis, respiration, partitioning of biomass to vegetative and reproductive organs, protein mobilization, and senescence. The model does not consider biological nitrogen fixation by chickpea and assumes soil fertility to be nonlimiting for crop growth. The model predicted flowering, pod initiation, beginning of seed growth, and physiological maturity within ±5 days of the observed values, except under extreme wet situations when the actual seed growth and physiological maturity of chickpea occurred later than the simulated dates. Leaf area index, total dry matter production, and partitioning to various plant organs under irrigated and water-stressed conditions were also predicted satisfactorily by the model. Predicted total dry matter and seed yields were significantly correlated with the observed data. The authors concluded that CHIKPGRO could be used to predict potential and water-limited yields of chickpea in the Indian plateau, but cautioned that further work requires inclusion of a soil fertility submodel and model testing over a wide range of environments.

Soltani et al. (1999) developed a simple mechanistic model which simulates crop phenology, development of leaves as a function of temperature, accumulation of crop biomass as a function of intercepted radiation, dry matter accumulation of grains as a function of time and temperature, and soil water balance. Phenology, leaf growth, and senescence and biomass production were made sensitive to soil water content. Tests of the model from a range of environments in Iran showed a good agreement between simulated and observed yield under both irrigated and rainfed conditions.

Other Crops

Wheat

Ritchie (1985) used the CERES-Wheat model to determine the expected yield variations that resulted from weather variations at a semiarid location in Oklahoma. All of the relationships used to calculate water balance components in CERES are quite empirical and the model has been tested with about 300 different measured data sets from several countries to demonstrate its generality (Ritchie and Otter, 1984). The simulation results showed that a early maturing genotype with an earlier floral initiation date is superior, simply because it has less chance of depleting the water supply before plant maturity. Bell and Fischer (1994) used the CERES-Wheat model to account for weather-based potential variation in wheat yield in the Yaqui Valley region of Mexico between 1978 and 1990, assuming no change in cultivar or management. They showed that the climatic potential yield declined because of increased temperature.

Aggarwal and Kalra (1994) used the wheat growth model, WTGROWS to evaluate the potential wheat yield at 138 sites across India. They concluded that yields, under optimal water and nitrogen availability, increased with higher latitude and with more inland sites, primarily because of variation in temperature. Predicted yields increased 428 kg ha^{-1} for each °C increase in temperature. They also showed that the yield gap between potential and actual yields of wheat at New Delhi is attributable in part to delayed sowing.

The CERES v3. model was used by Lal et al. (1998) for wheat and rice as incorporated in the DSSAT to examine the vulnerability of wheat and rice in the semiarid region of northwest India to projected climate change. Both the models were able to simulate observed year-to-year variations in yield over northwest India. Acute water shortage conditions combined with the thermal stress were projected to adversely affect both the wheat and more severely the rice productivity in northwest India even under the positive effect of elevated CO_2 in future.

Jamieson (2001) carried out a case study on modeling the response of wheat to drought using the wheat simulation model Sirius (Jamieson et al., 1998). In most of the treatments, observed and predicted responses were close. The exceptions were that leaf area index was overestimated late in the life of the crop when drought stress was severe and evapotranspiration was overestimated in the driest treatment and underestimated in the wettest treatment.

The Crop Growth Simulation System (CGMS) was developed by Vossen (1990) at the core of which is the crop growth simulation model WOFOST (van Keulen and Wolf, 1986). CGMS operates on grid cells of 50 x 50 km and for each grid cell the required inputs are daily weather data, soil characteristics and management practices. Using the crop growth simulation results, planted area to soft wheat and a trend function, Supit (1997) predicted wheat yield for 12 European countries, including Spain. The very long and severe dry spells in the semiarid regions of Spain in the early 1990s led to water shortages that were underestimated by CGMS.

O'Toole and Stockle (1987) employed a process oriented spring wheat model for analysing the temperature-related traits: root growth and grain filling. Simulation results indicated that tolerance of cereal roots to cool soil temperature deserves attention given the vital role of early root system development in the target environment. Similarly simulations with grain filling indicated that the functional form of grain filling rate versus temperature is relatively less important than limitations on grain filling duration in determining final grain weight. It was suggested that plant breeders and physiologists may benefit from more in-depth study of the physiological basis for stability in grain filling duration exhibited by adapted genotypes.

The APSIM-Wheat model was used to simulate above- and below-ground growth, grain yield, water and N uptake, and soil water and soil-N in wheat crops in Western Australia (Asseng et al., 1998). Grain yields were well-predicted, despite some underestimation during severe terminal droughts.

Rainfed Rice

For the drought-prone rainfed rice belt across the Gangetic Plains of Bihar and Uttar Pradesh in India, Jones and O'Toole (1987) used the ALMANAC model to simulate rainfed rice production for different production situations — adequate water supply and plant nutrition; near-optimum plant nutrition and limiting water supply; and suboptimal water and nutrient supply. Results showed that the models could be used to simulate growth of crops, and the effect of the variations in the supply of nutrients and water.

Soybean

Patron and Jones (1989) described the Tropical Soybean Production Model (TROSOY), which calculates the dry matter accumulation of soybean when water and nutrients are in optimal supply or when water is a constraint to production. They applied the model as a framework to synthesize and analyze the ecological structure of the soybean production system in southern Tamaulipas, Mexico.

Model results showed satisfactory agreement with values obtained by field experimentation and it was concluded that the model could be used, within reasonable limits, as an auxiliary research tool. The model was applied to assess potential and rainfed production levels of soybean in the region and to evaluate the prospect of using runoff farming as a strategy to stabilize and improve production.

Sinclair (1986) developed a simple and robust crop model for soybean using a phenomenological and physiological framework. Sinclair et al. (1987) generalized this modeling approach and used it to examine the yield potential and production risks of soybean, cowpea, and black gram under water deficits, and to assess the importance of different physiological traits in determining the productivity in these grain legumes. Coefficients of the relationships in the model describing leaf growth, carbon and nitrogen input, seed growth, and the water budget were obtained from literature and from glasshouse and field experimentation. The principal differences in input variables to model the growth of soybean, cowpea, and black gram were those describing leaf emergence rate, N fixation during seed filling, and the biochemical composition of the seeds. The relationship describing the response of leaf-area growth, radiation use efficiency, and nitrogen fixation to soil water content differed little among species.

The SOYGRO model simulates the effect of soil water on photosynthesis, leaf expansion and leaf senescence. The user-friendliness of the SOYGRO model, which was developed by Wilkerson in the early 1980s (Wagner-Riddle et al., 1997), contributed to its wide utilization.

Cowpea

Timsina et al. (1993) used MACROS.CSM, a mechanistic, noncrop-specific model developed for the simulation of crop growth and development as a function of weather and soil water, to estimate yields of partially irrigated and rainfed crops of early- and medium-maturing cowpea cultivars in various parts of the Philippines. The comparison of the measured and the simulated weights of the different crop parts as well as of the measured and calculated soil water content, showed that the mechanistic simulation modeling can be a useful tool to delineate the extrapolation domains of crops and cultivars.

Sunflower

Chapman et al. (1993) developed a sunflower simulation model, QSUN, with five interacting modules: grain yield, biomass accumulation, crop leaf area, phenology, and water balance. Using this model, Meinke et al. (1993) evaluated the production risks in a variable subtropical environment.

Modeling Crop Rotations and Intercropping

The diversity of crops grown is an important feature of small-holder agriculture in developing countries of the semiarid regions. Intercropping cereal crops with legume crops is a very common practice in the semiarid tropics and has been traditionally used by low-income farmers to make the most of their resources and to reduce risk. Similarly crop rotations of cereals with legumes are used to maintain soil fertility and ensure crop diversity in the risk-prone, semiarid regions. Following the success of simulating the growth and yield of sole crops, efforts have been made to simulate crop rotations and intercropping. Notable among such efforts is the APSIM package, followed by some other simple approaches such as GROWIT, as described in the next subsection.

APSIM

The Agricultural Production Systems Simulator (APSIM) is a software package that is structured around a central "engine" via which modules (crop growth, soil water, soil N and erosion) communicate with each other (McCown et al., 1996). Any module can, in principle, be incorporated in a simulation run, provided it uses the same variable as APSIM and the formats are recognized by APSIM. This flexibility allows simulation of crop rotations or mixtures by linking two or more crop growth modules to the engine. An innovation in APSIM, as compared to many other models, is that it considers the soil, instead of the crop, as the central entity (McCown et al., 1996). Turpin et al. (1992) used the APSIM model to simulate changes in soil nitrogen and effects on subsequent cereal crops (wheat and sorghum) after the soil organic matter content had been enhanced under a legume ley. Their results showed that the model was capable of producing sensible output for how soil organic matter and nitrogen behave after soil fertility has been raised by a ley and the contribution this makes to subsequent crops.

Jones et al. (1996) used the APSIM model with modules for the soil water balance (SoilWat V1.0), soil organic matter and N transformations (SoilN V1.0), surface residue dynamics (Residue V1.0), and for maize, sorghum, and Verano (Maize V1.0, CSSAT V0.1 and Stylo V1.0) to explore the benefit from legume pasture leys of Caribbean stylo (*Stylosanthes hamata* cv. Verano) to subsequent maize crops. The model was shown to adequately capture the principal effects, in terms of the extra nitrogen taken up by the maize crop after stylo leys compared with the grass leys, and the persistence of the effect of legume leys into at least the second maize crop. The authors concluded that with some further development, it should prove useful in examining the viability of pasture legume–cereal rotations in other environments and seasons.

The APSIM model was also used to simulate the performance of a hypothetical chickpea–wheat rotation on clay soils in Queensland, Australia (Probert et al., 1998). The legume effect was demonstrated by the soil nitrate available at the time of sowing of the next crop. The simulation results also showed that soil organic matter and nitrogen declined continuously in continuous wheat cropping for 25 years without the addition of nitrogen fertilizer. The inclusion of a legume crop in the rotation with wheat considerably reduced the decrease of soil fertility.

GROWIT

The millet/cowpea intercrop is well adapted to the poor soils and low rainfall of the Sahelian zone of West Africa. In Niger, over 70% of the cropland is intercropped and millet/cowpea intercrop is the primary crop association (Lowenberg-DeBoer et al., 1991). Cowpeas are typically planted 2 to 3 weeks after millet, though the exact planting date depends on rainfall. Lowenberg-DeBoer et al. (1991) adopted the generic plant growth model, GROWIT, in a spreadsheet format to simulate the growth and yield of the millet/cowpea intercrop. The GROWIT model calculates daily plant growth based on temperature, rainfall and soil characteristics. The structure of the GROWIT model was originally developed by Smith and Loewer (1983) to simulate a wide variety of crop species. It has been used extensively for perennial crops, but some annual crops have been simulated. In the case of the millet/cowpea intercrop model, competition between the two crops for available sunlight was simulated by assuming that millet was the dominant crop and cowpeas used residual space. Soil water budget was calculated assuming that both crops draw from a common soil water resource.

CropSys

Caldwell and Hansen (1993) used the CropSys model to assess the risks associated with an upland, rice-based cropping system involving rice and soybean. Their study showed an overall increase in yields when upland rice was preceded by nitrogen-fixing soybean.

Coupling CERES Models

Timsina et al. (1997) coupled CERES-Rice and CERES-Wheat models and used it to identify the causes for low, unstable yields and to quantify nutrient depletion rates in the intensely grown rice–wheat system now common in Indo-Gangetic Plains of India.

Singh et al. (1999) evaluated a soybean–chickpea sequencing model on Vertic Inceptisols in Patancheru, India, and used it to extrapolate the results over 22 years of historical weather records. Data on climate, crop, soil, and agronomic management were retrieved to create files needed for the execution of generic versions of SOYGRO and CHIKPGRO models available in DSSAT v3. Simulation results showed that in most years the broadbed-and-furrow (BBF) landform increased rainfall infiltration into the soil and had marginal effect on the yields of soybean and chickpea. The authors concluded that crop yields on Vertic Inceptisols can be further increased and sustained by adopting appropriate rainwater management practices for exploiting surface runoff and deep drainage water as supplemental irrigation to crops in a watershed setting.

NEEDS AND FUTURE PERSPECTIVES

The semiarid regions, home to a large majority of the rural poor in the world, are in general less-favored areas with low agricultural productivity and natural resource degradation. Due to the low and variable rainfall and the poor soils, crop production in most of the semiarid regions of the world is undertaken mainly for sustenance, with the exception of cash crops such as groundnut, cotton, and cowpea. Agriculture is risk prone and most of the subsistence farmers are reluctant to place large investments in the needed inputs such as improved varieties, fertilizers, and pesticides.

Given this ground reality in the semiarid regions, agricultural scientists and planners alike, are faced with the enormous challenge of ensuring continued increases in agricultural productivity to feed the growing populations under conditions of decreasing private and public investments to improve the drylands of the semiarid regions. With the growing realization that public investment in agricultural research is on a decline in many countries of the semiarid regions due to lack of adequate financial resources, and that the traditional practice of field experimentation to find solutions to the diverse questions concerning farming can no longer be supported as in the past, much hope was placed on the use of crop growth models to answer both strategic and tactical questions concerning agricultural planning as well as on-farm soil and crop management.

Although the potential for crop model applications covers at least three areas i.e., on-farm decision making, research and policy management, this review shows that in the semiarid regions much of the modeling applications to date have been mainly in the area of research. Many of the crop models have been used by the researchers to investigate possible outcomes of changes in planting dates, varieties, seeding rates, fertilizer inputs, etc. In addition, the current interest in the impacts of global warming on agriculture and forestry led researchers to investigate the different scenarios of the projected increases in green house gases using crop models and project likely outcomes. Some investigations also covered the issue of climate variability by combining the knowledge of SOI with output from crop growth models to allow derivation of yield probability distributions based on historical data.

How good is the current state of knowledge of crop growth modeling in the semiarid regions? A wide range of crops and cropping systems are used in the semiarid regions. Where farmers have access to irrigation, the natural preference is toward cash crops. The review of the current applications of models for different crops shows that considerable effort was devoted to modeling applications in maize and cotton. A beginning has been made for other cash crops such as groundnut and cowpea, but field applications as seen in the case of maize and cotton are lacking for these two crops. Sorghum is the predominant cereal crop in the semiarid region and some applications of sorghum modeling have been made. Wheat is a minor crop in the semiarid regions, but the large advances made in the modeling applications of wheat in the temperate regions carried some impact in such applications in the semiarid regions.

Although sole cropping is the major cropping pattern in the productive farm lands of the humid temperate regions, farmers in the semiarid regions, especially in the tropics, favor intercropping and crop rotations. Much of the crop modeling applications in the literature are based on sole cropping, but in the recent past, a beginning has been made for modeling the growth and yield of intercrops and crop rotations, particularly employing the APSIM package. A lot more work is needed in this very important area for the semiarid regions.

The disappointing reality is that few, if any, of the modeling applications in the semiarid regions are actually used by the farmers. Although the criticism that crop growth models have failed to simulate reality on the ground in the semiarid regions is understandable, it is important to recognize that there are many scientific challenges to enhancing crop modeling applications in the semiarid regions.

First, many of the crops grown by the resource-poor farmers of the semiarid regions have not received the kind of attention that is needed to understand fully the physical and physiological processes that govern their growth and yield under the variable soil and climatic conditions. This is a fundamental requirement to make progress in simulating the growth and yield of these crops.

Second, there is inadequate human resource capacity *in situ* to develop and validate simulation models in the semiarid regions.

Third, many of the scientific research institutions in the developing countries of the semiarid regions fail to promote the needed multidisciplinary research approach to develop and apply crop models, even in the research area. Hence, it is difficult to find comprehensive soil, crop and climate data sets to develop and validate crop models in the semiarid regions.

Fourth, the linkages between the research, teaching and extension departments in the developing countries of the semiarid regions are most often weak or are nonexistent, hence it is difficult to develop modeling applications that can be quickly disseminated to farmers.

Unless these challenges are addressed in earnest, it is difficult to see field applications of crop growth models in the semiarid regions in the near future. There is no doubt that research applications in crop modeling will continue, but such applications can not sustain long-term interest in crop modeling unless on-farm and policy applications are vigorously developed and applied.

REFERENCES

Adams, R.M., C. Rosenzweig, R.M. Peart, J.T. Ritchie, B.A. McCal, J.D. Glyer, R.B. Curry, J.W. Jones, K.J. Boote, and L.H. Allen. 1990. Global climate change and U.S. agriculture, *Nature,* 345:219–224.

Aggarwal, P.K. 1993. Agro-ecological zoning using crop growth simulation models: characterization of wheat environments in India. In *Systems Approaches for Sustainable Agricultural Development,* F.W.T Penning de Vries et al., Eds., Kluwer Academic Publishers, Dordrecht, The Netherlands, 97–109.

Aggarwal, P.K. and N. Kalra. 1994. Analyzing the limitations set by climatic factors, genotype, and water and nitrogen availability on productivity of wheat: II. climatically potential yields and management strategies, *Field Crops Res.,* 38:93–103.

Alberda, T., N.G. Seligman, H. van Keulen, and C.T. de Wit, Eds., 1992. Food from dry lands: an integrated approach to planning of agricultural development, Kluwer Academic Publishers, Dordrecht, The Netherlands.

Andrews, D.J. and K.A. Kumar. 1992. Pearl millet for food, feed and forage, *Advances in Agronomy,* 48:89–139.

Anon. 1981. *World Wheat Facts and Trends,* Report 1, International Maize and Wheat Improvement Center, Mexico.

ARS. 1993. *Communicating Research to the Farmer Using Modular Crop and Ecosystem Simulators and Expert Systems,* Agricultural Research Service, Washington, D.C.

Asseng, S., B.A. Keating, I.R.P. Fillery, P.J. Gregory, J.W. Bowden, N.C Turner, J.A. Palta, and D.G. Abrecht. 1998. Performance of the APSIM-Wheat model in western Australia, *Field Crop Res.,* 57:163–179.

Badini, O., C.O. Stockle, and E.H. Franz. 1997. Application of crop simulation modelling and GIS to agroclimatic assessment in Burkina Faso, *Agric. Ecosys. Env.,* 64:233–244.

Baez-Gonzalez, A.D. and J.G.W. Jones. 1995. Models of sorghum and pearl millet to predict forage dry matter production in semi-arid Mexico: I. Simulation Models, *Agric. Syst.,* 47:133–145.

Baker, D.N., J.R. Lambert, and J.M. McKinion. 1983. GOSSYM: A simulator of cotton crop growth and yield, *S.C. Agric. Exp. St. Bull.,* 1089, Clemson Univ., Clemson, SC.

Bationo, A., M.P. Sedogo, E.O. Uyovbiscere, and A.U. Mokwunye. 1993. Recent achievements on soil fertility management in the West African semi-arid tropics. Paper presented at the *International Workshop on Sustainable Lands, Environment Management in Tropical Africa,* organized by West and Central Africa Soil Science Associations (WACASS), December 6–10, Ouagadougou, Burkina Faso.

Bell, M.A. and R.A. Fisher. 1994. Using yield prediction models to assess yield gains: a case study for wheat, *Field Crops Res.,* 36:161–166.

Bindi, M., F. Ferrini, and F. Miglietta. 1992. Effect of CO_2-induced climatic change on the cultivated area of olive trees, *Agric. Meteorol.,* 122:41–44.

Boggess, W.G. and C.B. Amerling. 1983. A bioeconomic simulation analysis of irrigation environments, *S.J. Agric. Econ.,* 15:85–91.

Boote, K.J., J.W. Jones, G. Hoogenboom, and G.G. Wilkerson. 1987. PNUTGRO v 1.0, Peanut crop growth and yield model, technical documentation, Department of Agronomy and Agricultural Engineering, University of Florida, Gainesville, FL.

Boote, K.J., G. Bourgeois, and J. Goudriaan. 1988. Light interception and photosynthesis of incomplete hedgerow canopies of soybean and peanut, *Agron. Abstr.,* 1988:105.

Boote, K.J., J.W. Jones, and G. Hoogenboom. 1989. Simulating crop growth and photosynthesis response to row spacing, *Agron. Abstr.,* 1989:11.

Boote, K.J., J.W. Jones, and N.B. Pickering. 1996. Potential uses and limitations of crop models, *Agron. J.,* 88:704–716.

Boote, K.J., J.W. Jones, and P. Singh. 1992. Modelling growth and yield of groundnut. In *Groundnut — a Global Perspective. Proc. of an Int. Workshop, ICRISAT Center, Patancheru, India,* S.N. Nigam, Ed., November 25–29, 1991. International Crops Research Institute for the Semi-Arid Tropics, Patancheru, India.

Bouman, B.A.M. et al. 1996. The "School of de Wit" crop growth simulation models: a pedigree and historical overview, *Agric. Syst.,* 52:171–198.

Buringh, P., H.D.J. van Heemst, and G. Staring. 1979. Potential world food production. In *MOIRA, Model of International Relations in Agriculture, Contributions to Economic Annals,* Vol. 124, H. Linnemann, Ed., North Holland Publishing Company, Amsterdam, New York, 19–74.

Cabelguenne, M., P. Debaeke, and A. Bouniols. 1999. EPICphase, a version of the EPIC model simulating the effects of water and nitrogen stress on biomass and yield, taking account of developmental stages: validation on maize, sunflower, sorghum, soybean and winter wheat, *Agric. Syst.,* 60:175–196.

Cabelguenne, M., J. Puech, P. Debaeke, N. Bosc, and A. Hilaire. 1996. Tactical irrigation management using real time EPIC-phase model and weather forecast: experiment on maize. In *Irrigation Scheduling: From Theory to Practice, Proceedings,* Water Report No. 8. FAO, Rome, Italy, 185–193.

Caldwell, R.M. and J.W. Hansen. 1993. Simulation of multiple cropping systems with CropSys. In *Systems Approaches for Agricultural Development, Proc. of Int. Symp.,* December 2–6, 1991, Bangkok, Thailand, F.W.T. Penning de Vries et al., Ed., Kluwer Academic Publishers, Dordrecht, The Netherlands, 397–412.

Castrignanò, A., N. Katerji, F. Karam, M. Mastropirilli, and A. Hamdy. 1998. A modified version of CERES-Maize model for predicting crop response to salinity stress, *Ecological Modelling,* 111(2-3):107–120.

Cavero, J., I. Farre, P. Debaeke, and J.M., Faci. 2000. Simulation of maize yield under water stress with EPICphase and CROPWAT models, *Agron. J.,* 92:679–690.

Chapman, S.C., G.L. Hammer, and H. Meinke. 1993. A sunflower simulation model: I. Model Development, *Agron. J.,* 85:725–735.

Charreau, C. 1974. Soils of tropical dry and dry-wet climatic areas of West Africa and their use and management, *Agricultural Mimeograph* 74–76, Cornell University, Ithaca, New York.

Choudoury, B.J. 2001. Modeling radiation- and carbon-use efficiencies of maize, sorghum, and rice, *Agric. For. Meteorol.,* 106: 317–330.

Coop, L.B., B.A. Croft, C.F. Murphy, and S.F. Miller. 1991. Decision support system for economic analysis of grasshopper treatment operations in the African Sahel, *Crop Prot.,* 10:485–495.

Craufurd, P.Q., V. Mahalakshmi, F.R. Bidinger, S.Z Mukuru, J. Chantereau, P.A. Omanga, A. Qi, E.H., Roberts, R.H., Ellis, R.J., Summerfield, and G.L. Hammer. 1999. Adaptation of sorghum: characterisation of genotypic flowering responses to temperature and photoperiod, *Theor. Appl. Genet.,* 99:900–911.

de Jager, J.M., A.B. Potgieter, and W.J. van den Berg. 1998. Framework for forecasting the extent and severity of drought in maize in the Free State Province of South Africa, *Agric. Syst.,* 57(3):351–365.

de Reffye, P., F. Blaise, S. Chemouny, S. Jaffuel, T. Fourcaud, and F. Houllier. 1999. Calibration of hydraulic architecture-based growth models of cotton plants, *Agronomie,* 19:265–280.

De Wit, C.T. 1982. Simulation of living systems. In *Simulation of Plant Growth and Crop Production, Simulation Monographs,* F.W.T Penning de Vries and H.H. van Laar, Eds., PUDOC, Wageningen, The Netherlands.

De Wit, C.T. and J. Goudriaan. 1978. *Simulation of Assimilation, Respiration and Transpiration of Crops, Simulation Monographs,* PUDOC, Wageningen, The Netherlands.

De Wit, C.T., R. Brouwer, and F.W.T. Penning de Vries. 1970. The simulation of photosynthetic systems, In *Prediction and Measurement of Photosynthetic Productivity, Proc. of Int. Biological Program/Plant Production Technical Meeting,* Setlik, I., Ed., Trebon, PUDOC, Wageningen, The Netherlands.

De Wit, C.T. 1965. Photosynthesis of leaf canopies, *Agric. Res. Rep. No. 663,* PUDOC, Wageningen, The Netherlands.

De Wit, C.T. 1967. Photosynthesis: its relationship to overpopulation. In *Harvesting the Sun,* Academic Press, New York, 315–320.

Dhakhwa, G.B., C.L. Campbell, S.K. LeDuc, and E.J. Cooter. 1997. Maize growth: assessing the effects of global warming and CO_2 fertilization with crop models, *Agric. For. Meteorol.,* 87:253–272.

Dingkuhn, M., F.W.T. Penning de Vries, and K.M. Miezan. 1993. Improvement of rice plant type concepts: systems research enables interaction of physiology and breeding. In *Systems Approaches for Sustainable Agricultural Development,* F.W.T. Penning de Vries, P. Teng, and K. Metselaar, Eds., Kluwer Academic Publishers, Dordrecht, The Netherlands, 19–35.

Donald, C.M. 1968. The breeding of crop ideotypes, *Euphytica,* 17:385–403.

Donatelli, M., C. Stockle, E. Ceotto, and M. Rinaldi. 1997. Evaluation of CropSyst for cropping systems at two locations of northern and southern Italy, *Eur. J. Agron.,* 6:35–45.

Ferraris, R. and R.L. Vanderlip. 1986. Assessment of the SORGF/SORG5 model for predicting sweet sorghum growth. In *Proc. Australian Sorghum Conf.,* February 4–6, 1986, M.A. Foale and R.G. Henzell, Eds., Lawes, Queensland, Australia, 484–487.

Fournier, C. and B. Andrieu. 1999. ADEL-Maize: an L-system based model for the integration of growth processes from the organ to the canopy. Application to regulation of morphogenesis by light availability, *Agronomie,* 19:313–327.

Fraisse, C.W., K.A. Sudduth, N.R. Kitchen, P.C. Robert, R.H Rust, and W.E Larson. 1999. Evaluation of crop models to simulate site-specific crop development and yield, *Proc. of the Fourth Int. Conf. on Precision Agriculture,* Parts A and B, St. Paul, MN, July 19–22, 1998, American Society of Agronomy, Madison, WI, 1297–1308.

Fritz, J.O., R.L. Vanderlip, R.W. Heiniger, and A.Z. Abelhalim. 1997. Simulating forage sorghum yields with SORKAM, *Agron. J.,* 89:64–68.

Gijzen, H. and J. Goudriaan. 1989. A flexible and explanatory model of light distribution and photosynthesis in row crops, *Agric. For. Meteorol.,* 48:1–20.

Gommes, R. 1999. *Roving Seminar on Crop-Yield Weather Modelling,* World Meteorological Organization, Geneva, Switzerland.

Goudriaan, J. 1977. *Crop Micrometeorology: A Simulation Study, Simulation Monograph,* PUDOC, Wageningen, The Netherlands.

Goudriaan, J., H.H. van Laar, H. van Keulen, W. Louwerse. 1984. Simulation of the effect of increased atmospheric CO_2 on assimilation and transpiration of a closed crop canopy. In *Math-Nat. R.,* Wissenschafliche Zeitschrift Humboldt Universitaet Berlin, 33:352–356.

Heiniger, R.W., R.L. Vanderlip, and S.M. Welch. 1997a. Developing guidelines for replanting grain sorghum: I. Validation and sensitivity analysis of the SORKAM sorghum growth model, *Agron. J.,* 89:75–83.

Heiniger, R.W., R.L. Vanderlip, S.M. Welch, and R.C. Muchow. 1997b. Developing guidelines for replanting grain sorghum: II. Improved methods of simulating caryopsis weight and tiller number, *Agron. J.,* 89:84–92.

Heiniger, R.W., R.L. Vanderlip, S.M. Welch, and R.C. Muchow. 1997c. Developing guidelines for replanting grain sorghum: III. Using a plant growth model to determine replanting options, *Agron. J.,* 89:93–100.

Hodges, T., C. Botner, C. Sakamoto, and J.H. Haug. 1987. Using the CERES-Maize model to estimate production for the U.S. Corn Belt, *Agric. For. Meteorol.,* 40:293–303.

Hodges, T., E.T. Kanemasu, and I.D. Teare. 1979. Modeling dry matter accumulation and yield of grain sorghum, *Can. J. Plant Sci.*, 59:803–818.

Hoogenboom, G. 2000. Contribution of agrometeorology to the simulation of crop production and its applications, *Agric. For. Meteorol.*, 103(1–2):137–157.

Huda, A.K.S., S.M. Virmani, M.V.K Sivakumar, and J.G. Sekaran. 1980. Collaborative multilocation sorghum modeling experiment, *Proc. of the Cooperators' Meeting, April 2-4, Agroclimatology Progress Report No. 4,* ICRISAT, Patancheru, India.

Huda, A.K.S., S.M. Virmani, M.V.K Sivakumar, and J.G. Sekaran. 1982. Report of collaborative multilocation sorghum modeling experiment, *Agroclimatology Progress Report No. 7,* ICRISAT, Patancheru, India.

Hunt, L.A. 1993. Designing improved plant types: a breeder's view point. In *Systems Approaches for Sustainable Agricultural Development,* F.W.T. Penning de Vries et al., Eds., Kluwer Academic Publishers. Dordrecht, The Netherlands, 3–17.

IBSNAT. 1988a. Documentation for IBSNAT crop model input and output files, version 1.1 for the decision support system for agrotechnoloy transfer (DSSAT v2.1), *IBSNAT Tech. Rep. No. 5,* Dept. of Agronomy and *Soil Sci.*, College of Tropical Agric. and Human Resour., University of Hawaii, Honolulu, HI.

IBSNAT. 1988b. Experimental design and data collection procedures for IBSNAT, *IBSNAT Tech. Rep. No. 1,* 3rd ed., Dept. of Agronomy and *Soil Sci.*, College of Tropical Agric. and Human Resour., University of Hawaii, Honolulu, HI.

Jagtap, S.S., R.T. Alabi, and O. Adeleye. 1998. The influence of maize density on resource use and productivity: an experimental and simulation study, *African Crop Sci. J.*, 6(3):259–272.

Jallas, E. et al. 1999. COTONS, une nouvelle génération de modèles de simulation des cultures, *Agriculture et Développement,* 22:35–46.

Jamieson, P.D. 2001. Case study of modelling the response of wheat to drought. In *Agrometeorological Information Needs in Agricultural Production,* P.D. Jamieson et al., Eds., CAgM Report No. 85, Commission for Agricultural Meteorology, World Meteorological Organization, Geneva, Switzerland, 31–36.

Jamieson, P.D., M.A. Semenov, I.R. Brooking, and G.S. Francis. 1998. Sirius: a mechanistic model of wheat response to environmental variation, *Eur. J. Agron.*, 8:161–179.

Jones, C.A., P.T. Dyke, J.R. Williams, J.R. Kiniry, V.W. Benson, and R.H. Griggs. 1991. EPIC: an operational model for evaluation of agricultural sustainability, *Agric. Syst.*, 37:341–350.

Jones, C.A. and J.R. Kiniry. 1986. *CERES-Maize: A Simulation Model of Maize Growth and Development,* Texas A&M University Press, College Station, TX.

Jones, C.A. and J.C. O'Toole. 1987. Application of crop production models in agroecological characterization: simulation models for specific crops. In *Survey of Agroecological Methods,* Commonwealth Agricultural Bureau, Wallingford, U.K., 199–209.

Jones, J.W. 1993. Decision support systems for agricultural development. In *Systems Approaches for Agricultural Development, Vol. 2, Proc. Symp.,* F.W.T. Penning de Vries et al., Eds., Bangkok, Thailand, December 2–6, 1991, Kluwer Academic Publishers, Dordrecht, The Netherlands.

Jones, M.S. and A. Wild. 1975. *Soils of the West African Savanna,* Commonwealth Bureau of Soils, Technical Communication 55, Harpenden, Herts, U.K.

Jones, R.K., M.E. Probert, N.P. Dalgliesh, and R.L. McCown. 1996. Nitrogen inputs from a pasture legume in rotations with cereals in the semi-arid tropics of northern Australia: experimentation and modelling on a clay loam soil, *Aust. J. Exptl. Agric.*, 36:985–994.

Kiniry, J.R. and A.J. Bockholt. 1998. Maize and sorghum simulation in diverse Texas environments, *Agron. J.*, 90:682–687.

Kiniry, J.R., W.D. Rosenthal, B.S. Jackson, and G. Hoogenboom. 1991. Predicting leaf development of crop plants. In *Predicting Crop Phenology,* T. Hodges, Ed., CRC Press, Boca Raton, FL, 29–42.

Kropff, M.J., A.J. Haverkort, P.K. Aggarwal, and P.L. Kooman. 1995. Using systems approaches to design and evaluate ideotypes for specific environments. In *Eco-Regional Approaches for Sustainable Land Use and Food Production,* J. Bouma et al., Eds., Kluwer Academic Publishers, Dordrecht, The Netherlands, 417–435.

Lal, M., K.K. Singh, L.S. Rathore, G. Srinivasan, and S.A. Saseendran. 1998. Vulnerability of rice and wheat yields in NW India to future changes in climate, *Agric. For. Meteorol.*, 89:101–114.

Landivar, J.A. 1979. The application of cotton simulation model GOSSYM in genetic feasibility studies, M.Sc. thesis, Mississippi State University, Starkville, MS.

Lansigan, F.P., S. Pandey, and B.A.M. Bouman. 1997. Combining crop modelling with economic risk analysis for the evaluation of crop management strategies, *Field Crops Res.,* 51:133–145.

Linneman, H. 1979. *Moira, Model of International Relations in Agriculture,* North Holland Publishing Company, Amsterdam.

Lowenberg-De Boer, J. et al. 1991. Simulation of yield distributions in millet-cowpea intercropping, *Agric. Syst.,* 36:471–487.

Maas, S.J. 1988. Use of the remotely sensed information in agricultural crop models, *Ecological Modelling,* 41:247–268.

Maas, S.J. and G.F. Arkin. 1978. User's guide to SORGF: a dynamic grain sorghum growth model with feedback capacity, *Texas Agric. Exp. Stn. Res. Ctr. Program and Model Documentation No. 78–1,* Temple, TX.

Matthews, R.B. et al. 1995. *Modelling the Impact of Climatic Change on Rice Production in Asia,* CAB International, Wallingford, U.K.

McCown, R.L., G.L. Hammer, J.N.G. Hargreaves, D.P. Holzworth, and D.M. Freebairn. 1996. APSIM: a novel software system for model development, model testing, and simulation in agricultural systems research, *Agric. Syst.,* 50:255–271.

Meinke, H., G.L. Hammer, and S.C. Chapman. 1993. A sunflower simulation model: II. Simulating production risks in a variable sub-tropical environment, *Agron. J.,* 85:735–742.

Meinke, H. and G. L Hammer.1997. Forecasting regional crop production using SOI phases: an example for the Australian peanut industry, *Aust. J. Agric. Res.,* 48:789–793.

Miglietta, F. and J. Porter. 1992. The effects of CO_2-induced climatic change on development in wheat: analysis and modelling, *J. Exp. Bot.,* 43:1147–1158.

Monteith, J.L. 1996. The quest for balance in crop modelling, *Agron. J.,* 88:695–697.

Muchena P. and A. Iglesias. 1995. Vulnerability of maize yields to climate change in different farming sectors in Zimbabwe. In *Analysis of Potential International Impacts. ASA Special,* 59:229–239.

Muchow, R.C. 1985a. Phenology, seed yield and water use of grain legumes grown under different soil water regimes in a semi-arid tropical environment, *Field Crops Res.,* 11:81–97.

Muchow, R.C. 1985b. Canopy development in grain legumes grown under different soil water regimes in a semi-arid tropical environment, *Field Crops Res.,* 11:99–109.

Muchow, R.C. 1985c. Stomatal behaviour in grain legumes grown under different soil water regimes in a semi-arid tropical environment, *Field Crops Res.,* 11:291–307.

Muchow, R.C., G.L. Hammer, and R.L. Vanderlip. 1994. Assessing climatic risk to sorghum production in water-limited subtropical environments: II. Effects of planting date, soil water at planting, and cultivar phenology, *Field Crops Res.,* 36:235–246.

Muchow, R.C., T.R. Sinclair, and J.M. Bennett. 1990. Temperature and solar radiation effects on potential maize yield across locations, *Agron. J.,* 82:338–343.

Ng, N. and R.S. Loomis. 1984. *Simulation of Growth and Yield of Potato Crop, Simulation Monograph,* PUDOC, Wageningen, The Netherlands.

Nicholson, S.E. 1982. *The Sahel: A Climatic Perspective,* Club du Sahel, Organization for Economic Coop-eration and Development (OECD), Paris, France.

O'Toole, J.C. and C.O. Stockle. 1987. The role of conceptual and simulation modelling in plant breeding. Presented at the *Int. Symp. on Improving Winter Cereals under Temperature and Soil Salinity Stresses,* October 26–29, 1987, Cordoba, Spain.

Overman, A.R. and K. Robinson. 1995. Rational basis for the logistic model for forage grasses, *J. Plant Nutr.,* 18:995–1012.

Palanisamy, S., F.W.T Penning de Vries, S. Mohandass, T.M. Thiyagarajan, and A.A. Kareem. 1993. Simulation in pre-testing of rice genotypes in Tamil Nadu. In *Systems Approaches for Sustainable Agricultural Development,* F.W.T. Penning de Vries et al., Eds., Kluwer Academic Publishers, Dordrecht, The Netherlands, 63–75.

Pan, X., J.D. Hesketh, and M.G. Huck. 1997. An object-oriented and internet-based simulation model for plant growth. In *Seventh Int. Conf. on Computers in Agriculture,* USDA-ARS, Urbana, IL, 345–351.

Parry, M.L., J.H Porter, and T.R. Carter. 1990. Agriculture: climatic change and its implications, *Trends in Ecol. and Evol.,* 5:318–322.

Passioura, J.B. 1996. Simulation models: science, snake oil, education or engineering, *Agron. J.,* 88:690–694.

Patron R.S. and J.G.W Jones. 1989. Soybean production in the Tropics — a simulation case for Mexico, *Agricultural Syst.,* 29(3):219–231.

Paz, J.O. et al. 1999. Model-based technique to determine variable rate nitrogen for corn, *Agric. Syst.,* 61(1):69–75.

Penning de Vries, F.W.T. 1994. Computers, climate and risks. In *Proc. of the 5th Int. Computer Conf.,* London, U.K., 27.

Penning de Vries, F.W.T. 1977. Evaluation of simulation models in agriculture and biology: conclusions of a workshop, *Agricultural Syst.,* 2:99–105.

Penning de Vries, F.W.T., A.B. Brunsting, and H.H. van Laar, 1974. Products, requirements and efficiency of biological synthesis, a quantitative approach, *J. Theor. Biol.,* 45:339–377.

Penning de Vries, F.W.T., H. van Keulen, and R. Rabbinge. 1995. Natural resources and limits of food production in 2040. In *Eco-Regional Approaches for Sustainable Land Use and Food Production,* J. Bouma et al., Eds., Kluwer Academic Publishers, Dordrecht, The Netherlands, 65–87.

Phillips, J.G., M.A. Cane, and C. Rosenzweig. 1998. ENSO, seasonal rainfall patterns and simulated maize yield variability in Zimbabwe, *Agric. For. Meteorol.,* 90:39–50.

Probert, M.E., J.P. Dimes, B.A. Keating, R.C. Dalal, and W.M. Strong. 1998. APSIM's water and nitrogen modules and simulation of the dynamics of water and nitrogen in fallow systems, *Agric. Syst.,* 56:1–28.

Rabbinge, R. and F.H. Rijsdijk. 1983. EPIPRE: a disease and pest management system for winter wheat, taking account of micrometeorological factors, *EPPO Bull.,* 13:297–305.

Ram Niwas, C.V.S. Sastri, and O.P. Bishnoi. 1996. Testing of CERES millet simulation model using three pearl millet cultivars, *Haryana Agric.Univ. J. Res.,* 26:267–271.

Rasse, D.P., J.T. Ritchie, W.W. Wilhem, J. Wei, and E.C. Martin. 2000. Simulating inbred-maize yields with CERES-IM, *Agron. J.,* 922:672–678.

Reddy, K.R., H.F. Hodges, and J.M. McKinion. 1997. Crop modelling and application: a cotton example, *Adv. Agron.,* 59:225–290.

Ritchie, J.T. and G. Alagarswamy. 1989a. Simulation of sorghum and pearl millet phenology. In *Modelling the Growth and Development of Sorghum and Pearl Millet,* S.M. Virmani et al., Eds., Research Bulletin No. 12, International Crops Research Institute for the Semi-Arid Tropics, Patancheru, India, 24–26.

Ritchie, J.T. and G. Alagarswamy. 1989b. Genetic coefficients for the CERES models. In *Modelling the Growth and Development of Sorghum and Pearl Millet,* S.M. Virmani et al., Eds., Research Bulletin No. 12, International Crops Research Institute for the Semi-Arid Tropics, Patancheru, India, 27–28.

Ritchie, J.T. and G. Alagarswamy. 1989c. Simulation of growth and development in CERES Models. In *Modelling the Growth and Development of Sorghum and Pearl Millet,* S.M. Virmani et al., Eds., Research Bulletin No. 12, International Crops Research Institute for the Semi-Arid Tropics, Patancheru, India, 34–38.

Ritchie, J.T. and S. Otter. 1984. CERES-Wheat: A user-oriented wheat yield model. Preliminary documentation, AgRISTARS Publication No. YM-U3-04442-JSC-18892.

Ritchie, J.T. 1985. A user-oriented model of the soil water balance in wheat. In *Wheat Growth and Modeling,* Day, W. and Atkin, R.K., Eds., Plenum Press, New York, 293–305.

Ritchie, J.T., U. Singh, D.C. Godwin, and W.T Bowen. 1998. Cereal growth, development, and yield. In *Understanding Options for Agricultural Production,* G.Y. Tsujii et al., Eds., Kluwer Academic Publishers, Dordrecht, The Netherlands, 79–98.

Roberts, B.R. 1976. The potential for systems and modelling research in pasture science in South Africa, *Agric. Syst.,* 1:233–241.

Rosa, M., R. Genesio, B. Gozzini, G., Maracchi, and S. Orlandini. 1992. Plasmo: a computer program for grapevine downy mildew development forecast, *Comp. Electr. in Agric.,* 9:205–215.

Rosenthal, W.D., R.L. Vanderlip , B.S. Jackson, and G.F. Arkin. 1989. *SORKAM: A grain sorghum growth model, TAES Computer Software Demonstration Series,* MP-1669, Texas Agric. Exp. Stn., College Station, TX.

Rötter, R., H. van Keulen, and M.J.W. Jansen. 1997. Variations in yield response to fertilizer application in the tropics: I. Quantifying risks and opportunities for smallholders based on crop growth simulation, *Agric. Syst.,* 53(1):41–68.

Rötter, R. and H. van Keulen. 1997. Variations in yield response to fertilizer application in the tropics: II. risks and opportunities for smallholders cultivating maize on Kenya's arable land, *Agric. Syst.,* 53(1):69–95.

Seligman, N.G. and H. van Keulen. 1981. PAPRAN: A simulation model of annual pasture production limited by rainfall and nitrogen. In *Simulation of Nitrogen Behaviour of Soil-Plant Systems*, M.J. Frissel and J.A. van Veen, Eds., PUDOC, Wageningen, The Netherlands.

Shisanya, C.A. and H. Thuneman. 1993. The use of dynamic crop production simulation model in quantifying agricultural resources: an example from Machakos district, Kenya, *J. Eastern African Res. Dev.*, 23:176–191.

Sinclair, T.R. 1986. Water and nitrogen limitations in soybean grain production. I. model development, *Field Crops Res.*, 15:125–141.

Sinclair, T.R., R.C. Muchow, M.M. Ludlow, G.J. Leach, R.J. Lawn, and M.A. Foale. 1987. Field and model analysis of the effect of water deficits on carbon and nitrogen accumulation by soybean, cowpea and black gram, *Field Crops Res.*, 17:121–140.

Sinclair, T.R., R.C. Muchow, and J.L. Monteith. 1997. Model analysis of sorghum response to nitrogen in subtropical and tropical environments, *Agron. J.*, 89:201–207.

Sinclair, T.R. and N.G. Seligman. 1996. Crop modelling: from infancy to maturity, *Agron. J.*, 88:698–703.

Singh, P., G. Alagarswamy, G. Hoogenboom, P. Pathak, S.P. Wani, and S.M. Virmani. 1999. Soybean–chickpea rotation on Vertic Inceptisols: II. Long-term simulation of water balance and crop yields, *Field Crops Res.*, 63:225–236.

Singh, P. et al. 1994a. Evaluation of the groundnut model PNUTGRO for crop response to water availability, sowing dates and seasons, *Field Crops Res.*, 39:147–162.

Singh, P., K.J. Boote, and S.M. Virmani. 1994b. Evaluation of the groundnut model PNUTGRO for crop response to plant population and row spacing, *Field Crops Res.*, 39:163–170.

Singh, P. and S.M. Virmani. 1996. Modeling growth and yield of chickpea (*Cicer arietinum* L.), *Field Crops Res.*, 46:41–59.

Singh, U., P. Wilkens, V. Chude, and S. Oikeh. 1999. Predicting the effect of nitrogen deficiency on crop growth duration and yield. In *Proc. of the Fourth Int. Conf. on Precision Agriculture*, Parts A and B, St. Paul, MN, July 19–22, 1998, American Society of Agronomy, Madison, WI, 1379–1393.

Sivakumar, M.V.K. and J.S. Wallace. 1991. Soil water balance in the Sudano-Sahelian zone: need, relevance and objectives of the workshop. In *Soil Water Balance in the Sudano-Sahelian Zone, Proc. of an Int. Workshop*, M.V.K. Sivakumar et al., Eds., February 18–23, 1991, Niamey, Nigeria. IAHS Publication 199, IAHS Press, Institute of Hydrology, Wallingford, U.K., 3–11.

Sivakumar, M.V.K. 1989. Agroclimatic aspects of rainfed agriculture in the Sudano-Sahelian zone. In *Soil, Crop and Water Management Systems for Rainfed Agriculture in the Sudano-Sahelian Zone, Proc. of an Int. Workshop*, January 7-11, 1987, ICRISAT Sahelian Center, Niamey, Nigeria, ICRISAT No. 17, Patancheru, India.

Sivakumar, M.V.K. and J.L. Hatfield. 1990. Spatial variability of rainfall at an experimental station in Niger, West Africa, *J. Theor. and Appl. Climatology*, 42:225–237.

Smith, E.M. and O.J. Jr. Loewer. 1983. Mathematical logic to simulate the growth of two perennial grasses, *Trans. ASAE*, 26:878–883.

Soltani, A., K. Ghassemi-Golezani, F.R. Khooie, and M. Moghaddam. 1999. A simple model for chickpea growth and yield, *Field Crops Res.*, 62:213–224.

Steele, W.M., D.J. Allen, and R.J. Summerfield. 1985. Cowpea. In *Grain Legume Crops*, R.J. Summerfield and E.H. Roberts, Eds., Collins, London, 520–583.

Stockle, C.O. and R. Nelson. 1993. *CropSyst: Cropping System Simulation Model User's Manual*, Biological Systems Engineering Department, Washington State University, Pullman, Washington.

Stockle, C.O., S. Martin, and G.S., Campbell. 1994. CropSyst, a cropping systems model: water/nitrogen budgets and crop yield, *Agric. Syst.*, 46:335–359.

Supit, I. 1997. Predicting national wheat yields using a crop simulation and trend models. *Agric. For. Meterol.*, 88:199–214.

Swindale, L.D. 1982. Distribution and use of Arable Soils in the Semi-Arid Tropics. In *Managing Soil Resources, Trans. of the 12th Int. Congr. of Soil Science*, New Delhi, India, 67–100.

Thornton, P.K. et al. 1997. Estimating millet production for famine early warning: an application of crop simulation modelling using satellite and ground-based data in Burkina Faso, *Agric. For. Meterol.*, 83:95–112.

Thornton, P.K. and P.W. Wilkens. 1998. Risk assessment and food security. In *Understanding Options for Agricultural Production,* G.Y. Tsuji et al., Eds., Kluwer Academic Publishers. Dordrecht, The Netherlands, 329–345.

Timsina, J., F.W.T. Penning de Vries, and D.P. Garrity. 1993. Cowpea production in rice-based cropping systems of the Philippines — extrapolation by simulation, *Agric. Syst.,* 42:383–405.

Timsina, J., U. Singh, and Y. Singh, 1997. Addressing sustainability of rice-wheat systems: analysis of long-term experimentation and simulation. In *Application of Systems Approaches at the Field Level,* vol. 2, M.J. Kropff et al., Eds., Kluwer Academic Publishers. Dordrecht, The Netherlands, 383–397.

Trewartha, G.T. 1968. *An Introduction to Climate,* McGraw-Hill Book Company, New York.

Tsuji, G.Y., G. Hoogenboom, and P.K. Thornton, Eds., 1998. *Understanding Options for Agricultural Production, Systems Approaches for Sustainable Agricultural Development,* vol. 7, Kluwer Academic Publishers, Dordrecht, The Netherlands.

Tsuji, G.Y., G. Uehara, and S.S. Bala, Eds., 1994. *DSSAT v3,* University of Hawaii, Honolulu, HI.

Turpin, J.E., M.E. Probert, I.C.R. Holford, and P.L. Poulton. 1992. Simulation of cereal-legume rotations using APSIM, *Aust. J. Agric. Res.,* 43:334–338.

UNEP. 1992. *World Atlas of Desertification,* Edward Arnold, United Nations Environmental Programme, Nairobi.

Ungar, E.D. 1990. *Management of Agropastoral Systems in a Semiarid Region, Simulation Monographs,* PUDOC, Wageningen, The Netherlands.

UNSO. 1997. *Aridity Zones and Dryland Populations, an Assessment of Population Levels in the World's Drylands,* Office to Combat Desertification and Drought (UNSO) United Nations Development Programme (UNDP), 23pp. New York.

van Keulen, H. 1975. *Simulation of Water Use and Herbage Growth in Arid Regions, Simulation Monographs,* PUDOC, Wageningen, The Netherlands.

van Keulen, H. 1982. Crop production under semi-arid conditions, as determined by nitrogen and moisture availability. In *Simulation of Plant Growth and Crop Production,* F.W.T. Penning de Vries and H.H. van Laar, Eds., PUDOC, Wageningen, The Netherlands, 234–251.

van Keulen, H., F.W.T. Penning de Vries, and E.M. Drees. 1982. A summary model for crop growth. In *Simulation of Plant Growth and Crop Production, Simulation Monographs,* F.W.T. Penning de Vries and H.H. van Laar, Eds., PUDOC, Wageningen, Netherlands, 87–98.

van Keulen, H. and W. Stol. 1991. Quantitative aspects of nitrogen nutrition in crops, *Fertilizer Res.,* 27:151–160.

van Keulen, H. and J. Wolf, Eds., 1986. *Modelling of Agricultural Production: Weather, Soils and Crops, Simulation Monographs,* PUDOC, Wageningen, The Netherlands.

Vossen, P. 1990. Modèles agrométéorologiques pour le suivi des cultures et la prévision de rendements des grandes régions des Communautés Européennes. In *Proc. of the Conf. on the Application of Remote Sensing to Agricultural Statistics,* Varese, Italy, October 10–11, 1989. Publication EUR 12581 EN of the Office for Official Publications of the European Union, Luxembourg, 75–84.

Wagner-Riddle, C., T.J. Gillespie, L. A. Hunt, and C.J. Swanton. 1997. Modeling a rye cover crop and subsequent soybean yield, *Agron. J.,* 89:208–218.

Whisler, F.D., B. Acock, D.N. Baker, R.E. Fye, H.F. Hodges, J.R. Lambert, H.E. Lemmon, J.M. McKinion, and V.R Reddy. 1986. Crop simulation models in agronomic systems, *Adv. Agron.,* 40:141–208.

Wilkerson, G.G., J.W. Jones, K.J. Boote, K.T. Ingram, and J.W. Mishoe, 1983. Modeling soybean growth for crop management, *Trans. ASAE,* 26:63–73.

Wolf, J. 1993. Effects of climate change on wheat production potential in the European Community, *Eur. J. Agron.,* 2:281–292.

Zadoks, J.C., F.H. Rijsdijk, and R. Rabbinge. 1984. Epipre, a systems approach to supervised control of pests and diseases in wheat in Netherlands. In *Pest and Pathogen Control: Strategy, Tactical and Policy Models, International Series on Applied Systems Analysis,* G.R. Conway, Ed., John Wiley and Sons, New York, 34–351.

CHAPTER 10

Applications of Models with Different Spatial Scales

James R. Kiniry, Jeffrey G. Arnold, and Yun Xie

CONTENTS

INTRODUCTION

Simulation models integrate results from field research, providing valuable means of technology transfer. User-oriented models help agricultural producers, crop consultants, and policy makers make intelligent decisions based on current scientific knowledge and readily available soils and weather data. Such models integrate information from a wide range of sources into easily applied decision aids. The objective of this chapter is to describe some models of different scales, in such a way as to help users decide which is most appropriate for their situation.

Simulation models can be grouped into three categories based on spatial scale. Single-plant models simulate processes such as production of various yield components, leaf development, and reproductive development. They can be used to evaluate traits for optimizing yield production at different latitudes, in different rainfall zones, and on different soils. They can evaluate planting densities and planting dates as part of risk assessment in different environments.

Canopy-level, single-field models share some common applications with single plant models, but tend to use more conservative and more general approaches to simulating plants. Leaf growth can be simulated as leaf area index (LAI) and yield can be simulated as harvest index (HI). Although they are often not able to describe the detailed differences among cultivars of a crop, such models can be readily applied to several crops by deriving realistic crop parameters. Within crop species differences may be confined to maturity types for such a model. Single field models can simulate the impact of management systems (crop rotations, tillage, irrigation, manure and fertilizer management, and drainage) on edge-of-field sediments and pollutant loadings.

Basin scale models simulate crop growth in a more aggregated fashion, allowing reasonable leaf area index development and reasonable biomass production in order to simulate yields of water, sediment nutrients, and pesticides from sub-basins. Basin scale models can be used to assess off-site impacts such as channel erosion, reservoir sedimentation, wetlands, riparian zones, water supply, water transfer, and stream and reservoir water quality. The scale is such that plant parameters describe generic processes of crop growth and development.

This chapter describes three models developed by USDA-ARS at Temple, TX. CERES-Maize (Crop Environment Resource Synthesis) (Jones and Kiniry, 1986) is a maize (*Zea mays* L.) simulation model for individual plants. ALMANAC (Agricultural Land Management Alternatives with Numerical Assessment Criteria) (Kiniry et al., 1992) is a field-scale model that simulates a wide range of plant species and simulates competition among species. The SWAT model (Soil and Water Assessment Tool) (Arnold et al., 1998) simulates watersheds and subwatersheds and can also simulate many plant species. SWAT is an integral part of the HUMUS (Hydrologic Unit Model for the U.S.) (Srinivasan et al., 1993) hydrologic project. HUMUS combines SWAT with a geographic information system and with regional databases to simulate surface and subsurface water quantity and quality on a basin scale.

Several features shared by these models contribute to their widespread application. First, they were all developed with a high degree of cooperation with users and, since the models were developed by the USDA-ARS, they are available at no cost. The models, documentation, code, and example data sets can be obtained by contacting the authors. This has encouraged widespread application of the models and has increased feedback from users. Often users help decide which processes need to be simulated and what output is needed. As a result of this close cooperation, these models are easy to access and apply. They have been validated for a wide range of sites within the U.S. and throughout the world. Feedback from such users has been an important component of model improvement.

Second, the models rely on readily available daily weather data and on the extensive USDA-NRCS soils data. Commonly reported values of daily maximum and minimum temperatures, rainfall, and solar radiation are needed. This enables users to apply the models throughout the world by using data from the nearest weather station. In cases where weather data or portions of weather data are not available, realistic values can be generated, usually within the models themselves.

Third, the models use a daily time step, enabling rapid execution of multiple year runs. The models do not have iterative processes such as curve fitting or solving differential equations which can slow down execution. Users can make runs with several years of weather in a few minutes, enabling them to efficiently simulate an extensive range of management, crop, and soil scenarios.

Finally, the models share common features in their simulation of plant growth. The models simulate LAI, light interception with Beer's law, and potential daily biomass increase with a species-specific value of radiation use efficiency (RUE). The daily increases in LAI and biomass are reduced when plant available water in the current rooting depth is insufficient to meet potential evapotranspiration. Plant development is temperature driven, with duration of growth stages dependent on degree days. Each plant species has a defined base temperature and optimum temperature. Parameters for describing plant processes are easy to derive for a plant species or cultivar and easy to transfer among models.

Crops such as maize and sorghum (*Sorghum bicolor* L. Moench) are grown in a wide range of soils and climatic conditions and can be vulnerable to late-spring freezes, drought, and high temperatures during grain growth.This sequence seemed more logical during the growing season. Producers make decisions on planting date, maturity type, planting rate, and fertilizer rates, attempting to maximize profit and minimize risks associated with unpredictable weather conditions. Crop models offer hope as tools to optimize such management practices. A robust crop model can provide a quantitative means to predict crop yields under different environmental and climatic conditions. Crop consultants, using accurate soil information and updated weather data, can provide producers with realistic predictions on the outcome of various management alternatives. Likewise, crop advisory information can be linked to soil type and measurements of soil layer depths in individual fields.

ALMANAC and CERES-Maize were developed to simulate critical growth processes. ALMANAC was developed to simulate the impacts of various field-level management on the soil and water environment, and on crop yields. The crop model in ALMANAC was designed to simulate a wide range of plant species efficiently. CERES-Maize was developed to simulate phenological processes and yield components of maize and to describe accurately how different hybrids produce grain in different environments. Adapted versions of CERES-Maize accurately simulated dryland and irrigated maize yields in Kenya at one to nine plants m^{-2} (Keating et al., 1988) and reasonably simulated maize yields with variable planting density, sowing dates, and nitrogen rates in Kenya (Wafula, 1995). CERES-Maize was used to simulate maize yields in Kansas with weed and insect stresses (Retta et al., 1991). The model "gave excellent predictions of yield trends" when used to simulate variability within a field in Iowa, proving to be "a viable and powerful tool in developing and evaluating management prescriptions across a field" (Paz et al., 1999). The model was tested in the semiarid tropics under conditions with measured yields of 1.7 to 8.3 Mg/ha^{-1} (Carberry et al., 1989). CERES failed to simulate differences among data sets for high yielding conditions in Argentina when yields ranging from 11.7 to 16.7 Mg/ha^{-1}, but was the mean simulated yield was only 5% greater than the mean observed yield (Otegui et al., 1996). An adaptation of CERES-Maize to simulate sorghum was tested in Australia using data with measured yields ranging from 1.6 to 6.3 Mg/ha^{-1} (Birch et al., 1990). ALMANAC and CERES-Maize accurately simulated mean crop yields in nine states with diverse soils and climate (Kiniry et al., 1997) and at sites within Texas (Kiniry and Bockholt, 1998). ALMANAC accurately simulated spring wheat (*Triticum aestivum* L.) yields with different densities of competing oats (*Avena sativa* L.), oilseed rape (*Brassica napus* L.), and vetch (*Vicia sativa* L.) in France (Debaeke et al., 1997).

To be effective as tools, crop models must be capable of simulating crop yields in average rainfall years and in unusual rainfall years such as with drought or excess moisture. When applied to maize at eleven sites and sorghum at eight sites in Texas for the dry conditions of 1998, ALMANAC realistically simulated grain yields (Yun et al., 2001). In this study, the model demonstrated ability to simulate site-to-site differences in grain yields under dry climate conditions, showing it can be valuable for risk assessment of grain production.

ALMANAC is also capable of simulating grasses, both in monoculture and with multiple species growing together. Kiniry et al. (1996) successfully simulated Alamo switchgrass (*Panicum virgatum* L.) at several sites in Texas. In addition, ALMANAC realistically simulated range yields for 20 range sites representing the extremes of productivity for Texas (Kiniry et al., 2001b).

Crop models capable of accurately simulating long-term mean crop yields for diverse environments and capable of simulating annual crop yields in extreme climatic conditions would be valuable for risk assessment and management evaluation. Such models can greatly increase confidence in crop modeling. Of the models evaluated in this study, ALMANAC and SWAT simulate many crops by using different parameters, while CERES-Maize simulates individual maize hybrids with descriptive parameters.

DESCRIPTION OF ALMANAC AND CERES-MAIZE

The ALMANAC and CERES-Maize models simulate processes of crop growth and soil water balance including light interception by leaves, dry matter production, and partitioning of biomass into grain. A major difference between these models is their approach to simulating grain yields. ALMANAC simulates a grain yield based on HI, which is grain yield as a fraction of total aboveground dry matter at maturity. CERES simulates the seed number per plant (based on plant growth) and average mass per seed (based on potential seed growth rate).

CERES-Maize simulates phenology based on leaf development up to silking and on ear development thereafter. Leaf area is simulated on an individual leaf basis. Plants begin with six leaf primordia at seedling emergence and initiate an additional leaf for each 20 degree days base 8°C up to the date of tassel initiation. Prior to tassel initiation, plants are in the basic vegetative phase which is degree day dependent, the sum of which varies among hybrids. Plants are then in a photoperiod sensitive phase, which can be as short as 4 days in short days and is extended when photoperiods exceed 12.5 h. Hybrids differ in the sensitivity to photoperiod, with greater sensitivity causing greater delays in tassel initiation in long photoperiods. At tassel initiation, final leaf number is determined. The number of leaf tips that emerge from the leaf whorl requires 38 degree days base 8°C, after the second leaf. The first leaf is assumed to be present at seedling emergence and the last leaf emerges 20 degree days later. Silking is assumed to occur when the final leaf fully emerges. The degree days from silking to maturity is input as a hybrid-specific parameter. The effective filling period of the grain is assumed to be completed when 95% of these degree days have accumulated.

ALMANAC includes a generic LAI function. The maximum LAI of a crop species at high planting density is a parameter. This potential LAI is reduced as a function of planting density. The development of LAI as a function of fraction of seasonal degree day sum follows an "s" curve, with two input parameters defining the curve. Daily increments of LAI growth can be reduced by water stress. At a defined fraction of the seasonal degree days, grain growth is assumed to begin. A species specific value for HI defines the fraction of final above-ground biomass that is in grain. This potential HI can be reduced if drought stress occurs near anthesis (from 45 to 60% of the season degree days).

Recent improvements in the models include light extinction coefficients (k) based on row spacing for ALMANAC and a new seed number algorithm in CERES.

For ALMANAC, the extinction coefficient equation is a linear function of row spacing for maize and sorghum (Flénet et al., 1996):

$$k = 0.685 - 0.209 \ \ ROWS \tag{10.1}$$

where ROWS is the row spacing for maize and sorghum and k is the extinction coefficient. This function is not included in CERES-Maize because it reduced yield simulation accuracy.

The number of seeds per plant (SEEDS) for CERES is now estimated by a linear function of GROWTH (g plant 1 d 1) from silking to the beginning of grain growth (Kiniry et al., 2001a):

$$SEEDS = 90 \ GROWTH, \tag{10.2}$$

where SEEDS is constrained to not exceed a genotype-specific potential number of seeds per plant (G2). Although Andrade et al. (2000) and Otegui and Andrade (2000) described nonlinear seed number equations due to increased barrenness at abnormally high planting densities, we chose to use Eq. (10.2), which is similar to the function of Keating et al. (1988).

Since publication of CERES-Maize in 1986, some other studies have provided basic information about maize growth relationships described in the model. Improvements in the model based on these studies were described previously (Kiniry et al., 1997). The first change is that RUE is now reduced as mean daily vapor pressure deficit (VPD) exceeds 1.0 kPa (Stockle and Kiniry, 1990). Maize RUE is 4.33 g MJ^{-1} of intercepted photosynthetically active radiation for mean daily VPD less than 1.0 kPa and is reduced by mean daily VPD > 1.0 as:

$$RUE = 5.05 - 0.72 \text{ VPD} \tag{10.3}$$

The second change is that only 0.26 g of grain is produced for each g of carbohydrate lost from the stem and leaves (Kiniry et al., 1992b). Respiration, efficiency of conversion of glucose into grain, and translocation costs presumably are responsible for this being less than 1.0.

Critical for yield simulation in water-limited conditions is the simulated water demand. The three models calculate effects of soil water on crop growth and yield with similar functions. Potential evaporation (Eo) is calculated first, and then potential soil water evaporation (ES) and potential plant water transpiration (EP) are derived from potential evaporation and LAI. Based on the soil water supply and crop water demand, the water stress factor is estimated to decrease daily crop growth and yield, although some water balance equations differ between the two models. Each model has options on which technique is used to estimate Es, but for this study, Eo was estimated by the Penman method (1948) in ALMANAC, and by the Priestley–Taylor method (1972) in CERES-Maize. In ALMANAC, ES, and EP were estimated by:

$$E_P = E_o(LAI/3) \qquad 0 \leq LAI \leq 3.0 \tag{10.4}$$

$$E_P = E_o \qquad LAI > 3.0 \tag{10.5}$$

E_S is either $E_o \exp(-0.1BIO)$ or $E_o - E_p$, whichever is smallest, where BIO is the sum of the aboveground biomass and crop residue (Mg ha^{-1}). In CERES-Maize

$$E_P = E_o(1 - \exp(-LAI)) \qquad 0 \leq LAI \leq 3.0 \tag{10.6}$$

$$E_P = E_o \qquad LAI > 3.0 \tag{10.7}$$

$$E_S = E_o(1 - 0.43LAI) \qquad 0 \leq LAI \leq 1.0 \tag{10.8}$$

$$E_S = E_o \exp(-0.4LAI)/1.1 \qquad LAI > 1.0 \tag{10.9}$$

If $E_o < E_P + E_S$, then $E_P = E_o - E_s$.

Demonstration of CERES-Maize

CERES-Maize can simulate how changes in plant parameters affect grain yields in different weather conditions and on different soils. By evaluating the impact of changes in a plant parameter for a given set of conditions, users can efficiently determine how changes in hybrid characteristics can influence grain yields. These indicate the response of yield to changes in various plant characteristics. For this demonstration, we used a site near Ames, IA, on a Nicollet loam and a site near Temple, TX on a Houston Black clay, as described in Kiniry et al. (1997). Researchers used the weather data from 1983 to 1992 just as in the previous study and evaluated how changes in three traits altered grain yield.

Table 10.1 CERES-Maize Mean Simulated Grain Yields (Mg ha⁻¹) near Ames, Iowa, for 10 Years

P1 values	180	200	220	240	
	(100)[a]	(111)	(122)	(133)	
Mean yields	6.51	6.58	6.67	6.72	
	(100)	(101)	(102)	(103)	
G3 values	6	7	8	9	
	(100)	(117)	(133)	(150)	
Mean yields	5.23	5.99	6.72	7.33	
	(100)	(115)	(129)	(140)	
P5 values	550	600	650	700	750
	(100)	(109)	(118)	(127)	(136)
Mean yields	5.00	5.63	6.21	6.76	7.29
	(100)	(113)	(124)	(135)	(146)

Note: Crop parameters changed included the duration of the vegetative phase (P1, in GDD_8), the rate of grain filling (G3, mg seed⁻¹ d⁻¹), and the duration of grain filling (in GDD_8).

[a] Values in parentheses are relative percentages.

At 5 plants m⁻², degree days base 8°C (GDD_8) from silking to maturity of 685 GDD_8, and a grain filling rate of 7.8 mg per seed per day, the impact of change in number of leaves was measured by changing the heat units from seedling emergence to end of the juvenile phase. Each 20 GDD_8 increase in this "P1" causes an additional leaf primordia to be initiated and delays tasseling by 39 GDD_8. Values tested were 180, 200, 220, and 240 GDD_8. These allowed 9, 10, 11, and 12 leaves to be initiated during this stage resulting in final leaf numbers of 17, 18, 19, and 20 leaves.

The impact of changes in grain filling rate on final yield was evaluated next; rates of 6, 7, 8, and 9 mg seed⁻¹ d⁻¹ were tested, assuming 5 plants m⁻², 685 GDD_8 from silking to maturity, and a grain filling rate of 7.8 as in the original study.

The final trait studied was the duration of grain filling, tried at values of 550, 600, 650, 700, and 750 GDD_8 from silking to maturity. All other parameters were held constant.

The relative sensitivity of these changes differed between the two sites (Tables 10.1 and 10.2). The more drought-prone site in Texas tended to show less yield increases than the site in Iowa, due to the dominant influence of drought stress in Texas.

At Ames, increases in number of leaves (greater P1) gradually increased mean simulated yields up to a maximum increase of 3%. At the more drought-prone Temple site, mean yields decreased for the largest two P1 values.

Increases in grain filling rate (G3) caused increases in mean yields at Ames of up to 40%. At Temple, these increases were almost as large, the maximum being 39%.

Finally, increases in duration of grain filling (P5) caused increases up to 46% in Iowa. Temple mean yields also increased, but only up to a maximum of 36%.

Demonstration of ALMANAC

Farmers face a number of management decisions when growing dryland maize. They try to optimize their management based on past experiences and expected weather. Two known variables on which they can base management decisions at planting time are the depth of their soil, and thus their potential plant available water at field capacity, and how much of their soil profile has been refilled since last year's growing season. Researchers examined the effect of plant spacing on yields on a deep (2.0 m) Houston black clay soil (fine, montmorillonitic, thermic Udic Palusterts) with 9 years of Temple, TX measured weather. This was repeated with a 1.5 m and a 1.0 m deep soil. Next they looked at planting density effects on a 5 d earlier and 10 d earlier maturity maize hybrids

Table 10.2   CERES-Maize Mean Simulated Grain
Yields (Mg ha^{-1}) near Temple, Texas,
for 10 Years

P1 values	180	200	220	240	
	(100)[a]	(111)	(122)	(133)	
Mean yields	5.54	5.52	5.35	5.26	
	(100)	(100)	(97)	(95)	
G3 values	6	7	8	9	
	(100)	(117)	(133)	(150)	
Mean yields	4.39	5.07	5.63	6.12	
	(100)	(115)	(128)	(139)	
P5 values	550	600	650	700	750
	(100)	(109)	(118)	(127)	(136)
Mean yields	4.35	4.82	5.23	5.60	5.89
	(100)	(111)	(120)	(129)	(136)

Note: Crop parameters changed included the duration of
the vegetative phase (P1, in GDD$_8$), the rate of
grain filling (G3, mg seed^{-1} d^{-1}), and the duration
of grain filling (in GDD$_8$).

[a] Values in parentheses are relative percentages.

and finally simulated yields of different maturity hybrids and a sorghum hybrid when soil moisture was not entirely replenished.

For the first set of analyses, a 2.0 m deep Houston black clay soil that could hold 0.25 m of plant available water at field capacity was simulated. The three maturity types evaluated were normal maturity for this region (1600 GDD$_8$ from planting to maturity), 5 d earlier maturing (1500), and 10 d earlier maturing (1400). This range was based on the range of maturities measured at Temple, TX, for some hybrids of diverse maturity (Kiniry and Knievel, 1995). For each maturity type, investigators simulated four, five, six, and seven plants m^{-2} plant densities for years 1991 to 2000 at Temple.

The three statistics of interest were the average for the three lowest yielding years (as an indication of yields in dry years), the yields for the three greatest yielding years (as an indication of yield potential), and the average yields over the 10 years.

Results with different densities of different maturity types on a 2.0 m soil (Table 10.3) showed useful information on maturity type differences and grain yields. Optimum densities for greatest average yields were five plants m^{-2} for the normal maturity, six plants m^{-2} for the 3 d earlier hybrid, and seven plants m^{-2} for the 10 d earlier hybrid. For the normal maturity hybrid, decreasing planting density decreased yield potential but increased yield in the 3 driest years. Using the CV as an estimate of yield variability, CV values increased as population density increased above five plants m^{-2} for the earliest maturity results and above four plants m^{-2} for the other two. For any given density, earlier maturity caused a decrease in yield potential and an increase in yield stability (the CV decreased). The greatest yields in the 3 driest years were for the four plants m^{-2} density for the normal maturity, for the five plants m^{-2} density for the 5 d earlier maturity hybrid, and six plants m^{-2} for the 10 d earlier maturity hybrid.

Decreasing soil depth to less than 1.5 m decreased overall average yield and yield in the highest 3 years (Table 10.4). The change in soil depth from 2.0 m to 1.5 m had little or no effect on maize yields. These soil depths correspond to plant available water at field capacity of 250 mm, 206 mm, and 147 mm. The optimum planting density based on average yield was five plants m^{-2} for all three soil depths. Greater densities, although they had increased potential yields, had reduced values for the low yielding years and reduced yield stability (as indicated by large CV values).

Analysis of 89 years of Temple, TX, weather indicated the average rainfall during the period from maize harvest until the next year's planting was 483 mm. Ranking the 89 years for amount rainfall during this period, the average rainfall for the lowest 20% of these years was 254 mm. Our

Table 10.3 ALMANAC's Mean Simulated Grain Yields of Three Different Maturity Maize Hybrids on a 2.0-m deep Houston Black Clay with Temple, Texas, Weather Data from 1991 to 2000, with Different Planting Densities

	4 Plants m^{-2} 16,200 Plants Acre^{-1}	5 Plants m^{-2} 20,200 Plants Acre^{-1}	6 Plants m^{-2} 24,300 Plants Acre^{-1}	7 Plants m^{-2} 28,300 Plants Acre^{-1}
	(Mg ha^{-1})			
	Normal Hybrid			
Low 3 avg.	3.5	2.7	2.4	2.5
High 3 avg.	5.0	7.4	8.4	9.2
Avg.	4.4 (83)	5.2 (97)	4.7 (89)	5.1 (96)
CV (%)	16	37	53	55
	Early Hybrid			
Low 3 avg.	2.6	3.1	2.6	2.6
High 3 avg.	3.6	6.2	7.5	8.1
Avg.	3.2 (61)	4.9 (92)	5.1 (97)	4.9 (93)
CV (%)	15	27	39	46
	Very Early Hybrid			
Low 3 avg.	1.7	3.1	3.3	2.9
High 3 avg.	2.2	4.2	5.8	6.8
Avg.	2.0 (38)	3.8 (72)	4.8 (90)	5.0 (95)
CV (%)	15	15	24	32

Note: The latter two maturity types reached maturity 5 d earlier and 10 d earlier than the common maturity type for the region. The value in parentheses is the yield in bushels per acre.

Table 10.4 ALMANAC's Mean Simulated Grain Yields for Three Soil Depths of a Houston Black Clay for a Common Maturity Maize with Temple, Texas, Weather Data from 1991 to 2000 with Different Planting Densities

	4 Plants m^{-2} 16,200 Plants Acre^{-1}	5 Plants m^{-2} 20,200 Plants Acre^{-1}	6 Plants m^{-2} 24,300 Plants Acre^{-1}	7 Plants m^{-2} 28,300 Plants Acre^{-1}
	(Mg ha^{-1})			
	2.0 m Soil Depth			
Low 3 avg.	3.5	2.7	2.4	2.5
High 3 avg.	5.0	7.4	8.4	9.2
Avg.	4.4 (83)	5.2 (97)	4.7 (89)	5.1 (96)
CV (%)	16	37	53	55
	1.5 m Soil Depth			
Low 3 avg.	3.4	2.6	2.4	2.5
High 3 avg.	5.0	7.5	8.4	9.2
Avg.	4.4 (83)	5.2 (98)	4.7 (89)	5.1 (96)
CV (%)	16	38	54	55
	1.0 m Soil Depth			
Low 3 avg.	2.8	2.3	2.3	2.4
High 3 avg.	5.0	6.7	7.6	7.7
Avg.	4.2 (80)	4.8 (91)	4.5 (84)	4.6 (86)
CV (%)	22	38	50	48

Note: The value in parentheses is the yield in bushels per acre.

Table 10.5   ALMANAC's Mean Simulated Grain Yields Following 254 mm of Rainfall during the Previous Fallow Period, for Three Different Maturity Maize Hybrids Simulated on a 2.0-m Deep Houston Black Clay with Temple, Texas, Weather Data from 1991 to 2000 with Different Planting Densities

	4 Plants m^{-2} 16,200 Plants Acre^{-1}	5 Plants m^{-2} 20,200 Plants Acre^{-1}	6 Plants m^{-2} 24,300 Plants Acre^{-1}	7 Plants m^{-2} 28,300 Plants Acre^{-1}
	(Mg ha^{-1})			
	Normal Hybrid			
Low 3 avg.	1.1	1.3	1.4	1.4
High 3 avg.	3.8	3.6	3.6	3.9
Avg.	2.4 (44)	2.4 (45)	2.5 (47)	2.7 (51)
CV (%)	49	44	40	40
	Early Hybrid			
Low 3 avg.	1.2	1.2	1.3	1.3
High 3 avg.	3.5	3.7	4.0	3.7
Avg.	2.4 (46)	2.4 (44)	2.5 (47)	2.5 (47)
CV (%)	40	49	51	42
	Very Early Hybrid			
Low 3 avg.	1.7	1.2	1.2	1.2
High 3 avg.	2.2	3.8	3.7	3.7
Avg.	2.0 (38)	2.5 (47)	2.3 (43)	2.3 (44)
CV (%)	15	44	50	50
	Grain Sorghum (25 plants)			
Low 3 avg.	1.6			
High 3 avg.	4.0			
Avg.	2.8			
CV (%)	38			

Note: The average rainfall for this period for 89 years was 483 mm. The latter two maturity types reached maturity 5 d earlier and 10 d earlier than the common maturity type for the region. The value in parentheses is the yield in bushels per acre.

Grain sorghum results for a common pllanting density are included for comparison.

two scenarios for looking at management of maize following low winter rainfall were 254 mm and an intermediate value of 381 mm during the period. Fallow season rainfall was adjusted accordingly, using the growing season weather for 1991 to 2000 at Temple, as described previously.

With the lowest winter soil recharge (254 mm), there did not appear to be a benefit of reducing planting density but sorghum showed promise as having superior yields to maize (Table 10.5). Yields of the normal maturity maize hybrid were low, averaging 2.4 to 2.7 Mg/ha^{-1}. Again looking at planting densities of four to seven plants m^{-2}, the highest average yields were at seven plants m^{-2} for the normal maturity maize hybrid, at six to seven for the early hybrid, and at five for the very early hybrid. The sorghum average yield exceeded all of the maize average yields. Sorghum yields were more stable than those of maize, as indicated by the smaller CV values of sorghum.

With an intermediate amount of winter soil recharge (381 mm), optimum density of maize was reduced and maize average yields were greater than sorghum yields (Table 10.6). The optimum planting rates to achieve maximum average yields were four plants m^{-2} for the normal maturity maize, and five plants m^{-2} for the early and very early hybrids. With such soil moisture recharge, there appeared to be sufficient soil moisture to take advantage of reduced planting density. Yields in the 3 years with wettest growing season conditions were greatest for these low densities. Sorghum was not as competitive as it was with the 254 mm winter rainfall.

Table 10.6   ALMANAC's Mean Simulated Grain Yields Following 381 mm of Rainfall during the Previous Fallow Period for Three Different Maturity Maize Hybrids Simulated on a 2.0 m Deep Houston Black Clay with Temple, Texas, Weather Data from 1991 to 2000 with Different Planting Densities

	4 Plants m^{-2} 16,200 Plants Acre^{-1}	5 Plants m^{-2} 20,200 Plants Acre^{-1}	6 Plants m^{-2} 24,300 Plants Acre^{-1}	7 Plants m^{-2} 28,300 Plants Acre^{-1}
	(Mg ha^{-1})			
	Normal Hybrid			
Low 3 avg.	2.3	1.8	1.9	2.1
High 3 avg.	4.9	6.0	5.1	5.6
Avg.	3.8 (71)	3.6 (68)	3.4 (63)	3.6 (69)
CV (%)	32	48	46	47
	Early Hybrid			
Low 3 avg.	2.7	2.2	1.9	2.0
High 3 avg.	3.6	5.6	5.3	5.0
Avg.	3.3 (61)	3.9 (73)	3.3 (62)	3.2 (61)
CV (%)	15	37	48	46
	Very Early Hybrid			
Low 3 avg.	2.8	2.0	1.8	2.0
High 3 avg.	3.8	5.7	4.5	4.9
Avg.	3.4 (64)	3.9 (74)	3.0 (56)	3.2 (61)
CV (%)	15	38	44	46
	Grain Sorghum (25 plants)			
Low 3 avg.	2.2			
High 3 avg.	4.3			
Avg.	3.3			
CV (%)	28			

Note: The average rainfall for this period for 89 years was 483 mm. The latter two maturity types reached maturity 5 d earlier and 10 d earlier than the common maturity type for the region. The value in parentheses is the yield in bushels per acre.

Grain sorghum results for a common pllanting density are included for comparison.

DESCRIPTION OF THE SWAT MODEL AND THE HUMUS PROJECT

The SWAT model simulates water quantity and quality in large, complex basins. SWAT predicts the impact of topography, soils, land use, management and weather on water, sediment, nutrient (nitrogen and phosphorus), and agricultural chemical yields for large watersheds with an insufficient number of gages. To meet the design criteria SWAT:

1. Does not require calibration (which is impossible on ungaged watersheds).
2. Uses inputs that are readily available for large areas.
3. Efficiently simulates hundreds of interacting sub-basins using a daily time step.
4. Simulates hundreds of years in a continuous time model to assess long-term impacts.

The command structure routes water, nutrients and chemicals through streams and reservoirs and inputs measured data for point sources of water and nutrients (Figure 10.1). Basins are subdivided into grid cells or subwatersheds to increase input and output detail.

Model sub-basin components consist of components of hydrology, weather, sedimentation, soil temperature, crop growth, nutrients, pesticides, and agricultural management. The model simulates hydrologic processes including surface runoff estimated from daily rainfall using the USDA-NRCS

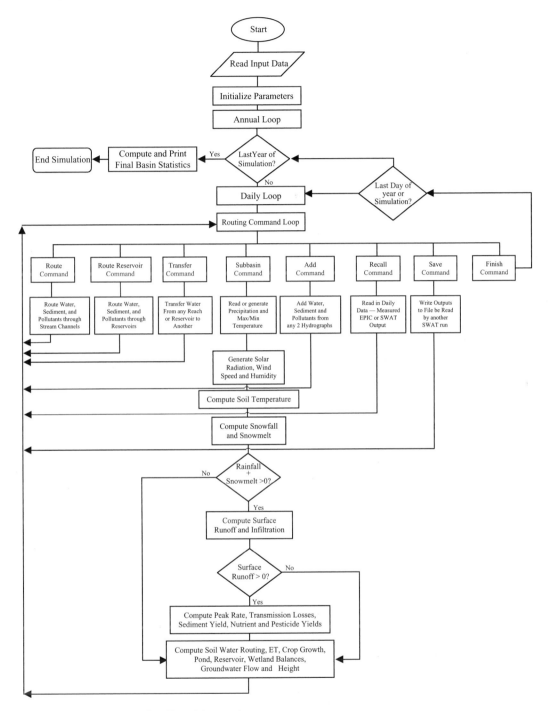

Figure 10.1 Flowchart of SWAT model operation.

curve number; percolation modeled with a layered storage routing technique combined with a crack flow model; lateral subsurface flow; groundwater flow to streams from shallow aquifers, potential evapotranspiration by the Hargreaves, Priestley–Taylor or Penman–Monteith methods; snowmelt; transmission losses from streams; and water storage and losses from ponds.

Daily precipitation, maximum and minimum air temperatures, solar radiation, wind speed, and relative humidity drive the hydrologic model. A weather generator simulates variables based on monthly climate statistics derived from long-term measured data. Weather data can differ among sub-basins.

SWAT computes sediment yield for each sub-basin with the modified universal soil loss equation. Soil temperature is updated daily for each soil layer as a function of air temperature; snow, plant and residue cover; damping depth; and mean annual temperature.

The model simulates crop growth with a daily time step using a simplification of the EPIC crop model which predicts phonological development based on daily accumulation of degree days, harvest index for partitioning grain yield, a radiation use efficiency approach for potential biomass, and adjustments for water and temperature stress. Both annual and perennial crops are simulated using crop-specific input parameters.

SWAT simulates nitrate losses in runoff, in percolation and in lateral subsurface flow. The model simulates organic nitrogen losses from soil erosion and an enrichment ratio. A nitrogen transformation model modified from EPIC includes residue mineralization, soil humus, mineralization, nitrification, denitrification, volatilization, fertilization and plant uptake. Phosphorus processes include residue and humus, mineralization, losses with runoff water and sediment, fertilization, fixation by soil particles and plant uptake. Pesticide transformations are simulated with a simplification of the GLEAMS model (Leonard et al., 1987) approach and include interception by the crop canopy; volatilization; degradation in soils and from foliage; and losses in runoff, percolation, and sediment.

The model simulates agricultural management practices such as tillage effects on soil and residue mixing, bulk density and residue decomposition. Irrigation may be scheduled by the user or applied automatically according to user-specified rules. Fertilization with nitrogen and phosphorus can also be scheduled by the user or applied automatically. Pesticide applications are scheduled by the user. Grazing is simulated as a daily harvest operation.

SWAT simulates stream processes including channel flood routing, channel sediment routing, nutrient and pesticide routing, and transformations modified from the QUAL2E model (Brown and Barnwell, 1987). Components include algae as chlorophyll-a, dissolved oxygen, organic oxygen demand, organic nitrogen, ammonium nitrogen, nitrite nitrogen, organic phosphorus, and soluble phosphorus. In-stream pesticide transformations include reactions, volatilization, settling, diffusion, resuspension, and burial.

The ponds and reservoirs component includes water balance, routing, sediment settling, and simplified nutrient and pesticide transformation routines. Water diversions into, out of, or within the basin can be simulated to represent irrigation and other withdrawals from the system.

HUMUS was designed to improve existing technologies for making national and river basin scale water resource assessments, considering both current and projected future climatic characteristics, water demands, point-sources of pollution, and land management affecting non-point pollution. The project was implemented as part of the U.S. Resources Conservation Act Assessment completed in 1997. The major cooperators in the HUMUS project were the U.S. Department of Agricultural Research Service and the Texas Agricultural Experiment Station, part of the Texas A&M University System.

The major components of the HUMUS system were:

1. The basin-scale SWAT to model surface and sub-surface water quantity and quality
2. A geographic information system to collect, manage, analyze, and display the spatial and temporal inputs and outputs of SWAT
3. Relational databases used to manage nonspatial climate, soil, crop and management data required as input to and generated as output from SWAT

A SWAT/GRASS input interface (Srinivasan and Arnold, 1994) was used in this project. The Geographic Resource Analysis Support System-Geographic Information System (GRASS) (U.S.

Army, 1987) is a GIS system developed by the U.S. Army Corps of Engineers. The interface project manager is used to extract, aggregate, view, and edit model inputs. This manager helps the user collect, prepare, edit and store basin and sub-basin information to be formatted into a SWAT input file. Most of the SWAT input data are derived from GRASS map layers. The data collected by the interface include basin attributes such as area of the basin, its geographic location, and soil attributes needed for SWAT. These are extracted from the STATSGO (USDA-SCS, 1992) database. Topographic attributes include accumulated drainage area, overland field slope, overland field length, channel dimensions, channel slope, and channel length. Land use attributes include crop name, planting and harvesting date based on heat unit scheduling, and weather station information for the weather generator.

Digital Elevation Model (DEM) Topographic Attributes

The overland slope and slope length were estimated for each polygon using the 3-arc second DEM data for each state. Measuring the slope using the neighborhood technique (Srinivasan and Engel, 1991) for each cell within a sub-basin, a weighted average based on area for the entire sub-basin was then calculated. The USLE slope length factor was computed using the standard table from the USDA Handbook 537 (Wischmeier and Smith, 1978) and the estimated overland slope.

Land Use Attributes

The USGS-LUDA data were used to develop crop inputs to SWAT. The land use with the greatest area was selected for each sub-basin and the crop parameter database characterized each crop (Williams et. al, 1990). The broad classification categories used in the LUDA were urban, agriculture/pasture, range, forest, wetland, and water. Planting date of a land use was calculated with a heat unit scheduling algorithm using latitude and longitude of the sub-basin, monthly mean temperatures of the sub-basin, and land use type. This automated approach also identifies other operations associated with a cropping system. For this study maize as used in the agricultural areas because it is the most prevalent crop in many parts of the U.S.

Soil Attributes

The STATSGO-soil association map was used to select soil attributes for each sub-basin. Each STATSGO polygon contains multiple soil series and the areal percentage of each. The soil series with the largest area was selected by the GIS interface. The interface then extracted the physical properties of the soil series for SWAT from a relational data structure and wrote them to SWAT input files. The runoff curve number (CN) was assigned to each sub-basin based on the type of land use and the hydrologic condition of the soil series using a standard CN table (USDA-SCS, 1972).

Irrigation Attributes

This study used the STATSGO database to identify locations using irrigation. In the STATSGO "yldunits" table, irrigated crop yield is reported. Hence, if a STATSGO polygon had irrigated crop yield for any crop in this table, and if the sub-basin's land use (from USGS-LUDA) was agriculture, then that sub-basin was simulated as irrigated agriculture. Using the irrigation map layer, the interface created input parameters for automated irrigation application for each sub-basin. The model automatically irrigated a sub-basin by replenishing soil moisture to field capacity when crop stress reached a user-defined level.

Weather Attributes

The SWAT model accessed data from 1130 weather stations in the U.S. The input interface assigned the closest weather station for each sub-basin. The interface also extracted and stored the monthly weather parameters in a model input file for each sub-basin.

Once the data were gathered for all the sub-basins for each state, the SWAT model was executed for a 20-year simulation run. Using the SWAT/GRASS output interface, average annual output were created as layers, which included rainfall, water yield, actual ET, potential ET, biomass, grain production, water surplus (rainfall minus actual ET), and irrigation applied.

Demonstration of SWAT

The U.S. Environmental Protection Agency reported nutrient enrichment as the major cause for impairment of lakes and other water bodies in the U.S. (USEPA, 1994). EPA's water quality inventory of 1996 indicated that forty percent of the surveyed rivers, lakes, and estuaries were polluted relative to their designated uses (USEPA, 1998). To restore the quality of these water bodies, the Total Maximum Daily Load (TMDL) process was established by Section 303(d) of the Clean Water Act. A TMDL quantifies pollutant sources, and maximum allowable loads of contributing point and nonpoint sources so that water quality standards are attained for uses such as for drinking water and aquatic life (USEPA, 1998). Once necessary pollutant reduction levels are identified through the establishment of TMDLs, control measures such as best management practices are implemented. The USEPA Office of Science and Technology has developed a framework for states to analyze impaired water bodies called BASINS (Better Assessment Science Integrating point and Non-point Sources). BASINS consists of five components:

1. National databases
2. Assessment tools
3. Utilities
4. Watershed models
5. Post-processing and output tools

SWAT and its associated GIS interface have been integrated into BASINS and is being used in several states for TMDL analysis.

The SWAT model was applied to the 4277 km^2 Bosque River watershed in central Texas. This river flows into Lake Waco, which is the source of drinking water for the city of Waco, TX. The watershed is mostly range and pasture in the upper portion while cropland is widespread in the lower portion. Manure from the 41,000 dairy cows in this watershed is applied on an area of 9450 ha. There is a strong positive correlation between elevated levels of phosphorus, the number of cows and the total acreage of manure application fields (McFarland and Hauck, 1999). Other sources of pollution include runoff from cropland and urban areas and effluent from wastewater treatment plants.

SWAT was calibrated and validated at two USGS gaging stations in this watershed, at Hico and Valley Mills (Santhi et al., 2001). After the model was validated, several management practices were simulated to see which practices would reduce phosphorus concentrations in the river below water quality standards.

The calibrated model was used to study the long-term effects of various BMPs related to dairy manure management and municipal wastewater treatment plant loads in this watershed. Among several scenarios studied, four scenarios are discussed in this paper. Detailed description of the BMPs can be found in Santhi et al. (2002). The existing condition scenario simulates the watershed with the present dairy herd size, the present waste application fields, the average manure application rate of 13 Mg ha^{-1}yr^{-1}, the present discharge volumes from waste water treatment plants (WWTPs)

Table 10.7   Comparison of SWAT Corn Yields vs. Ag Census and National Ag Statistics Corn Yields (Mg ha^{-1})

State	FIPS-id	AGCENSUS (1987)	NASS (20 yr avg)	SWAT Yield
Illinois	17	6.6	6.0	6.7
Indiana	18	6.5	6.2	6.2
Iowa	19	6.6	6.2	6.6
Kansas	20	5.3	5.9	5.5
Kentucky	21	4.5	4.5	4.9
Michigan	26	4.0	4.5	2.9
Minnesota	27	5.1	4.5	3.6
Missouri	29	4.8	4.5	5.0
Nebraska	31	6.2	6.6	3.7
North Carolina	37	3.1	4.0	3.0
Ohio	39	5.7	5.6	5.8
Pennsylvania	42	4.9	4.9	2.6
South Dakota	46	3.3	3.1	3.6
Wisconsin	55	5.3	5.2	3.6

with the current median concentrations for nutrients and present urban and cropland areas (Table 10.7). The future condition scenario reflects the projected conditions of the watershed in year 2020 with a projected dairy herd size of 67,000 cows, manure application in waste application fields at the crop N requirement rate of 46 Mg of N ha^{-1}yr^{-1}, waste application field area calculated at N rate requirement, maximum permitted discharge volumes from WWTPs using nutrient concentrations defined by current median values, urban area increased by 30% to reflect the projected population growth in 2020, and cropland area at current levels (due to no increase in cropland over last two decades) (Table 10.7). Three additional WWTPs with 1 mg/L concentration of total P were input into the model as point sources along the North Bosque River to account for possible industrial future growth outside existing communities.

Several management practices on dairy manure and WWTP effluents were simulated to study the impact in reducing the mineral P loadings. Imposed dairy management practices included hauling solid manure from the watershed, applying manure at crop P requirement rate (P rate) of 6.3 Mg ha^{-1}yr^{-1} (because the N rate allows more applied P than crops require), and reducing the dairy diet P to 0.4% (resulting in a 29% reducvtion in dairy manure P content as suggested by Keplinger, 1999). The concentrations of total P in WWTP effluents were reduced to 1 mg/L^{-1}. Scenario E was a modification of the existing condition scenario with additional conditions imposed on manure application rate (P rate), hauling off 38% of the manure, P diet reduction in animal feed, and 1 mg/L^{-1} limits of P in WWTPs (Table 10.8). Scenario F was a modification over the

Table 10.8   Assumptions of BMP Scenarios in the Bosque Watershed

	WWTP Flow Period	WWTP P Limit	Dairy Manure Application Rate	Reduced P in Diet	Haul-Off Manure
Existing scenario	1997–1998 (actual)	Median concentration	Btw N&P rate	No	No
Future scenario	2020 (permitted)	Median concentration	N rate	No	No
Scenario E	1997–1998	All WWTPs at Median concentration and Stephenville WWTP — 1mg/L	P rate	Yes	Yes
Scenario F	2020	All WWTPs with loads equal to Scenario E and Stephenville WWTP — with load equal to 1mg/L of future	P rate	Yes	Yes

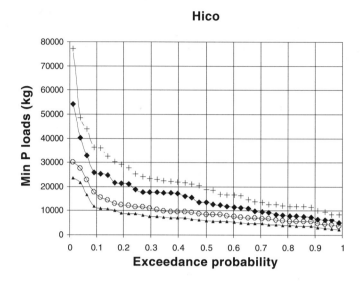

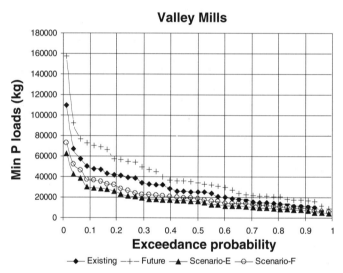

Figure 10.2   Exceedence probability of phosphorus loadings for various BMPs in the Bosque River.

future scenario with manure applied at P rate, hauling off 38% of the manure, P diet reduction, and 1 mg/L^{-1} P limits on all WWTPs.

Mineral P loadings are displayed as probability exceedance plots to analyze the effectiveness of BMPs. In these exceedance plots, annual mineral P loadings (y-axis) for the simulation period (1960 through 1998) were ordered and plotted with their associated exceedance probability values (x-axis) for Hico and Valley Mills (Figure 10.2). These plots provide information on the probability of achieving a particular load of mineral P through a BMP at a particular location. Mineral P loading curves for the scenarios varied from 10,000 kg to 40,000 kg at 10 probability at Hico whereas it varied from 20,000 kg to 80,000 kg at Valley Mills. These curves showed loadings within 10,000 kg at Hico at 90% probability and they showed loadings within 20,000 kg at Valley Mills for the same probability.

In general, the loading curves were wider at lower probabilities and become closer as they reach higher probabilities. The mineral P loadings were increased by about 27% at Hico and 29%

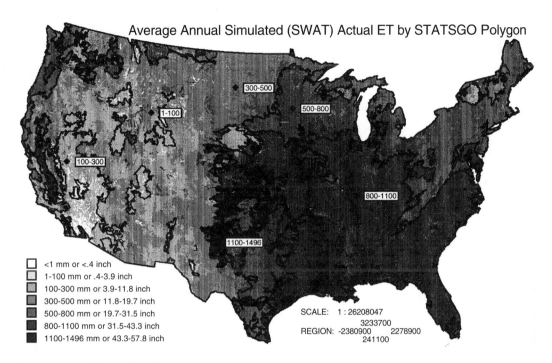

Figure 10.3 ET from HUMUS project.

at Valley Mills in the future condition scenario as compared to the existing condition scenario. These increases were predominantly caused by projected conditions for dairy and WWTPs in the future scenario (Table 10.7). Scenario E showed reduction in mineral P loadings of about 67% at Hico and 57% at Valley Mills from the future scenario. With scenario F, mineral P loadings were reduced 54% at Hico and 48% at Valley Mills from the future scenario. Scenario E indicated that with existing conditions, implementation of the BMPs would come closest to achieving the desired water quality goals; however, with year 2020 growth (future) conditions, more stringent controls will be required to meet the water quality goals.

Demonstration of HUMUS

Various hydrologic and crop growth outputs from the SWAT model simulation for the entire U.S. for the HUMUS project are given in Arnold et al. (1998). Penman-Monteith ET methodology was used in the simulation. Average annual ET generated from 20-year SWAT model simulations had highs and lows in parts of Kansas and Nebraska (Figure 10.3). These were due to the irrigation database used in this study. The high actual ET in most of Kansas was because the STATSGO database showed most of the state as irrigated land. With irrigation automatically triggered when plant available soil water was 50% of plant demand, irrigation of the agricultural cropland areas were greatest in parts of California, Kansas, and eastern New Mexico (Figure 10.4). The average annual biomass production (Figure 10.5) of irrigated cropland areas ranged from 25 to 32 Mg ha^{-1}. For non-irrigated cropland areas this ranged from 21 to 25 Mg ha^{-1}. For forest land areas values ranged from 16 to 21 Mg depending on their spatial and temporal distributions. Grains yields for irrigated land ranged from 9 to 11 Mg ha^{-1}, for non-irrigated land ranged from 6 to 9 Mg ha^{-1} in Midwest U.S. and 3 to 6 Mg in other grain production areas. These grain yields agreed reasonably well with state averages (Table 10.8).

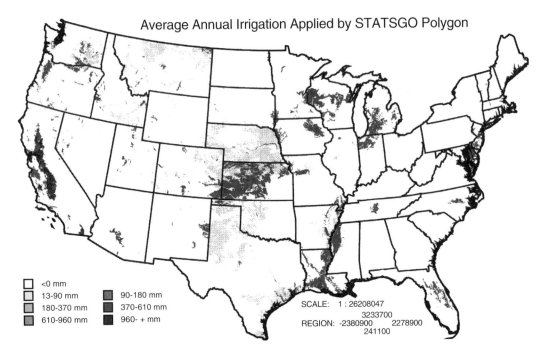

Figure 10.4 Irrigation applied from HUMUS.

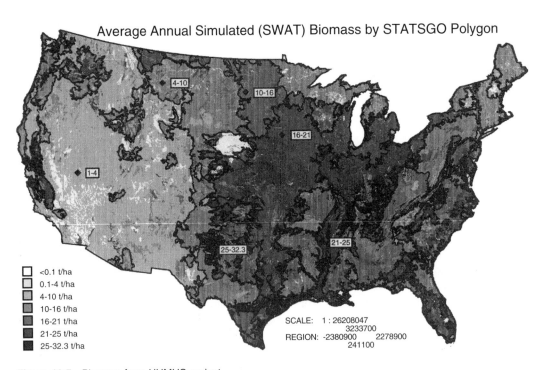

Figure 10.5 Biomass from HUMUS project.

CONCLUSIONS, RECOMMENDATIONS, AND AREAS FOR FUTURE STUDY

Model development and improvement are evolving processes, driven by users' needs while providing direction to basic research to fill knowledge gaps on key processes. Future work on model improvement is needed in several areas. We need to address limitations to model inputs, including availability of input data and problems with scale. Likewise, coding improvements can make models easier to understand and more modular. Interface tools for models and inputs of different scale are also needed. Especially challenging is the improvement in quantification of processes and process interactions in models. Finally, models need to be tested in environments distinct from the ones used for model development.

Limitations to inputs often become obvious when testing a model in a diverse group of sites. Precipitation data can be a problem because rain-gage density is insufficient to describe the spatial variability for accurate hydrologic simulation. Radar data (NEXRAD) can provide spatially detailed precipitation data for use in large scale models. Solar radiation data is often difficult to find, forcing model users to rely on weather stations several kilometers from field sites. Soil profile description can be derived from USDA-NRCS soil surveys, but actual description of layer depths within a field requires labor intensive soil sampling.

Making models modular allows portions to be easily transferred to new models. Once a model has been sufficiently validated and applied by many users, others may want to use only a portion, such as the plant growth. Thus, easily read code and favorable modularity become important. Often model developers, committed to working with users to apply models and develop reasonable inputs, may not have the resources and time to rewrite model code to make it modular. Such efforts may require outside funding and a special programmer to make the model code more object-oriented and user-friendly.

Interface tools are another area of promise for future work on modeling. GIS interfaces have been developed to automate spatial inputs and spatially display outputs of basin scale models. More research is needed to determine better basin discretization schemes (how to subdivide a basin, such as by sub-basins, on a grid scale, or by landuse overlays) and to assist users in developing management scenarios.

Functions within a model that quantify processes are usually the best available approximation at the time the model was developed and often can be improved by additional basic research. An example, for basin scale modeling, is the simulation of surface/groundwater interaction. Since groundwater can be a significant portion of stream flow at large scales, accurate simulation of groundwater flow and surface interaction (recharge) is essential. Likewise, functions to simulate bacteria fate and transport are needed for some basin scale models. Numerous TMDLs across the U.S. involve bacteria and basin scale bacteria processes which are not well understood or simulated. For single plant models there is a need for better description of many plant processes such as stress effects on plant phenology. For field scale and single plant models, there is a need for better description of root:shoot dynamics with and without drought or nutrient stress. Future research on the phenology and growth of perennial woody species in temperate and tropical environments will benefit application of these models to many areas.

The ultimate goal for process-based models is realistic simulation in a wide range of environments, not just those used for model development. Applying crop models developed in temperate conditions to new regions in the tropics can cause phenological simulations to fail. Maize leaf appearance rate as a function of degree days is much slower near the equator than in temperate zones. Careful analysis of simulations under high evaporative demand environments can identify weaknesses in soil water balance simulation and in plant responses to drought stress. Riparian zones and buffers are becoming important management tools with much need for accurate simulation. Realistic simulation of such zones is critical for many applications of large scale models.

ACKNOWLEDGMENTS

The SWAT model can be downloaded at http://www.brc.tamus.edu/swat/index.html. The other models are available directly from the authors.

The authors express their appreciation to the National Key Basic Research Special Foundation Project (G2000018605) in China for their support.

REFERENCES

Andrade, F.H., M.E. Otegui, and C. Vega. 2000. Intercepted radiation at flowering and kernel number in maize, *Agron. J.,* 92:92–97.

Arnold, J.G., Srinivasan, R., Muttiah, R.S., and Williams, J.R. 1998. Large area hydrologic modeling and assessment part I: model development, *J. American Water Resour. Assoc.,* 34(1):73–89.

Birch, C.J., P.S. Carberry, R.C. Muchow, R.L. McCown, and J.N.G. Hargreaves. 1990. Development and evaluation of a sorghum model based on CERES-Maize in a semi-arid tropical environment, *Field Crops Res.,* 24:87–104.

Brown, L.C. and T.O. Barnwell, Jr. 1987. The enhanced water quality models QUAL2E and QUAL2E-UNCAS documentation and user manual, EPA document # EPA/600/3-87/007, Cooperative Agreement #811883, Environmental Research Laboratory, U.S. Environmental Protection Agency, Athens, GA.

Carberry, P.S., R.C. Muchow, and R.L. McCown. 1989. Testing the CERES-Maize simulation model in a semi-arid tropical environment, *Field Crops Res.,* 20:297–315.

Debaeke, P., J.P. Caussanel, J.R. Kiniry, B. Kafiz, and G. Mondragon. 1997. Modeling crop:weed interactions in wheat with ALMANAC, *Weed Res.,* 37:325–341.

Flénet, F., J.R. Kiniry, J.E. Board, M.E. Westgate, and D.C. Reicosky. 1996. Row spacing effects on light extinction coefficients of corn, sorghum, soybean, and sunflower, *Agron. J.,* 88:185–190.

Jones, C.A. and Kiniry, J.R, Eds., 1986. *CERES-Maize: A Simulation Model of Maize Growth and Development,* Texas A&M Univ. Press, College Station, TX.

Keating, B.A., B.M. Wafula, and R.L. McCown. 1988. Simulation of plant density effects on maize yield as influenced by water and nitrogen limitations. In *Proc. Int. Congr. Plant Physiol.,* New Delhi, India. February 15–20, 1988, Soc. for Plant Physiol. and Biochem., New Delhi, India, 547–559.

Keplinger, K.O. 1999. *Cost Savings and Environmental Benefits of Dietary P Reductions for Dairy Cows in the Bosque River Watershed,* PR99-09, Texas Institute for Applied Environmental Research, Tarleton State University, Stephenville, TX.

Kiniry, J.R. and A.J. Bockholt. 1998. Maize and sorghum simulation in diverse Texas environments, *Agron. J.,* 90:682 687.

Kiniry, J.R. and D.P. Knievel. 1995. Response of maize seed number to solar radiation intercepted soon after anthesis. *Agron. J.* 87:228–234.

Kiniry, J.R., J.R. Williams, P.W. Gassman, and P. Debaeke. 1992a. A general, process-oriented model for two competing plant species, *Trans. ASAE,* 35:801–810.

Kiniry, J.R., C.R. Tischler, W.D. Rosenthal and T.J. Gerik. 1992b. Nonstructural carbohydrate utilization by sorghum and maize shaded during grain growth. *Crop Sci.* 32:131–137.

Kiniry, J.R., M.A. Sanderson, J.R. Williams, C.R. Tischler, M.A. Hussey, W. R. Ocumpaugh, J.C. Read, G. VanEsbroeck, and R.L. Reed. 1996. Simulating Alamo switchgrass with the ALMANAC model, *Agron. J.,* 88:602–606.

Kiniry, J.R., J.R. Williams, R.L. Vanderlip, J.D. Atwood, D.C. Reicosky, J. Mulliken, W.J. Cox, H.J. Mascagni,Jr., S.E. Hollinger, and W.J. Wiebold. 1997. Evaluation of two maize models for nine U.S. locations, *Agron. J.,* 89:421–426.

Kiniry, J.R., Y. Xie, and T.J. Gerik. 2002a. Similarity in maize seed number for a diverse set of sites. Agronimie: *Agric. and Envir.* (in press).

Kiniry, J.R., H. Sanchez, J. Greenwade, E. Seidensticker, J.R. Bell, F. Pringle, G. Peacock Jr., and J. Rives. 2001. Simulating range productivity on diverse sites in Texas, *J. Soil and Water Conserv.,* (in press).

Leonard, R.A., W.G. Knisel, and D.A. Still. 1987. GLEAMS: groundwater loading effects of agricultural management systems, *Trans. ASAE,* 30(5):1403–1418.

McFarland, A.M.S. and L.M. Hauck. 1999. Relating agricultural landuses to in-stream stormwater quality, *J. of Environmental Qual.,* 28:836–844.

Muttiah, R.S., Allen, P.M., Arnold, J.G. and Srinivasan, R. 1994. Baseflow day characteristics of Texas. In *Toxic Substances and the Hydrologic Sciences,* American Institute of Hydrology, Austin, TX, 212–219.

Otegui, M.E., R.A. Ruiz, and D. Petruzzi. 1996. Modeling hybrid and sowing date effects on potential grain yield of maize in a humid temperate region, *Field Crops Res.,* 47:167–174.

Otegui, M.E. and F.H. Andrade. 2000. New relationships between light enterception, ear growth, and kernel set in maize. In *Physiology and Modeling Kernel Set in Maize.* M.E. Westgate and K. Boote, Eds. Crop Science Society of American special Publivcation No. 29, CSSA/ASA, Madison, WI. 89–102.

Paz, J.O., W.D. Batchelor, B.A. Babcock, T.S. Colvin, S.D. Logsdon, T.C. Kaspar, and D.L. Karlen.1999. Model-based technique to determine variable rate nitrogen for corn, *Agric. Syst.,* 61:69–75.

Penman, H.L. 1948. Natural evaportranspiration from open water, bare soil and grass, *Proc. R. Soc. London Ser. A,* 193:120–145.

Priestley, C.H.B. and R.J. Taylor. 1972. On the assessment of surface heat flux and evaporation using large-scale parameters, *Mon. Weather Rev.,* 100:81–92.

Retta, A., R.L. Vanderlip, R.A. Higgins, L.J. Moshier, and A.M. Feyerherm. 1991. Suitability of corn growth models for incorporation of weed and insect stresses, *Agron. J.,* 83:757–765.

Santhi, C., J.G. Arnold, J.R. Williams, W.A. Dugas, and L. Hauck. 2001a. Validation of the SWAT model on a large river basin with point and nonpoint sources, *J. Amer. Water Resour. Assoc.* 37(5):1169–1188.

Santhi, C., J.G. Arnold, J.R. Williams, W.A. Dugas, and L. Hauck. 2002. Application of a watershed model in TMDL analysis, *Trans. ASAE,* (accepted).

Srinivasan, R., Arnold, J.G., Muttiah, R. S., Walker, C., and Dyke, P.T. 1993. Hydrologic Unit Model for United States (HUMUS). In *Proc. Advances in Hydro-Science and Engineering,* University of Mississippi, Mississippi State, MS, 451–456.

Srinivasan, R. and Arnold, J.G. 1994. Integration of a basin-scale water quality model with GIS, *Water Resour. Bull.,* (30)3:453–462.

Srinivasan, R. and Engel, B.A. 1991. A knowledge-based approach to extract input data from GIS, *ASAE Paper No. 91-7045,* ASAE Summer Meeting, Albuquerque, NM. July 1991.

Stockle, C.O. and J.R. Kiniry. 1990. Variability in crop radiation use efficiency associated with vapor pressure deficit, *Field Crops Res.,* 21:171–181.

U.S. Army. 1987. *GRASS Reference Manual,* U.S. Army Construction Engineering Research Laboratory, Champaign, IL.

USDA-SCS. 1972. *National Engineering Handbook,* Hydrology Section 4, U.S. Government Printing Office, Washington, D.C., chapters 7–10.

USDA-SCS. 1992. States Soil Geographic Database (STATSCO) Data User's Guide, Publ. No. 1492, U.S. Government Printing Office. Washington D.C.

USEPA. 1994. The quality of our nation's water: executive summary of the national water quality inventory, report to United States Congress, EPA841-R-97-008, U.S. Environmental Protection Agency, Washington D.C.

USEPA. 1998. National water quality inventory: 1996 report to United States Congress, EPA841-R-97-008. U.S. Environmental Protection Agency, Washington D.C.

Wafula, B.M. 1995. Application of crop simulation in agricultural extension and research in Kenya, *Agric. Syst.,* 49:399–412.

Williams, J.R, J.R., P.T. Dyke, W.W. Fuchs, V.W. Benson, O.W. Rice, and E.D. Taylor. 1990. *EPIC — Erosion/Productivity Impact Calculator: 2. User Manual,* A.N. Sharpley and J.R. Williams, Eds., USDA Tech. Bull. No. 1768.

Wischmeier, W.H. and Smith, D.D. 1978. Predicting rainfall erosion losses: a guide to conservation planning, *Agric. Handbook No. 537,* U.S. Dept. Agric.

Yun, X., J.R. Kiniry, V. Nedbalek, and W.D. Rosenthal. 2001. Maize and sorghum simulations with CERES-Maize, SORKAM, and ALMANAC under water-limiting conditions for Texas in 1998, *Agron. J.,* 93:1148–1155.

CHAPTER 11

Modeling Crop Growth and Nitrogen Dynamics for Advisory Purposes Regarding Spatial Variability

Kurt Christian Kersebaum, Karsten Lorenz, Hannes I. Reuter, and Ole Wendroth

CONTENTS

INTRODUCTION

Technical development of agricultural machinery, combined with a global positioning system (GPS) and yield monitoring, allows farmers to consider the heterogeneity within fields for their crop management. Site-specific crop management provides a better efficiency of applied nutrients combined with lower emissions of agrochemicals. Especially for nitrogen, it is important that the nutrient supply closely matches the demand of the crops because a surplus can be easily leached into the groundwater. Until now, fertilizer recommendations for nitrogen for entire fields have usually been based on measurements of soil mineral nitrogen content in early spring (Wehrmann and Scharpf, 1986) supported sometimes by measurements of the crop nitrogen status by optic sensors (Leithold, 2000). Nevertheless, all measurements are just snapshots of a present situation which enlighten neither the reason for an observed phenomenon nor the probable future development. Regarding the spatial and temporal variability of soil mineral nitrogen within fields, it is doubtful that frequent and dense soil sampling might be a realistic approach under practical conditions for a site-specific fertilizer application. Therefore, methods are required to estimate easily the local nitrogen demand considering the spatial variability of soil nitrogen supply and crop yield potentials. Agricultural system models provide a tool to transfer the spatial heterogeneity of time-stable soil and terrain attributes, which have to be estimated once for a field, into a temporal dynamic of the relevant state variables of the soil-crop nitrogen dynamics (Figure 11.1).

To evaluate the capability of such a model-based approach, the authors investigated the spatial structure of some relevant soil characteristics, crop yield, and mineral nitrogen content on two fields in Germany. Spatial estimation techniques were tested for their capability to improve or to facilitate the determination of spatial patterns. The relatively simple agricultural system model HERMES was applied to simulate crop growth and nitrogen dynamics using different levels of spatial information. The model has been applied successfully to calculate fertilizer recommendations for cereals for entire fields or to estimate nitrogen leaching on a regional scale (Kersebaum and Beblik, 2001). The aim was to determine if the model approach was sensitive enough to reflect the temporal and spatial variability of crop growth and soil mineral nitrogen within fields and, therefore, be feasible for site-specific fertilizer recommendations. Furthermore, the model will be applied to compare results of different fertilizer strategies between field plots.

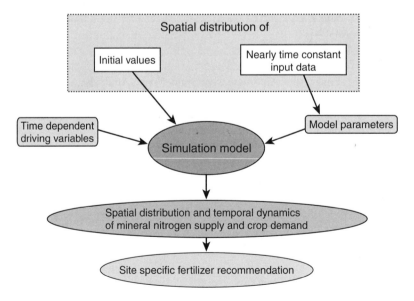

Figure 11.1 Concept to estimate site-specific dynamics from stable spatial structures.

METHODS

Simulation Model

The model HERMES (Kersebaum and Richter, 1991; Kersebaum, 1995) was developed for practical purposes, which implies that relative simple model approaches were chosen to operate under restricted data availability. The model and the concept for model-based fertilizer recommendations has been described in detail by Kersebaum and Beblik (2001); therefore, only a brief characterization of the fundamentals will be given here.

The vertical one-dimensional model operates on a daily time step using daily precipitation, temperature, vapor pressure deficit, and global radiation. The main processes considered are nitrogen mineralization, denitrification, transport of water and nitrogen, crop growth, and nitrogen uptake. A capacity approach is used to describe soil water dynamics. The capacity parameters required by the model are attached to the model by external data files that are consistent with the German soil texture classification and their capacity parameters. The basic values are modified by organic matter content, bulk density, and hydromorphic indices. The latter are derived for the investigated fields from the digital elevation model (DEM) as a topographic scaled wetness index.

Modeling of nitrogen net mineralization simulates the release of mineral N from two pools of potentially decomposable nitrogen that are derived from soil organic matter and amounts of crop residues related to the yield of the previous crop. Therefore, the spatial distribution of soil organic carbon and the yield map of the previous crop have been used as input variables. Daily mineralization is simulated depending on temperature and soil moisture. Denitrification is simulated for the top soil using a Michaelis–Menten kinetic modified by reduction functions dependent on nitrate content, water filled pore space, and temperature.

The submodel for crop growth was developed on the basis of the SUCROS model (van Keulen et al., 1982). Driven by the global radiation and temperature, the daily net dry matter production by photosynthesis and respiration is simulated. Dry matter production is partitioned according to crop development stages calculated from a thermal sum (°C days) modified by day length and vernalisation. The yield was estimated at harvest from the weight of the ears and nitrogen recycling with crop residues is calculated automatically from the simulated crop uptake. Crop growth is limited by water and nitrogen stress. Temporary limitation of soil air by water logging is considered through reducing transpiration and photosynthesis according to Supit et al. (1994). Water and nitrogen uptake is calculated from potential evaporation and crop nitrogen status, depending on the simulated root distribution and water and nitrogen availability in different soil layers. The different crops within a rotation are considered using crop specific parameter sets from an external file.

Data required by the model can be separated into three parts: weather data, soil information, and management data. The model operates on a daily weather data basis. Input data required are precipitation, average air temperature, global radiation or alternatively sunshine duration, and the vapor pressure deficit at 2 a.m.

Soil information is required at a resolution of 10 cm for the profile. For the plowing layer, the organic matter content and its C/N ratio should be given. The soil texture classes of A.G. Boden (1994) are the most important soil information required. Information of groundwater depth and soil texture are used to calculate capillary rise if applicable. Bulk density and percentage of stones are used to modify capacity parameters.

Mandatory data for management are crop species, dates of sowing, harvest and soil tillage, nitrogen fertilizer, and water application (kind of fertilizer, quantity and date of application/incorporation). Measured vertical distribution of mineral nitrogen content and soil moisture are used as initial values for the simulation.

Investigation Sites

The investigations for spatial variability and its effect on crop yield and nitrogen dynamics were carried out at two locations in Germany. For both locations, the relevant meteorological data were recorded by automatic weather stations.

Site 1

The 20-ha field "AUTOBAHN" is located in the northwestern part of Germany in Beckum/North Rhine-Westphalia. The texture varies from slightly loamy sand in the north to silty and clay loam with underlying calcareous marl coming up into the root zone at the southern part of the field. The elevation ranges from 96 to 102 m and a deciduous forest is located in the center of the field. During the investigation period of 1999/2000, winter wheat was grown following the previous crop of oilseed rape.

Site 2

The second investigation site is located in the southeastern hummocky loess region of Luette-witz/Saxony on a farm of the Suedzucker Company. The silty aeolian sediment of the investigated 30-ha field "SPORTKOMPLEX" varies only slightly in texture. The elevation ranges from 246 to 276 m above sea level with maximum slopes of 13°. The crop rotation under investigation was spring barley (1997/1998), winter rye (1998/1999), and winter wheat (1999/2000).

Spatial Data Acquisition

Measurements

Site 1

Auger sampling from 0 to 90 cm was performed in September 1999 at 60 locations in a 50×50-m grid mixing five auger samples at each location. Nested sampling was done at two locations with 16 points in a 5×5-m grid to investigate the small-scale variability. Samples were analyzed for three 30-cm depth increments for soil texture, soil moisture, mineral nitrogen, and organic carbon/nitrogen (only 0 to 30 cm). Sampling was repeated at the same locations in August 2000 after harvest of winter wheat and in early Spring 2000 on 49 selected points representing the soil map units of the field as a base for fertilizer recommendations. To evaluate the capability of remote sensing techniques to improve soil map accuracy and to reduce laborious grid mapping, electro-magnetic induction measurements were performed by Durlesser and Sperl (Soilinvest Company) with an EM-38 equipment (Durlesser and Stanjek, 1997). Geo-referenced grain yields were obtained from a combine harvester and aggregated for model validation using a 10-m radius around the sampling locations eliminating values beyond three standard deviations (Jürschik et al., 1999).

Site 2

After harvest of triticale in August 1997, auger samples were taken for 0 to 90 cm depth on 225 locations of a 28×28-m grid with five samples mixed at each location. At each grid point, the initial soil moisture and mineral nitrogen content in layers of 30-cm thickness was analyzed. Small-scale variability of mineral nitrogen was measured by nested sampling at five locations with 25 points in a 6×6-m grid. Basic soil information for soil texture and organic carbon content were obtained for every second sample of the standard grid. Soil mineral nitrogen was measured at selected standard grid cells at different time intervals during the growing period 1998 and 1999 and after harvest of spring barley in 1998 and winter rye in 1999. Results had to be transferred in

1999 to a new shifted 54 × 54-m grid that was the basic cell size to perform an experiment with different fertilizer applications in the growing season 2000. For these grid cells, mineral nitrogen content in the root zone (0 to 90 cm) was measured after harvest of winter rye as initial values for nitrogen simulation, for selected cells in early spring as a base for the conventional fertilizer recommendation and at all grid cells after harvest in August 2000. Grain yields were mapped every year by a combine harvester (Claas Agromap). The raw yield data were processed using the GIS ArcInfo according to a method described by Jürschik et al. (1999).

Terrain Analysis

Spatial variability of crop growth or soil properties often has a topographic background. Therefore, researchers obtained topographical information for the study sites to derive spatial patterns of model input variables such as solar irradiation or soil characteristics. For field AUTO-BAHN (site 1) a digital elevation model was digitized from a topographical map (Scale 1:10,000). Although the field shows no steeper slopes, the forest in the center of the field was included in the elevation model to estimate the shading effect on the arable crops. Therefore, we include the forest patch using an estimated height of the trees and a time-variable light permeability to consider the seasonal fluctuations of the canopy (Erhardt et al., 1984).

For SPORTKOMPLEX, a digital elevation model was obtained by laser altimetry with a horizontal resolution of 1 × 1 m and a vertical resolution of 10 to 15 cm. In this elevation model, the shape of trees and buildings are included. A local filter was applied to remove small-scale noise and resampling to a cell size of 5 × 5 m was performed.

The spatial variation of the incoming long- and short-wave radiation was estimated using the SRAD modul from the TapesG-package (Gallant and Wilson, 1996). Irradiation for each month was calculated for 3 days, 10 days apart in time steps of 12 minutes due to computational limitations. Measurements of the weather station beside the field were used for parametrization. The results were aggregated by averaging to a cell size of the 28 × 28-m grid and normalized to the computed radiation of the weather station. For simulation, we used an average monthly correction for each grid cell. A more detailed description of the procedure is given by Reuter et al. (2001).

Additionally, the elevation model of SPORTKOMPLEX was used to derive the topographic wetness index according to Moore et al. (1993). The specific catchment area (A_s) and the slope (D8-method) were derived using the GRID modul of ArcInfo. The topographic wetness index was scaled to be used as a hydromorphic modification factor for the field capacity parameter in the HERMES model.

Spatial Estimation Techniques

Cokriging

Soil textural information is available or can be determined experimentally at spatial resolutions that are relatively coarse, only. In order to obtain a picture about soil textural distribution at higher resolutions, spatial estimation techniques can be applied. Values for the variable of interest can be estimated at unsampled points. The estimation uncertainty is reduced, the more closely the "expensive" variable is spatially correlated to a variable that is easier to sample, and therefore available at a high resolution. For the AUTOBAHN site, clay content was determined at sampling locations shown in Figure 11.4, with point separation distances of 5 m, 50 m, and 100 m. Electrical conductivity values were determined "on the go" at irregular separation distances of 1 m or less within the management direction and approximately 20 m perpendicular to this direction.

The objective of the spatial estimation was to compare interpolation using kriging based on clay content measurements only with a coregionalization technique (cokriging), where spatial clay estimation was supported by electrical conductivity values, and their spatial covariance (see, e.g., Alemi et al., 1988; Deutsch and Journel, 1992; Zhang et al., 1999).

First, the spatial covariances of clay content in the upper 30-cm soil layer and of electrical conductivity were determined with univariate semivariance by:

$$\gamma_z(h) = \frac{1}{2N(h)} \sum_{i=1}^{N(h)} [Z(x) - Z(x+h)]^2 \tag{11.1}$$

where semivariance $\gamma_Z(h)$ is calculated for a number of $N(h)$ observations $Z(x)$ that are separated by a lag distance h. In order to quantify the bivariate spatial covariance behavior, the cross-semivariance of two variables Y and Z was calculated according to:

$$\Gamma_{YZ}(h) = \frac{1}{2N(h)} \sum_{i=1}^{N(h)} [Y(x) - Y(x+h)][Z(x) - Z(x+h)] \tag{11.2}$$

As can be seen from Figure 11.4, clay contents were determined at 79 locations, whereas electrical conductivity was determined for approximately 6400 locations. Sampling locations for both variables were not identical. For covariogram calculation, for each of the 79 sampling locations the electrical conductivity values obtained for the surrounding 16 m were averaged.

Variogram models were fitted to the calculated $\gamma_Z(h)$ and $\Gamma_{YZ}(h)$ relations, respectively. Both variogram models and the cross-variogram model are comprised of different so-called nested structural components. These components need to exhibit the same spatial scales of correlation structure, i.e., the range parameters of different structures have to be identical for the different variograms and the covariogram, and each of the sill parameters represents the specific covariance structure. In this study, variogram models were composed from an exponential part with (Deutsch and Journel, 1992):

$$\gamma(h) = c \left[1 - \exp\left(-\frac{3h}{a} \right) \right] \tag{11.3}$$

and a spherical model with:

$$\gamma(h) = c \left[1.5 \frac{h}{a} - 0.5 \left(\frac{h}{a} \right)^3 \right], \quad \text{if } h \leq a$$

$$\gamma(h) = c, \qquad\qquad\qquad \text{if } h > a \tag{11.4}$$

In Eqs. 11.3 and 11.4, c denotes the sill, and a is the range parameter. The kriged value for an unsampled location was estimated from observations in the spatial neighborhood, by (Yates and Warrick, 2002):

$$\hat{Z}(x) = \sum_{i=1}^{N} \lambda_i Z(x_i) \tag{11.5}$$

where each observation contributes with a weight λ_i, which depends on the number of neighborhood observations considered and their respective distance to the unsampled location, as well as the semivariance associated with the respective distance. For a particular kriging estimation, the sum of respective weights equals 1. The minimization of the variance of errors is obtained from the Lagrange multiplier α. The kriging estimation variance is calculated via:

$$\sigma_k^2 = \alpha + \sum_{i=1}^{N} \lambda_i \gamma (x_i - x) \tag{11.6}$$

An unsampled value is estimated by cokriging via:

$$\hat{Z}(x) = \sum_{i=1}^{N} \lambda_i Z(x_i) + \sum_{k=1}^{M} \omega_k Y(x_k) \tag{11.7}$$

where λ_i and ω_i are respective weights for Z(x) and Y(x). Accordingly, the kriging estimation variance is obtained from:

$$\sigma_k^2 = \alpha + \sum_{i=1}^{N} \lambda_i \gamma_{11}(x_i - x) + \sum_{k=1}^{M} \omega_k \gamma_{12}(x_k - x) \tag{11.8}$$

The effect of using an auxiliary variable that could be sampled with higher density, while being spatially correlated to clay content, should be evaluated by comparing the maps of kriging and cokriging standard deviations.

First-Order Autoregressive State–Space Analysis

The spatial association between two variables j and k is quantified in the cross-covariance function $C_{jk}(h)$ with (Shumway, 1988)

$$C_{jk}(h) = \frac{1}{N} \sum_{i=1}^{N-h} [Z_j(x+h) - \mu_j][Z_k(x) - \mu_k] \tag{11.9}$$

with means μ_j and μ_k, respectively. The cross-correlation function as the normalized cross-covariance function is

$$r_{jk}(h) = \frac{C_{jk}(h)}{\sqrt{s_j^2 s_k^2}} \tag{11.10}$$

with variances s_j^2 and s_k^2, respectively.

The state–space model consists of two equations. The first is the model equation, in this case a first-order autoregressive equation relating the state vector X_s to the state vector at location X_{s-1} via a matrix G of transition or autoregression coefficients.

$$X_s = G(X)_{s-1} + \omega_s \tag{11.11}$$

The term ω_s denotes the model uncertainty.

The state vector is embedded in an observation equation, because the observed vector Y_s is affected by observation and calibration uncertainty v_s.

$$Y_s = M_s X_s + v_s \tag{11.12}$$

For further details, the reader is referred to Shumway (1988), and Wendroth et al. (1992). For numerical stability in the state–space estimation, observed series $Z(x)$ are normalized $\Theta(x)$ with respect to their mean and standard deviation by (Wendroth et al., 2001):

$$\Theta(x) = \frac{Z(x) - (\mu_Z - 2\sigma_Z)}{4\sigma_Z} \qquad (11.13)$$

In order to evaluate whether an additional variable improves the model result, the corrected form of the Akaike Information Criterion AICc is employed (Shumway and Stoffer, 2000):

$$\text{AICc} = \ln \hat{\sigma}_k^2 + \frac{n+k}{n-k-2} \qquad (11.14)$$

with $\hat{\sigma}_k^2$ denoting the average of the sum of squared differences between observation and prediction by a model with k regression coefficients and n as the number of observations.

Fertilization Strategies and Scenarios

On the two sites different fertilization strategies were applied. On the field AUTOBAHN, a fertilizer recommendation scheme was derived by an algorithm of Wenkel et al. (2001), which considers the mineral nitrogen content measured in early spring at different soil map units of the field, soil type, and related yield potential estimations. This site-specific recommendation was applied on the east side of the field AUTOBAHN, whereas on the other side of the field an average of the recommended doses was applied uniformly over the field.

On the field SPORTKOMPLEX, four different fertilizer strategies and a zero fertilization were applied for the winter wheat in 2000. The conventional strategy uses the average of the measured soil mineral nitrogen content in early spring on the field to calculate the first nitrogen application according to the N_{min}-method (Wehrmann and Scharpf, 1986). The next two applications were estimated site-specifically by the online optic Hydro-N-sensor (Leithold, 2000). The HERMES uniform"strategy used the average of the model-based fertilizer recommendation for all grid cells of the field (simulations used only the N_{min} observations of August 1999), while on the HERMES site-specific plots, the grid-cell-specific recommendation was applied. The HERMES + 30% strategy adds 30% to the previously-mentioned HERMES uniform recommendation. Details of the investigation plan are given by Schwarz et al. (2001).

The observations and simulations of soil mineral nitrogen residues after harvest and the yields of the different strategies were compared. To eliminate the effects of different site conditions on the results, scenario calculations with the model were carried out assuming the strategies would have been applied to the same sites. For the field AUTOBAHN, the site-specific application was simulated for the uniformly treated field section and vice versa. For SPORTKOMPLEX, the uniform strategies are simulated for the site-specifically treated grid cells while the other way was not realized because the estimations of fertilizer demand by the model and by the sensor were dependent on the already applied fertilizer and the actualized weather data.

RESULTS

Spatial Distribution of Input Data

On field SPORTKOMPLEX, texture varies little within the field. Although the clay content ranges between 12 and 23%, the texture is dominated by silt. Only two texture classes exist, and

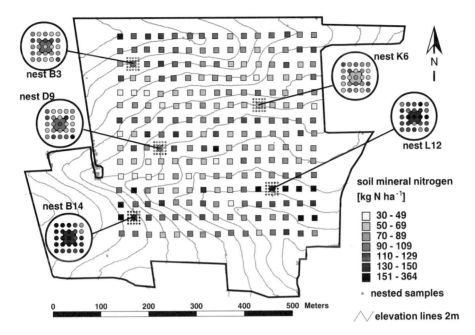

Figure 11.2 Spatial distribution of soil mineral nitrogen in the root zone of field SPORTKOMPLEX in August 1997 for standard grid and nested sampling grid.

they are very similar in their capacity parameters. The organic matter content ranges from 0.9 to 1.7% and is correlated to the clay content. Nevertheless, soil mineral nitrogen at the beginning of the investigations showed a very high variability (30–364 kg N ha^{-1}) within the field (Figure 11.2). Nested sampling gave different results for the northern and southern part of the field. While the semivariogram of the nests B3 and K6 fits well into the variogram of the standard grid, the small distance variations within the nests B14, D9, and L12 exceed the sill of the standard grid. Because the variability of the standard model inputs was small compared to the observed yield and soil mineral nitrogen variation, additional information for the model runs was derived from the DEM. Figure 11.3 shows the digital elevation model and the resulting spatial pattern for solar irradiation and the wetness index scaled from 0 (no wetness) to 3 (strong wetness). Irradiation is considered in the model for photosynthesis while wetness might cause growth reduction by soil air shortage.

The spatial distribution of the main standard model input variables on field AUTOBAHN are shown in Figure 11.4. Soil texture is very clearly structured (Figure 11.4a). The map units reflect the average texture in 0 to 90 cm excluding stones according to the German texture classification system (corresponding signatures according to the American soil taxonomy in brackets). Organic matter (Figure 11.4b) is strongly correlated to the fine texture fractions and ranges from 0.9 to 3.6% C_{org}. Remaining mineral nitrogen content after harvest of oilseed rape in 1999 is quite high. Most of the observations were in the range of 120 to 200 (Figure 11.4c). Measurements were used as initial values for simulation.

Supporting Soil Information by Geoelectrical Measurements

For the 79 observations of clay content in the 0 to 30 cm soil depth, and for approximately 6400 on-the-go measurements of electrical conductivity, variograms were calculated for lag intervals of 10 m. The resulting variograms for clay content and electrical conductivity and their crossvariogram are shown in Figure 11.5. As described earlier, variogram models with a common set of two different structures were fitted according to Eq. 11.4 and the variogram and covariogram model parameters are shown in Table 11.1. The exponential part of the model exhibits a range of 50 m

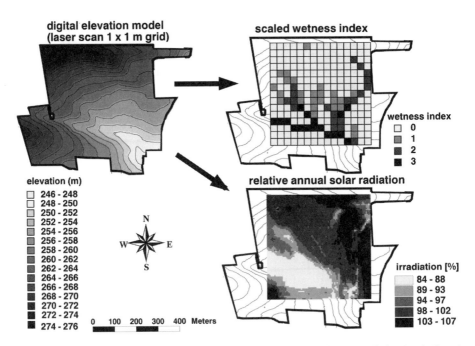

Figure 11.3 Digital elevation model of field SPORTKOMPLEX and derived wetness index (scaled) and spatial distribution of relative solar radiation in 1998.

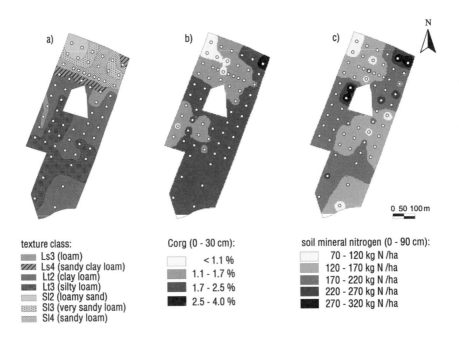

Figure 11.4 Spatial distribution of some relevant model inputs on field AUTOBAHN nesting sampling location not shown: (a) average texture class in the upper 90 cm (German soil texture classification and corresponding texture of soil taxonomy); (b) C_{org}-content in 0–30 cm; and (c) soil mineral nitrogen (N_{min}) in 0–90 cm (initial observation September 15, 1999).

Table 11.1   Variogram Model Parameters for Variograms in Figure 11.5

	c_e	a_e	c_s	a_s
Variogram: clay content	0.63	50 m	43.28	150 m
Cross-variogram: clay — electr. conductivity	-3.74	50 m	55.05	150 m
Variogram: electr. conductivity	22.07	50 m	70.00	150 m

Note: The parameters for the exponential and spherical part of the model are denoted by indices of *e* and *s*, respectively.

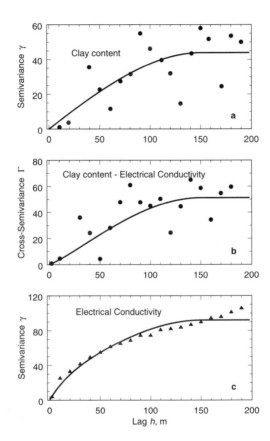

Figure 11.5 Variograms and covariogram for clay content and electrical conductivity for kriging and cokriging analyses at the field site AUTOBAHN in Beckum, Germany.

whereas the range in the spherical model is 150 m. Based on the variogram in Figure 11.5a, clay contents were kriged. The resulting map is displayed in Figure 11.6a. The map of kriging standard deviation is shown in Figure 11.6c. It is obvious that close to sampling points, the kriging standard deviation decreases. On the other hand, in the southern part of the field, where observation density is lower than in the northern part, kriging standard deviation generally increases.

Clay contents were cokriged with electrical conductivity measurements, based on both variograms and the covariogram shown in Figure 11.5. The pattern of cokriged clay contents (Figure 11.6b) becomes more pronounced compared to the kriged clay contents (Figure 11.6a) due to the close spatial association of electrical conductiviy and clay content. Moreover, the cokriging standard deviation of clay content (Figure 11.6d) is lower in general compared to the kriging standard deviation (Figure 11.6c). The lower standard deviation around the sampling locations for clay content ranges over a wider sphere than in the kriging case. Due to the coarser sampling of clay content in the southern part of the field, highest cokriging standard deviations appear in this

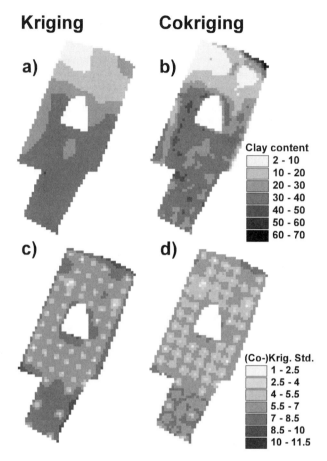

Figure 11.6   Kriged and cokriged map of clay distribution (Figures 11.6a and 11.6b) and the respective (co-)kriging standard deviations (Figures 11.6c and 11.6d) at the field site AUTOBAHN in Beckum, Germany.

region; however, the area with high cokriging standard deviations was smaller and limited to narrow bands at half distance between sampling locations.

Hence, due to the relatively coarse sampling of clay content associated with a relatively high analytical effort and the indirect, relatively cheap measurements of electrical conductivity taken across the entire field at a high resolution, the estimation of the spatial pattern of clay content could be improved and its accuracy increased.

Spatially Distributed Simulation

Model runs, with their individual combination of input data, were performed for each grid cell. Validity of the model is proved using observed soil moisture, mineral nitrogen contents, and crop yield mapping. For SPORTKOMPLEX, the model runs from August 1997 to August 1999 without any reset to observed data. Crop residues for the second simulation period were generated automatically at harvest from the simulated yield and N-uptake of the spring barley. Figure 11.7 demonstrates the time course of the simulated mineral nitrogen in the root zone of selected grid cells compared to the corresponding observations. Until May 1999 the simulation reflects the temporal dynamics of the mineral nitrogen within the standard deviations of the observations. Remaining mineral nitrogen after harvest in 1999 was underestimated while crop yield was overestimated (observed: 6.4 t ha^{-1}, simulated: 6.8 t ha$^{-1)}$.

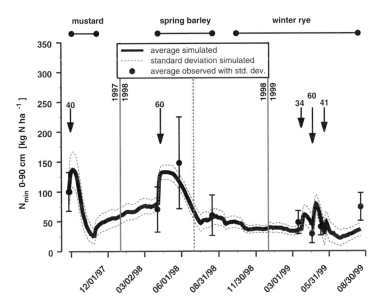

Figure 11.7 Simulated and observed soil mineral nitrogen content in 0–90 cm of all measured grid points including their standard deviation at field SPORTKOMPLEX from August 1997 to August 1999 (arrows indicate date and amount of nitrogen fertilization in kg N ha^{-1}).

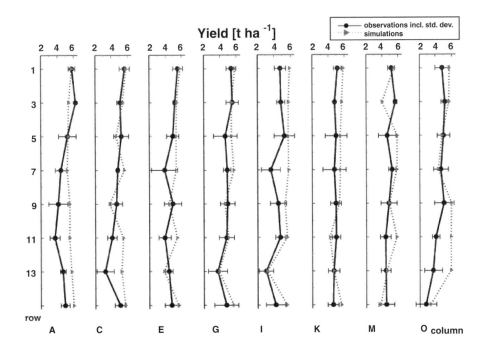

Figure 11.8 Simulated and observed (including standard deviation) yields (dry matter) of spring barley in August 1998 on a double-spaced grid (56 × 56 m) on field SPORTKOMPLEX.

In Figure 11.8, simulated and observed yields of the spring barley in 1998 are compared for every second column/row of the standard grid. The error bars indicate the standard deviation of the recorded yields within the 10-m radius circle around the grid points. On average, the coefficient of variation of all grid points was 0.15, which is composed of small-scale spatial variability and

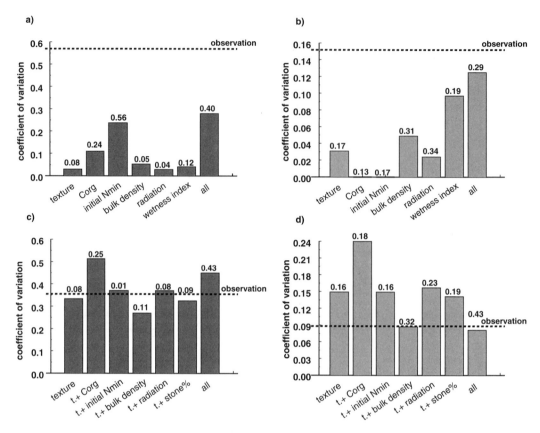

Figure 11.9   Separated effects of different spatial variable input parameters on the variation of (a) soil mineral nitrogen after harvest 1998 on SPORTKOMPLEX; (b) yields of spring barley in 1998 on field SPORTKOMPLEX; (c) soil mineral nitrogen after harvest 2000 on AUTOBAHN; and (d) yields of winter wheat in 2000 on field AUTOBAHN (bar labels show correlation coefficients).

the technical estimation error. Due to this high variability within the grid cells compared to the total field variability, only very low coefficients of determination ($r^2 = 0.09$) were achieved. Of the simulations 53% (1999: 92%) were within these error bars. While the yield depression in row 13, which corresponds to the hollow across the field, is well reflected in most cases, there is a bigger area of overestimation in the southeast (O10 to O14) which can also be seen in the 1999 yield data of the winter rye. This might be due to a former pasture area that can be identified on older field maps.

Figures 11.9a and 11.9b show the effects of different spatially variable inputs on the variation of the model output for soil mineral N (Figure 11.9a) and crop yield (Figure 11.9b) after a one-year simulation at harvest 1998 on field SPORTKOMPLEX. Also shown are the corresponding correlation coefficients between simulations and observations. For this analysis, the spatial variability of only one input parameter was considered, while averages of all the others were used. From the results, it can be seen that texture has only a minor effect on soil mineral nitrogen within this field as can be expected from its low variability. Organic matter content and initial soil mineral nitrogen had no effect on crop yield due to the high level of nitrogen supply but the highest effects on the residual nitrogen contents in the next year. In contrast, bulk density class and wetness index, both influencing the amount of air filled pores at field capacity, had the largest effects on yield but not on mineral N content. The variable solar radiation does not influence soil mineral N. The effect on yield variability is in the same order of magnitude as from texture but shows the highest correlation. The effect of solar radiation on crop yield of winter rye is somewhat higher because

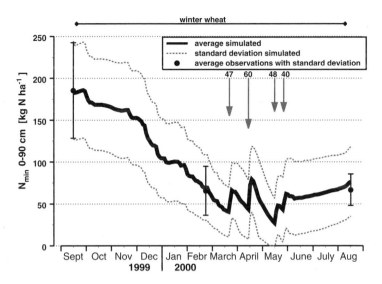

Figure 11.10   Simulated and observed soil mineral nitrogen 0–90 cm of all measured grid points including their standard deviations from September 1999 to August 2000 on field AUTOBAHN (arrows indicate date and average amount of nitrogen fertilization in kg N ha⁻¹).

the main differences in irradiation occur during the end of autumn and beginning of winter. The last column shows the combination effect of all input variables on the output variation.

A similar procedure was done for field AUTOBAHN (Figures 11.9c and 11.9d), but researchers did not take an average of the texture for the analysis of the other inputs because this would have caused artificial effects in some combinations, e.g., if the initial water content on a clay loam site exceeded the water holding capacity of the averaged texture class. So investigators used the variability of the texture as basic information, adding one of the other inputs for simulation each time. The pictures in Figures 11.9c and 11.9d are completely different from that of SPORTKO-MPLEX. Texture and organic matter content play the major roles for the variation of mineral nitrogen (Figure 11.9c) and yield (Figure 11.9d); the topographic wetness index has no influence, and irradiation only affects yield at two grid cells in the direct neighborhood of the forest. Leaching is generally higher on AUTOBAHN, which limits the effect of initial soil mineral variability on the variation of the following year. The small effect on yield indicates that the winter wheat is sufficiently supplied with nitrogen. The higher stone content in the subsoil of the southern part of the field reduces the differences between the water holding capacity of the different textures and the corresponding variability of yield and mineral N. Again, bulk density has the highest effect altering the texture-induced variance of yield and soil mineral nitrogen.

Figure 11.10 shows the simulation of the temporal dynamics of soil mineral nitrogen in the root zone of field AUTOBAHN for 49 grid points in comparison to the observations. Starting from a very high level after harvest, the mineral N content decreases until spring mainly due to nitrate leaching. Absolute variation also decreases until spring. Then, the site-specific fertilization on one side of the field leads to an increase of the variability in the simulation results again. On average, the model results agree well with the observations.

Due to the distinct differences in soil texture and the clear spatial structure of the field, relationships between measured and observed state variables are generally closer on field AUTOBAHN. The high differences in water holding capacity lead to a narrow spatial relationship (Figure 11.11a) between measured and simulated soil moisture contents of the root zone in summer 2000 ($r^2 = 0.79$). The correlation between observed and simulated grain yield ($r^2 = 0.2$) is much lower than for moisture content. Regarding the variability of the yield detection around the grid points (Figure 11.11b) this low correlation is not surprising. Nevertheless, 63% of the simulated yields

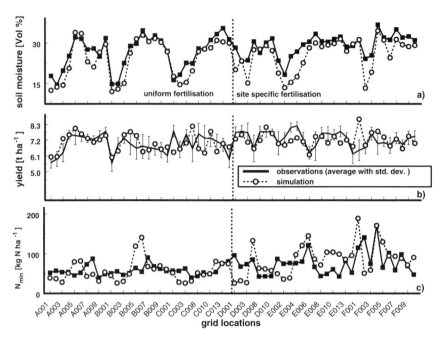

Figure 11.11   Observations and simulations in August 2000 on field AUTOBAHN of (a) soil moisture content (0–90 cm); (b) yield of winter wheat (error bars indicate standard deviations of recorded yields within 10 m radius); and (c) soil mineral nitrogen content (N_{min}) in 0–90 cm.

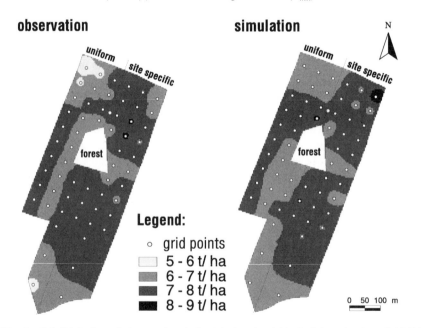

Figure 11.12   Spatial distribution of observed and simulated grain yield of winter wheat on field AUTOBAHN in August 2000.

are within one standard error of the measurement. The simulation of grain yield reflects the general trend between the observations at different locations, which explains the similarity of observed and simulated yield maps (Figure 11.12). Although the temporal dynamics of mineral nitrogen is well described by the model, high deviations occur at single locations (Figure 11.11c). Still, the trends between the differently managed field sides can be described by the model (see Table 11.2).

Table 11.2 Comparison of Observed and Simulated Average Grain Yields of Winter Wheat and Residual Soil Mineral Nitrogen in the Root Zone (0–90 cm) in August 2000 for Differently Fertilized Field Plots on Fields SPORTKOMPLEX and AUTOBAHN, and Scenario Calculations to Compare Fertilizer Schemes for Identical Areas

Scenario / Field	Strategy (average nitrogen fertilization)	No. of Cells	Actual Management Yield Dry Matter (t ha⁻¹) meas	sim	Actual Management N_min (kg N ha⁻¹) 0–90 cm meas	sim	Scenario Calculations HERMES Yield Dry Matter (t ha⁻¹) sim	HERMES N_min (kg N ha⁻¹) 0–90 cm sim	HERMES + 30% Yield Dry Matter (t ha⁻¹) sim	HERMES + 30% N_min (kg N ha⁻¹) 0–90 cm sim
SPORTKOMPLEX Zero		8	4.6	5.2	39	17	6.2	37	6.2	69
Conventional	(179 kg N ha⁻¹)	8	6.9	6.7	60	73	6.6	39	6.6	72
HERMES	(136 kg N ha⁻¹)	8	6.8	6.6	48	36			6.6	66
HERMES + 30%	(178 kg N ha⁻¹)	8	7.0	6.5	65	67	6.4	36		
Site-specific	(139 kg N ha⁻¹)	32	6.8	6.5	44	44	6.7	44	6.7	77

For AUTOBAHN the Scenario Calculations sub-headers are **Uniform** (in place of HERMES) and **Site-Specific** (in place of HERMES + 30%).

Scenario / Field	Strategy (average nitrogen fertilization)	No. of Cells	Actual Management Yield Dry Matter (t ha⁻¹) meas	sim	Actual Management N_min (kg N ha⁻¹) 0–90 cm meas	sim	Scenario Calculations Uniform Yield Dry Matter (t ha⁻¹) sim	Uniform N_min (kg N ha⁻¹) 0–90 cm sim	Site-Specific Yield Dry Matter (t ha⁻¹) sim	Site-Specific N_min (kg N ha⁻¹) 0–90 cm sim
AUTOBAHN Uniform	(185 kg N ha⁻¹)	30	6.9	7.0	58	56			7.0	70
Site-specific	(204 kg N ha⁻¹)	30	7.3	7.2	77	86	7.2	71		
Field average		60	7.1	7.1	68	71	7.1	64	7.1	78

Comparison of Fertilization Strategies

Table 11.2 shows the results of measured and simulated yields and residual soil mineral nitrogen after harvest of winter wheat in August 2000 for both fields under investigation. For AUTOBAHN, the measurements show that the site-specific management results in slightly higher yield than the uniform application. This is also indicated by the simulation if the real management is considered. Using the model to transfer both management strategies virtually to the other side, it is apparent that the observed differences in yield between both sides were mainly caused by site effects. Moreover, the site-specific management scheme seems to leave behind higher amounts of mineral nitrogen after harvest, which is at risk to be leached out during the next winter period. Regarding the high spatial variability and the uncertainty of the model calculations, the differences in residual soil mineral nitrogen between both sides of AUTOBAHN cannot be stated as significant.

On field SPORTKOMPLEX, the application of different fertilization rates, especially the zero fertilization plots, led to an increase of yield variability (Figure 11.13) which improved the correlation ($r^2 = 0.4$) between observations and simulations. For 89% of the grid cells, yield estimation deviates less than 15% from the observations. Again, the correlation between measured and simulated mineral nitrogen is poor ($r^2 = 0.17$) due to the very high small-scale variability within the grid cells (average coefficient of variation = 0.3).

Looking at the averages of the fertilization variants (Table 11.2), the model results reflect the yields and residual mineral nitrogen content satisfactorily. For the zero fertilized plot, the earlier senescense of the winter wheat due to nitrogen deficiency was not sufficiently considered, which led to an overestimation of yield and nitrogen uptake and a corresponding underestimation of mineral nitrogen content at harvest. The observations as well as the simulations indicate that the yields of the fertilization strategies are not significantly different. Instead, there is a slight tendency to higher residual mineral nitrogen contents after harvest for the higher fertilized strategies

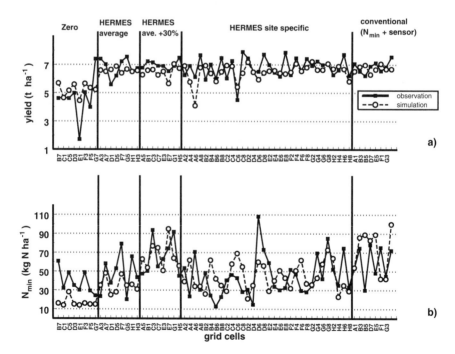

Figure 11.13   Simulated and observed values on field SPORTKOMPLEX at harvest of winter wheat in August 2000 at differently fertilized grid cells for a) grain yield (dry matter) and b) soil mineral nitrogen 0–90 cm.

HERMES + 30% and conventional, which received about 40 kg N ha^{-1} more than the two other strategies. Transferring the uniform application strategies to the other grid cells show that there is nearly no effect on residual mineral nitrogen after harvest between uniform and site specific application if a similar amount (strategy HERMES) would have been fertilized, but an increase of remaining nitrogen if the higher amount (HERMES + 30%) would have been applied to the cells of the variants HERMES and HERMES site-specific. Scenario calculations for the following winter period indicate that at single locations, especially in the hollow area where crop growth is limited by site conditions, a reduction of nitrogen leaching with site specific fertilization can be achieved (Kersebaum et al., 2001). This shows that the concept of site-specific fertilization has the potential to reduce nitrogen leaching although the average results do not reflect this due to the small contribution to the area of the investigated fields.

Spatial Crop Yield Estimation with Autoregressive State–Space Models

A set of different variables observed in the year 1998 in the field site SPORTKOMPLEX should be evaluated, i.e., to what extend different variables supported spatial estimation of crop yield. For this estimation, these variables were crop yield itself (Yi), normalized difference vegetation index (NDVI), crop-nitrogen status for May 15, 1998 simulated according to Kersebaum (1995) (Crop-N), and solar radiation for the month of April 1998 as mentioned above (Sol-Rad). The normalized difference vegetation index (Baret, 1995) is assumed to integrate patterns of soil properties and biomass that are related to the yield at the end of the growing season. The simulated crop-nitrogen status was chosen as a variable because, recently, sensors have been developed that allow for chlorophyll activity mapping, underlying nitrogen fertilizer recommendation records. Under field conditions, such a variable would not have to be simulated but would be available from a sensor. The solar radiation and its variation in space and time proved to have a significant impact on the deterministic modeling in this study. Therefore, it should be evaluated for the empirical autoregressive approach in the state–space analysis. These four variables were measured along a 15 by 15 grid with separation distances of 28 m. Because this methodology considers only a one-dimensional array of data, measurements were treated as if they were taken across a transect in a string-type arrangement, i.e., values were taken along a row in a west–east direction followed by the next row in an east–west-direction, and so on.

The spatial distributions of the four variables are shown in Figure 11.14. The measured yield and NDVI exhibit patterns with local fluctuations smaller than general variation indicating structured variation (Figure 11.14a). For calculated crop-nitrogen status (Crop-N), many locations yield the same value (Figure 11.14b). Lower crop-nitrogen levels are calculated at those locations where mineralization and growing conditions were unfavorable. The spatial pattern of solar radiation reflects the surface relief and exposure properties.

Crosscorrelation functions for yield and the other three variables are shown in Figure 11.15. The crosscorrelation coefficient at lag $h = 0$ equals the ordinary correlation coefficient. Although $r_{xy}(0)$ are close to or below 0.5, the structured variation and spatial association between the respective variable pairs is manifested by the smooth decay in the crosscorrelation functions.

Spatial process of crop yield was modeled in three scenarios with different combinations of variables (Table 11.3). In each of the scenarios, the input information of crop yield itself was varied, i.e., with all yield observations considered in the estimation (a), with every other yield value considered in the estimation (b), and with every fourth value, only (c).

The transition or autoregression coefficients are shown in Table 11.3. In all models except for scenario 3c, the yield value at the previous location has the highest weight in the estimation. The impact of the other variables changes for the respective scenarios. For the same number of observations considered in the estimation, the AICc criterion provides information on the suitability of the model. If all yield observations are considered, the four variables combined in scenario 3 yield

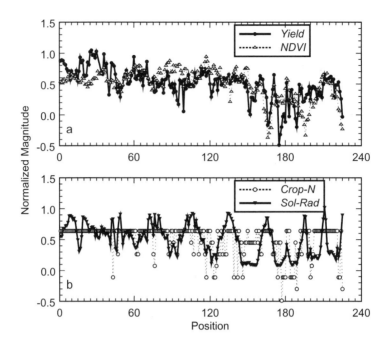

Figure 11.14 Measured spring barley (Yield, a), (NDVI, a), simulated crop nitrogen status at 15 May 1998 (Crop-N, b), and calculated solar radiation in April 1998 (Sol-Rad, b) at the field site SPORT-KOMPLEX in Luettewitz, Germany.

the best description; however, if the number of observations considered is reduced to 50% and 25%, respectively, scenario 2 with crop yield, NDVI, and Crop-N status yields the most promising result. For this analysis, the AICc was based on all 225 values, regardless of how many were considered in the estimation, in order to reflect how well the model reflected the values at those locations that were not considered in the estimation. The spatial yield process for the three cases in scenario 2 is shown in Figure 11.16. From case a to b, the 95% confidence range increased considerably; however, reducing input information of yield to 25% did not increase the confidence range strongly. Fluctuations in the yield series are conserved by the model only when extreme yield observations are included in the considered data. For example, the low values at positions 175 to 177 are still kept moderately in scenario 2b. In scenario 2c, however, the low level of yield values at these locations is ignored. The similarity in the description of local fluctuations between scenario 2b and 2c is manifested in the similar state covariance between Yi_i and Yi_{i-1} (Table 11.3). In the other scenarios, the state covariance decreases when observation frequency is reduced from 50% to 25%.

CONCLUSIONS

The investigations on two fields in different landscapes have shown that grain yield formation is determined by various factors differing in their relevance from site to site. To benefit from the potential of precision farming technologies in terms of increasing nitrogen fertilization efficiency, it is important to consider spatial variability of the most relevant influences of a specific site. This requires more detailed and accurate spatial information and a better knowledge of the interactions between site conditions and yield formation. Quality of traditional soil maps is often insufficient and needs to be improved by investigations of higher resolution. Grid sampling data to be used for geostatistical interpolation methods are too time-consuming and expensive for practical purposes. Results of easy on-the-go electric conductivity measurements with a high spatial resolution can be

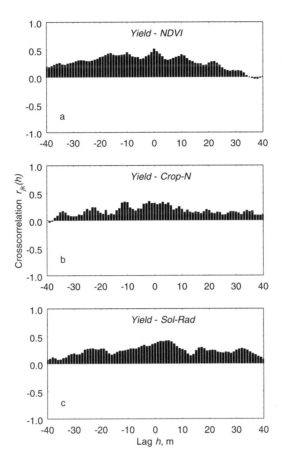

Figure 11.15 Cross-correlation function for variable pairs: (a) spring barley grain yield and normalized difference vegetation index; (b) yield and crop nitrogen status on 15 May 1998; and (c) yield and solar radiation integrated for the month of April 1998 in the field site SPORTKOMPLEX in Luettewitz, Germany.

combined with information of a coarse grid sampling by cokriging to improve the spatial resolution and accuracy of soil maps if texture varies enough within a field.

For the site AUTOBAHN, most of the grain yield variation can be reflected by the deterministic simulation using the spatial variation of basic soil analysis data. For the field SPORTKOMPLEX, soil texture variation is small and topographical impact on yield formation seems to be much higher. Although we use some topographical information to alter the model input, we have to consider that yield variability can only be partly explained by the simulation. This might be caused on one side by lateral processes, which cannot be considered by the chosen approach yet and, on the other side, by secondary effects, e.g., weed competition, which enhances yield depression caused by primary stress factors. It is interesting to note that solar radiation was very helpful to improve the deterministic model calculations when it was embedded in a set of bio-geochemical and physical equations. If, on the other hand, it was used in an empirical fashion in the autoregressive state–space analysis, solar radiation pattern cumulated for the month of April did not improve the model result. Moreover, the results of the autoregressive state–space analysis indicate that measurements of the crop-nitrogen status in May and of NDVI might be useful to predict the final spatial yield distribution.

For both sides, it has to be concluded that model validation based on yield mapping is limited by the small-scale variability of detected grain yields induced by various technical errors (Blackmore and Moore, 1999) and the inherent spatial variability. Although sensitivity of the model has to be

Table 11.3   Model Evaluation for Different Scenarios of Spatial Yield (Yi) Estimation

	Yi_{i-1}	$NDVI_{i-1}$	$Crop\text{-}N_{i-1}$	$Sol\text{-}Rad_{i-1}$	AICc
Covariance of normalized original data					
Yi_i	0.0457	0.0317	0.0193	0.0234	—
	+	+	—	+	—
Scenario 1a: all observations considered					
Autoregr. Coeff: Yi_i	0.7307	0.1276	—	0.1318	−3.569
State Cov.: Yi_i	0.0174	0.0056	—	0.0001	
Scenario 1b: 50% of observations considered					
Autoregr. Coeff: Yi_i	0.8924	0.0412	—	0.0625	−2.711
State Cov.: Yi_i	0.0050	0.0029	—	−0.0002	
Scenario 1c: 25% of observations considered					
Autoregr. Coeff: Yi_i	0.9623	−0.0252	—	0.0611	−2.306
State Cov.: Yi_i	0.0013	0.0021	—	−0.0009	
	+	+	+	—	—
Scenario 2a: all observations considered					
Autoregr. Coeff: Yi_i	0.6880	0.0874	0.2233	—	−3.577
State Cov.: Yi_i	0.0174	0.0052	−0.0011	—	
Scenario 2b: 50% of observations considered					
Autoregr. Coeff: Yi_i	0.7633	0.0641	0.1783	—	−2.882
State Cov.: Yi_i	0.0091	0.0033	−0.0008	—	
Scenario 2c: 25% of observations considered					
Autoregr. Coeff: Yi_i	0.7906	0.0111	0.2070	—	−2.564
State Cov.: Yi_i	0.0089	0.0054	−0.0020	—	
	+	+	+	+	—
Scenario 3a: all observations considered					
Autoregr. Coeff: Yi_i	0.6794	0.0959	0.1759	0.0476	−3.584
State Cov.: Yi_i	0.0174	0.0052	−0.0013	−0.0000	
Scenario 3b: 50% of observations considered					
Autoregr. Coeff: Yi_i	0.7901	0.0531	0.1526	0.0088	−2.817
State Cov.: Yi_i	0.0075	0.0032	−0.0009	0.0001	
Scenario 3c: 25% of observations considered					
Autoregr. Coeff: Yi_i	0.2249	0.1505	0.8891	−0.2182	−2.280
State Cov.: Yi_i	0.0030	0.0014	0.0009	0.0041	

Note: NDVI denotes normalized difference vegetation index; Crop-N denotes the simulated nitrogen in the crop on May 15; and Sol-Rad denotes the solar radiation calculated for the month of April.

improved for site-specific fertilization recommendations, the results indicate that the general concept to derive fertilizer recommendations works quite well. It is evident that a beneficial effect in terms of higher efficiency and lower emissions can only be achieved if fertilizer supply is just at the threshold of yield reduction and meets the crop demand very closely. Agricultural systems models play an important role to achieve this aim.

ACKNOWLEDGMENTS

The authors thank the German Research Foundation and the Federal Ministry for Research and Technology for their financial support, the Suedzucker AG for their financial and practical engagement, the Claas Company and the Amazone Company for their financial and technical support, and the farmers, Steigerwald and Luedeke. The authors also thank Michael Baehr, Norbert Wypler and Michael Heisig for their technical and sampling assistance.

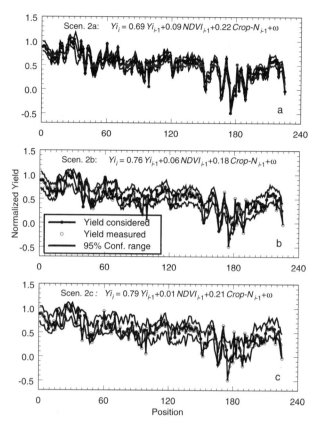

Figure 11.16 Autoregressive state–space analysis of spring barley grain yield in the year 1998 for the field site SPORTKOMPLEX in Luettewitz, Germany. The scenarios 2a, 2b, and 2c are explained in the text and in Table 11.3.

REFERENCES

AG Boden. 1994. *Bodenkundliche Kartieranleitung*, 4th ed., Schweizerbarth, Stuttgart.

Alemi, M.H., M.R. Shahriari, and D.R. Nielsen. 1988. Kriging and Cokriging of soil water properties, *Soil Technol.*, 1:117–132.

Baret, F. 1995. Use of spectral reflectance variation to retrieve canopy biophysical characteristics. In *Advances in Environmental Remote Sensing,* F.M. Danson and S.E. Plummer, Eds., John Wiley and Sons, Chichester, 33–51.

Blackmore, B.S. and M.R. Moore. 1999. Remedial correction of yield map data, *Precision Agric.*, 1, 53–66.

Deutsch, C.V. and A.G. Journel. 1992. *GSLIB, Geostatistical Software Library and User's Guide,* Oxford University Press, New York.

Durlesser, H. and H. Stanjek. 1997. Capability and limits of a DGPS supported EM38 survey for the fast estimation of the spatial variation of clay and water contents of soil. In *Field Screening Europe,* J. Gottlieb et al., Eds., Kluwer Academic Publishers, Dordrecht, The Netherlands, 73–76.

Erhardt, O. and F.P. Riedinger. 1984. Zum Strahlungshaushalt eines Buchenwaldes, *Proc. Int. Symp. in Memory of F. Sauberer,* Vienna, Universität für Bodenkultur, Vienna, Oct. 23–25, 1984. 125–127.

Jürschik, P., A. Giebel, and O. Wendroth. 1999. Processing of point data from combine harvesters for precision farming. In *Proc. 2nd Europ. Conf. Precision Agric.,* J.V. Stafford, Ed., Odense, Sheffield Academic Press, 297–307.

Gallant, J.C. and J.P. Wilson. 1996. TAPESG: a terrain analysis program for the environmental sciences, *Computers and Geosciences,* 22:713–722.

Kersebaum, K.C. 1995. Application of a simple management model to simulate water and nitrogen dynamics, *Ecol. Modelling,* 81:145–156.

Kersebaum, K.C. and A.J. Beblik. 2001. Performance of a nitrogen dynamics model applied to evaluate agricultural management practices. In *Modeling Carbon and Nitrogen Dynamics for Soil Management,* M. Shaffer, L. Ma, and S. Hansen, Eds., CRC Press, Boca Raton, 551–571.

Kersebaum, K.C., K. Lorenz, O. Wendroth, H.I. Reuter, J. Schwarz, and P. Jürschik. 2001. Effects of site-specific nitrogen fertilization on nitrogen leaching–comparison of different strategies in arable fields based on observations and simulations. In *Proc. 3rd European Conf. on Precision Agric.,* June 18–20, 2001. G. Grenier and S. Blackmore, Eds., Montpellier, 683–688.

Kersebaum, K.C. and J. Richter. 1991. Modelling nitrogen dynamics in a soil-plant system with a simple model for advisory purposes, *Fertilizer Res.,* 27:273–281.

Leithold, P. 2000. Der Hydro N Sensor bestimmt den Stickstoffbedarf von Getreide, *Neue Landwirtschaft,* 1:56–57.

Moore, I.D.. 1993. Soil attribute prediction using terrain analysis, *Soil Sci. Soc. Am. J.,* 57:443–452.

Reuter, H.I., O. Wendroth, K.C. Kersebaum, and J. Schwarz. 2001. Solar radiation modelling for precision farming — a feasible approach for better understanding variability of crop production. In *Proc. 3rd European Conf. on Precision Agric.,* June 18–20, 2001. G. Grenier and S. Blackmore, Eds., Montpellier, 845–850

Schwarz, J., K.C. Kersebaum, H.I. Reuter, O. Wendroth, and P. Jürschik. 2001. Site specific fertilizer application with regard to soil and plant parameters. In *Proc. 3rd European Conf. on Precision Agric.,* G. Grenier and S. Blackmore, Eds., Montpellier, 713–718.

Shumway, R. H. 1988. *Applied Statistical Time Series Analysis,* Prentice Hall, Englewood Cliffs.

Shumway, R.H. and D.S. Stoffer. 2000. *Time Series Analysis and Its Applications,* Springer, New York.

Supit, I., A.A. Hooijer, and C.A. van Diepen. 1994. System description of the WOFOST 6.0 crop simulation model implemented in CGMS, Vol. 1: Theory and Algorithms, EC Publication EUR 15956, Luxemburg.

van Keulen, H., F.W.T. Penning de Vries, and E.M. Drees. 1982. A summary model for crop growth. In *Simulation of plant growth and crop production,* F.W.T. Penning de Vries and H.H. van Laar, Eds., PUDOC, Centre of Agricultural Publishing and Documentation, Wageningen, The Netherlands, 87–97.

Wehrmann, J. and H.C. Scharpf. 1986. The N_{min}-method — an aid to integrating various objectives of nitrogen fertilization, *Z. Pflanzenernaehr. Bodenk,* 149:428–440.

Wendroth, O.., A.M. Al-Omran, C. Kirda, K. Reichardt, and D.R. Nielsen. 1992. State-space approach to spatial variability of crop yield, *Soil Sci. Soc. Am. J.,* 56:801-807.

Wendroth, O., P. Jürschik, K.C. Kersebaum, H. Reuter, C. Van Kessel, and D.R. Nielsen. 2001. Identifying, understanding, and describing spatial processes in agricultural landscapes — four case studies. In Special issue: landscape research — exploring ecosystem processes and their relations at different scales in space and time, C. Van Kessel and O. Wendroth, Eds., *Soil Till. Res.,* 58:113-128.

Wenkel, K.-O., S. Brozio, R.I.B. Gebbers, K.C. Kersebaum, and K. Lorenz.2001. Development and evaluation of different methods for site specific nitrogen fertilization of winter wheat. In *Proc. 3rd European Conf. on Precision Agric.,* June 18–20, 2001. G. Grenier and S. Blackmore, Eds., Montpellier, 743–748.

Yates, S. R. and A.W. Warrick. 2002. Geostatistics. In *Methods of Soil Analysis, ASA Monograph,* 3rd ed., Agron. Monogr. ASA and SSSA, G.C. Topp and J.H. Dane, Eds., Madison, WI.

Zhang, R., P. Shouse, and S. Yates. 1999. Estimates of soil nitrate distributions using cokriging with pseudo-cross-variograms, *J. Environ. Qual.,* 28:424–428.

CHAPTER 12

Addressing Spatial Variability in Crop Model Applications

E. John Sadler, Edward M. Barnes, William D. Batchelor, Joel Paz, and Ayse Irmak

CONTENTS

INTRODUCTION

The topic of this chapter, addressing spatial variability in crop model applications, comes from the logical combination of two trends in agricultural research. The first trend dates to the 1950s when Monsi and Saeki (1953) published the first known application of physical models to explain processes important to agricultural production. During the succeeding years, many models have been developed for research purposes, including those discussed later in this chapter. The second trend is the more-recent development of site-specific or precision agriculture, the formal start of which is often attributed to the granting of a patent for a device to apply dry granular fertilizer on a site-specific basis (Ortlip, 1986), followed soon by the development of combine-mounted yield monitors (Vansichen and De Baerdemaeker, 1991).

The ability to custom-apply fertilizer immediately generated the need for site-specific fertilizer recommendations, and the yield variation observed in yield maps immediately suggested a need for explanations of causes of variation. For both theoretical and empirical reasons, however, traditional statistical methods are not well-suited to address spatial problems. Seeking new tools to meet both these needs, researchers logically embraced process-oriented crop models, although for reasons discussed below, applying existing models proved not to be a simple task. Handling spatial variability in models, which were usually one-dimensional (1-D) in the vertical soil profile direction, embodied accounting for two additional horizontal dimensions. This added at least an order of magnitude greater complexity than simply increasing the resolution in the vertical dimension.

The remainder of this chapter describes the current state of the art in applying models to site-specific agricultural problems, using three approaches. The first is an extension of conventional methods, the second applies remote sensing tools to provide input data, and the third employs inverse modeling to generate spatially distributed inputs that produce the best description of the spatially distributed yield. Before starting on the spatial modeling, a discussion of temporal variability and how it is handled in dynamic models is useful.

During development, a model's structure depends upon the modeler's compromises between the objective and what knowledge can be encoded into the model. In simple terms, these are what can be predicted and what can be described. Although lack of suitable input data can constrain the choices, for the most part, the objective defines the time basis for the prediction. For example, predicting canopy temperature during cloud passage requires a time basis ranging from seconds to minutes, while predicting organic matter contents under decades or centuries of conservation tillage may require a time basis ranging from months to years. Common examples of several varying temporal scales include the Root Zone Water Quality Model (RZWQM, Ahuja et al., 2000) at sub-hourly time steps (for hydrology), the CERES (Jones and Kiniry, 1986) and CROPGRO models (Hoogenboom et al. 1994; Boote et al., 1998a) at daily time steps, and the CENTURY model (Parton et al., 1992) at monthly time steps. The remainder of this chapter discusses daily time step models, often using the CERES-Maize or CROPGRO-Soybean models as examples.

If one concludes that increasing the temporal resolution of a model requires a smaller time step, then it is possible that this will eventually require alterations in the model structure. This happens if empirical approximations break down under a smaller time step. For instance, a daily time step model cannot, by definition, handle diurnal patterns except by using approximations based on daily averages, ranges, or other statistical descriptions. In most such cases, the empirical approximation must be replaced by a module somewhat more mechanistic in nature to describe the time-sensitive processes at the smaller time step.

Often, temporal and spatial problems, and the programming solutions to them, are linked. In the case of the soil water balance, many models, including the DSSAT suite, currently use the SCS Curve Number method (USDA-SCS, 1972) to compute runoff and infiltration, which is desirable because the temporal scale is daily, corresponding to widely available daily total rainfall data. In order to adequately simulate runoff and redistribution within a field in a two- or three-dimensional (2-D or 3-D) soil water balance model, more accurate predictions of runoff and surface flow are needed. Better methods are available, but require shorter time steps and intra-day (or even sub-hour) rainfall data. As described in the next section, this topic poses yet another challenge to spatial modeling.

SPATIAL MODELING

Prior to widespread use of spatial tools, models were (usually):

1. Dynamic, meaning they accounted for temporal variability
2. One-dimensional in the vertical soil profile dimension
3. Sensitive to large differences in cultivars, soils, and weather
4. Validated with plot averages

To apply to spatially variable applications, however, they must retain their dynamic nature, as well as:

1. Add two horizontal dimensions.
2. Account for subtle differences in, primarily, soils, with secondary differences in weather.
3. Predict the variance as well as the mean.

These additional requirements deserve considerable thought. Within a field, the soil texture and chemical component (nutrients, salinity, etc.) variation might be significant, but certainly less so than differences across states, countries, or continents. Weather variability is very much reduced (and in some models, not a spatial input at all at the field scale, despite many meteorological parameters being known to have significant spatial variation), and cultivar characteristics are usually constant across a field. Despite the apparent reduction in these known sources of variation, the yield variability within a single field can be as large or larger than the yield variability measured between fields or counties within a state. Even though the soil types may not vary as much within a field as from field to field or county to county, there is still tremendous variability within the field that these models must address. Furthermore, if these models are to be reliable for evaluating variable rate decisions within a field, they must not only be able to provide good predictions of mean yield, but also of within-field variation in yield, as a response to highly spatially variable factors that affect yield. Mathematically stated, to use the results in an optimization algorithm, not only must the mean be predicted well, but also the partial differential with respect to all the important inputs. These criteria for success are severely stringent.

For a model to be successful, all important variations must be reflected in the model processes and associated variables established as model inputs. For some situations where mixed results have been obtained, variations in the observed data may not have been reflected in either one or both of the model's processes or of the model's inputs. For instance, if a model predicts phenology as a function of air temperature collected at a weather station several kilometers away, it is not likely that the observed spatial variation in canopy temperature, and hence energy balance and associated processes, will be modeled correctly within the field. Another example comes from the use of plot averages during model validation — observed high-yield spots within a field have been observed to exceed the model-allowed limit for harvest index, which had been chosen based on plot averages (Paz et al., 2001). These are clearly cases where simply adding finer-scale input data will not guarantee success. On the other hand, if a model were to have sufficient detail, but require correspondingly higher resolution of spatial soils data, then increasing the resolution of inputs might be productive, but it will almost certainly be expensive. The increase in expense can be easily proven — doubling the spatial resolution of a measurement means that the number of samples is doubled in both directions, with four times the cost for sampling.

One can speculate on what model processes and inputs would be necessary to fully characterize spatial yield variation in typical cropped fields, but both the availability of data and knowledge of the basic processes are usually quite severely limited (e.g., Robert, 1996). For instance, high-resolution spatial and temporal variation in soil physical and chemical properties would be prohibitively expensive to characterize. Some progress has been made in using terrain analysis and hydrologic modeling to predict within-field redistribution of runoff (e.g., Simmons et al., 1989; Moore et al., 1993; Kaspar et al., 2001), although transient effects of spatially variable evapotranspiration on soil water content and the several feedbacks into crop water stress, future infiltration, and eventual crop yield appear to be significant (Sadler et al., 2000a). Beyond these effects within the framework of the isolated monoculture are an entire litany of "external" factors including weeds, insects, nematodes, diseases, and other landscape-level ecological factors that so far have not been integrated completely into many crop models.

In our collective experiences, we have modeled spatially variable crop growth using three general approaches. The first approach is essentially a brute force method, acquiring inputs and running the model conventionally at multiple points in space (e.g., Sadler et al., 1998, 1999, 2000b). A second approach (e.g., Barnes et al., 1997, 2000; Jones and Barnes, 2000) used remote sensing to either augment inputs or test outputs and change state variables iteratively. The third approach (e.g., Irmak et al. 2001; Paz et al., 2001) employed objective parameterization, using optimization routines or database searches, to solve for spatially variable inputs that minimize errors between simulated and measured yield across seasons. In all three cases, the models used were 1-D models repeated in space rather than fully 3-D models of crop growth and yield. These latter, although desired, are not yet available.

Conventional Method

The conventional method approach was designed simply to acquire input data at more points than in an otherwise traditional way, either on grids or in management zones, and then to run the model conventionally at each point. The state of this art in 1995 was catalogued by Sadler and Russell (1997), who described approximately 20 such efforts, including AEGIS (Papajorgji et al., 1993; Engel et al., 1995) and other model-running shell programs (e.g., Han et al., 1995). Depending on the circumstances and how far removed the simulation was from typical conditions, these efforts suggested two things. First, success was mixed. Some of the work provided acceptable results, but other results were disappointing (see Sadler and Russell, 1997; Sadler et al., 1998, 1999, 2000b). Second, acquiring the extensive input data encouraged the search for more efficient procedures.

The usual procedure has been to obtain input values for soil parameters from soil surveys and typical pedon descriptions at the county level (~1:24000) or from similar techniques employed at a fine scale (~1:1200 – 1:5000). These have been supplemented occasionally with physical property measurements for profiles on transects and grids. In nearly every case, however, there existed additional variations not captured in the soil data collected (e.g., Sadler et al., 1998, 1999, 2000b). Despite the amount of data employed, it did not appear to be sufficient. Increasing the resolution using standard survey techniques appeared to be neither feasible nor productive, because even the finer scale approaches have not met with unambiguous success. Making the extensive measurements deemed necessary has been attempted at considerable effort in research settings, but it is not generally considered economically feasible in production settings.

The foregoing has dealt with traditional data that has had location added to it. There exists a data type that is acquired literally *en masse* (such as via photography), or practically so (such as with a scanning sensor in remote sensing or an on-the-go yield monitor). One characteristic that distinguishes such data from the traditional data mentioned above is that where the above is usually data-starved, these inherently spatial data sets are data-rich. This distinction allows several additional uses, some of which are worthwhile either in isolation or as a contribution to other spatial modeling efforts.

Spatial sensors were cited as one of the primary research needs to help solve the lack-of-data problem at several of the early Precision Agriculture Conferences (Schueller, 1993; Robert, 1996), and this may still be the primary bottleneck. Where such data have been obtained, they have been applied in modeling applications in one of three general ways. The first use has usually been to define areas where variation occurs in soil properties and crop development, illustrating areas that need to be managed or, in this context, modeled separately. Where one is fortunate enough to have spatial data for outputs of models, using them for validation of models is quite valuable. Where the observations are intermediate or state variables in models, in-season adjustments can improve the performance of models under certain conditions. Where the observations correspond to model inputs, these can be considered traditional data collected much more efficiently in space. Examples of such data include depth to clay layer, organic matter, plant population, and topography and terrain attributes (see review by Sudduth et al., 1997). One particular type of such data is the basis of the second approach for modeling spatial variation.

Remote Sensing Methods

The particular example of inherently spatially variable data is by remotely sensed (RS) observations, usually corresponding to intermediate variables. One of the most commonly cited uses of RS to provide a linkage with crop models has been to estimate leaf area index (LAI; Weigand et al., 1979). One method is to relate LAI and RS data with a radiative transfer model (RTM; e.g., Asner and Wessman, 1997) used in either direction. An RTM can be used to calculate LAI based on the radiometric characteristics of the field for comparison to the model, or the LAI output from the model can be used to calculate the scene reflectance for direct comparison to the RS data (Guérif and Duke, 2000). An advantage of this method is that it is not extremely dependent on site-specific relationships between the crop and RS data. A disadvantage is that the input data requirements for some RTM models are themselves quite severe. A second method is to use a locally determined, empirical relationship between the RS data and variable of interest (e.g., a simple linear regression with LAI as the independent variable and vegetation index, as in Jones and Barnes, 2000). Other crop parameters that have been estimated from RS data and incorporated into crop models include crop water status (Barnes et al., 2000), evapotranspiration rates (Moran et al., 1995), and canopy chlorophyll content (Weiss et al., 2001).

Exactly how the link is made between RS data and a model has varied according to the objectives of the researchers involved, but can be grouped by method. In a review on the topic, Moulin et al. (1998) placed methods to integrate RS data and models into four categories:

1. Inputting a variable estimated from RS data
2. Updating a state variable in a model from an RS estimate
3. Adjusting model's initial conditions
4. Calibrating parameters to produce better agreement between RS estimates and model predictions during the season

A fifth category uses remotely sensed data to identify areas where crop development is significantly different from surrounding areas and, thus, requires independent simulation (Jones and Barnes, 2000).

For category 1, it is theoretically possible to build a model that accepts the state variable as an input rather than as a computation from other inputs. This requires that a sufficiently intensive time series of spatial data could be obtained for a state variable. There are no known practical examples of such an application using remotely sensed data directly at daily time steps; however, estimates of state variables have been interpolated between image acquisitions to derive daily values to drive a model (examples cited in Moulin et al., 1998).

An example of updating a state variable (category 2) in CERES-Wheat is taken from Barnes et al. (1997), who modified the LAI predicted by the model based on remotely sensed estimates. LAI was replaced by a RS estimate when an estimated LAI was available for a particular day and the model's predicted LAI was outside of a predefined tolerance from the RS estimate. If the prediction was outside of the tolerance, the model's predicted LAI was set to the RS estimate by adjusting the model's prediction of accumulated green leaf area and leaf weight to match the RS estimate. The simulation then would continue until the next RS observation or end of the simulation. This approach is illustrated in Figure 12.1a, which shows a ratio vegetation index (RVI = ratio of

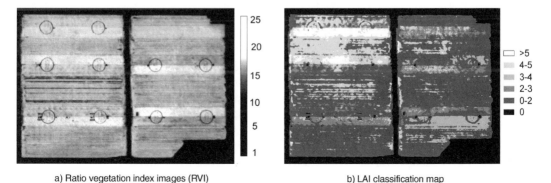

a) Ratio vegetation index images (RVI)   b) LAI classification map

Figure 12.1   Maps of a wheat field derived from March 31, 1966, image data (a) RVI and (b) LAI classification.

Table 12.1   Predicted and Observed[a] Wheat Yields Corresponding to the LAI Classes of Figure 12.1b

LAI Class in Figure 12.1b	Yield (kg ha⁻¹)		% Difference
	Observed	Predicted	
>5	8000	7454	6.8
4–5	7500	7417	1.1
3–4	7000	7366	5.2
2–3	6500	7008	7.8
0–2	5700	6045	6.0

[a] Observed is the approximate yield determined for the various treatments during the 1995–1996 experiment, which has been assigned to an LAI class based on the LAI of that treatment during the time the image was acquired.

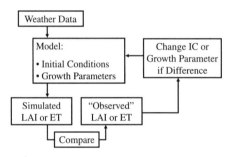

Figure 12.2   Schematic of approach developed by Maas. (*Agron. J.,* 85:354-358, 1993.)

near-infrared to red reflectance) image acquired from an aircraft on March 31, 1996, near the time of anthesis. In the image, the bands of increased RVI running left to right correspond to a high nitrogen treatment. The circles apparent in the image were from pipes that were used to inject carbon dioxide (see Kimball et al., 1999, for a description of the experiment). CERES-Wheat was used to simulate the field conditions, assuming adequate nitrogen and water were present. On the date the image was acquired, the LAI classes from the RVI image (Figure 12.1b) were input to the model and then the simulation was resumed, still assuming adequate nitrogen and water. Reasonable yield predictions were obtained with this particular image, because it was near the time of anthesis (see Table 12.1); however, this approach is subject to several limitations. Accuracy decreased for LAI modifications more than ~10 days before or after anthesis. This method also did not work as well if the "base" model run was underpredicting LAI (i.e., it was easier to lower the model's predictions than to raise them). Difficulties obtaining near-real-time data limit the application of this particular technique for real-time farm management, and the need for data near the time of anthesis significantly limits the amount of corrective action available to a farm manager.

An example that uses a combination of categories 3 and 4 is the approach used by Maas (1988, 1993) to calibrate model parameters initially based on LAI. This approach was later expanded by Moran et al. (1995) to consider RS estimates of evapotranspiration (ET). A flow diagram of their approach is illustrated in Figure 12.2. The model's initial condition of water content and field capacity were adjusted based on the difference between RS-estimated and model-predicted ET. To match RS estimates of LAI, leaf span or biomass partitioning was changed through adjustments of the model's calibration parameters. The approach provided accurate simulation for growth and yield of grain sorghum, corn, spring wheat (Maas, 1993) and alfalfa (Moran et al., 1995).

Objective Parameterization Methods

The third approach described in this chapter uses an inversion of modeling, which was developed more recently, and which, because of the complexity of the method, requires somewhat more explanation than the prior two approaches. It uses the relationships embodied in models to simultaneously derive the spatial array of input values that produces the best match to the observed data. The impetus to this work is that specific values for some critical spatial model inputs, especially soil properties and rooting depth, are not available at the desired spatial resolution within a field to adequately predict yield variability. Often, these properties are available only at the soil type scale, estimated years ago using techniques that provide typical ranges of values within the soil type. Using values estimated or measured at this larger resolution introduces unacceptable error for precision farming applications.

To refine the spatial estimates of the selected critical inputs, this approach uses the model with a range of the chosen critical inputs to predict yield or some other factor of interest, such as temporal soil water content, and minimizes the error between the set of predicted and measured values. The idea is that, if these critical parameters are estimated correctly, the model should perform well across seasons (temporally). Typically, this approach is applied to small homogeneous areas within a field, and the analysis is conducted independently for each area.

This method, objective parameterization, has been approached in two ways. Both require an objective function be defined, usually to minimize error between simulated and measured yield. One method links a classical optimization algorithm, such as Simulated Annealing (Goffe et al., 1994) or the AMOEBA method (Nelder and Mead, 1965; Press et al., 1992), to the model. Then, the optimizer runs the model multiple times while incrementally varying the chosen critical inputs within a reasonable range, and searches for the values of the input parameters that satisfy this objective function. The second approach constructs a database by running the model with the selected spatial inputs varied in a linear fashion over the expected range of variation and searching the database for combinations of inputs that minimize error according to the objective function.

The result from both approaches is a field of spatial inputs calibrated, or fine-tuned, to improve model performance. The key to success for both is to correctly identify a limited number of key spatial inputs that are uncertain, and calibrate those inputs within a realistic range. All other important inputs must be known with reasonable certainty.

The first example of objective parameterization is outlined in Paz et al.(2001). The goal of this work was to use the CROPGRO-Soybean model (Hoogenboom et al., 1994; Boote et al., 1998b) to determine causes of spatial soybean yield variability and to estimate the impact of different yield-limiting factors on yield variability for a field in Central Iowa. In this example, they identified water stress, soybean cyst nematodes (SCN), and weeds as the major yield-limiting factors. They built on previous work with modifications of the model to account for SCN damage (Fallick et al., 2001), incorporated the effects of tile drainage and nutrient movement (Shen et al., 1998), and then added the effects of weed damage using a separate model. They divided the 50-ha field into 77 grids and developed the appropriate model inputs for each grid for three seasons. Next, they linked the simulated annealing algorithm (Goffe et al., 1994) to the model for parameter estimation. Finally, they solved for the values of tile spacing, saturated hydraulic conductivity of the impermeable layer, and root depth distribution using the simulated annealing process. They were able to explain approximately 80% of the spatial yield variability over the 3-year period (Figure 12.3) caused by water stress, weeds, and SCN.

Once calibration was completed, the model could be used to study the relative effects of the different yield-limiting factors. They used the calibrated parameters in the model to calculate the yield loss caused by SCN, weeds, and water stress for one year. Figure 12.4 shows the predicted yield potential for each grid when all stresses were turned off. Each data point represents the

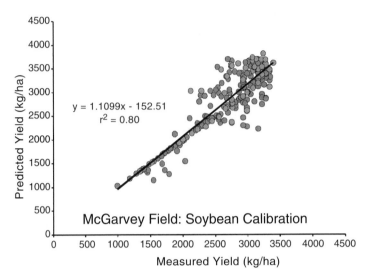

Figure 12.3   Predicted vs. measured yield calibrated for 3 seasons for the McGarvey field near Perry, Iowa. (Paz et al. 2001. A modeling approach to quantify the effeds of spatial soybean yield-limiting factors. *Trans. of the ASAE* 44(5):1329–1334. With permission.)

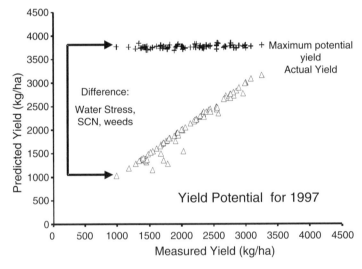

Figure 12.4   Example for effect attributed to water stress, weeds, and soybean cyst nematode, using 1997 yield data. (Paz et al. 2001. A modeling approach to quantify the effeds of spatial soybean yield-limiting factors. *Trans. of the ASAE* 44(5):1329–1334. With permission.)

predicted yield potential for a grid. The yield potential differs in each grid because of differences in measured soybean plant population for this season. A sequence of model runs was made by turning off each stress individually to predict the yield reduction due to water stress, SCN, and weeds in each individual grid. Table 12.2 shows a summary of the results when averaged over all grids. Water stress caused approximately 709 kg ha^{-1} in yield loss averaged over all grids. Some grids had large yield reductions due to water stress, while other grids experienced small yield reductions. Similarly, SCN and weeds caused average yield reductions of 119 and 20 kg ha^{-1}, respectively. The interaction among the three yield-limiting factors caused an additional 93 kg ha^{-1} of yield loss over the field.

Table 12.2   Predicted Yield Loss Due to Water Stress, Soybean Cyst Nematodes and Weeds for the Mcgarvey Field in 1997

Stress	Yield Loss (kg ha⁻¹)
Water stress	709
SCN	119
Weeds	20
Interactions	93
All stresses	941

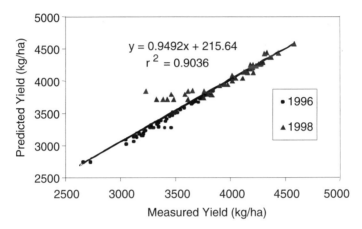

$$y = 0.9492x + 215.64$$
$$r^2 = 0.9036$$

Figure 12.5 Results of case 4 scenario. (Irmak et al., 2001, Estimating spatially variable soil properties for application of crop models in precision agriculture, *Trans. of the ASAE* 44(5):1343–1353. With permission.)

The second example used the database search method outlined by Irmak et al. (2001). The goal of this work was to use the CROPGRO-Soybean model to determine causes of yield variability in a 30-ha field in eastern Iowa. Similar to the previous method, they divided the field into 48 grids, and developed the appropriate crop model input files for each grid for a 2-year period. Upon analyzing the data, they concluded that water stress and soil fertility were the likely causes of yield variability. Thus, their focus was to calibrate several uncertain model inputs dealing with the soil water balance and soil fertility for each grid. They used the same modified version of the CROP-GRO-Soybean model used by Paz et al.(2001), adapted for tile drainage conditions found in the Midwest. They ran the model for all possible combinations of saturated hydraulic conductivity of the impermeable layer, a soil productivity factor that reduces yield based on soil fertility and other unknown factors (Paz et al., 2001), maximum rooting depth, tile spacing, and SCS runoff curve number, all within accepted ranges of values for each parameter. They conducted this analysis for each of the 48 grids and two seasons of data, totaling nearly 75,000 model runs. Each combination of model parameters and corresponding predicted yield was entered into a large database. An objective function was developed and the database was searched to determine the combination of parameters that minimized the error between predicted and measured yields in each grid over the 2-year period. The database search procedure took about half the time required by the simulated annealing approach.

Figure 12.5 shows the predicted and measured yield for the five-parameter calibration for 1996 and 1998. Each data point represents yield in a single grid. Using this approach, they were able to explain more than 90% of the yield variability within the field. Figure 12.6 shows the spatial pattern

Spatial Results - Soil Characteristics

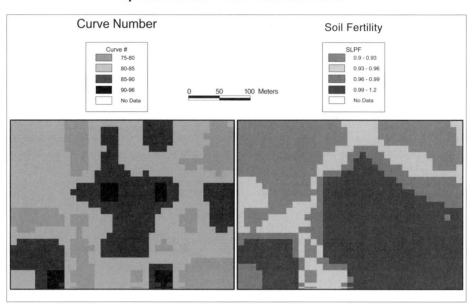

Figure 12.6   Example of spatial inputs from the database search routine. (Irmak et al., 2001, Estimating spatially variable soil properties for application of crop models in precision agriculture, *Trans. of the ASAE* 44(5):1343–1353.

for the SCS curve number and soil fertility factor produced by this technique. The model estimates of these five parameters consisted of a realistic spatial structure, further adding credibility to this approach.

CONCLUSIONS

Several conclusions can be drawn from the foregoing. First, the three approaches, both individually and collectively, contributed significantly to the body of knowledge about applying models to spatial applications. Conventional methods have helped define both strengths and critical gaps in basic knowledge to be incorporated into models. Remote sensing methods have employed high-resolution data to refine model estimates and in some cases, to reset the model during a season's run. Objective parameterization has shown how multiple years of data can be analyzed to provide the best set of spatial inputs for a field, and also to calibrate the model for those conditions. In all cases, the success of the models illustrated the potential for either further use or refinement of the models.

The choices of critical inputs to be measured, evaluated, or solved for in these approaches collectively illustrate the opinions and the conclusions reached by the researchers involved. These critical variables included plant population, LAI, fertility, rooting depth, and several soil physical properties, particularly surface runoff characteristics, water holding capacity, and hydraulic conductivity. One case, the simulated annealing example, also included tile spacing, weeds, and soybean cyst nematode infestation. These latter studies indicate one modeling need — pests are known to affect yield, and spatial variation in pests would then need to be accounted for. Nonetheless, for all examples, the importance of soil water and crop water stress is evident.

Extending that thought suggests future directions for improvement in spatial modeling might profitably concentrate on implementing 3-D modeling of water flow, particularly runoff and lateral subsurface flow. Such transfers of water in the horizontal directions are well known to occur under common weather, soil, and terrain conditions. As such, they are difficult to handle with 1-D models, even used at multiple points in space as described herein. Judging from the importance attributed to water-related soil parameters in the several modeling studies described here, the additional effort to implement and the additional computer resources to run 3-dimensional models may be justified.

REFERENCES

Ahuja, L.R. ., K.W. Rojas, J.D. Hanson, M.J. Shaffer, and L. Ma, Eds., 2000. *The Root Zone Water Quality Model,* Water Resources Publications LLC, Highlands Ranch, CO.

Asner, G.P. and C.A. Wessman. 1997. Scaling PAR absorption from the leaf to landscape level in spatially heterogeneous ecosystems, *Ecological Modelling,* 103(1):81–97.

Barnes, E.M. et al. 1997. Modification of CERES-Wheat to accept leaf area index as an input variable, Paper No. 97-3016, Am. Soc. of Agricultural Engineers, St. Joseph, MI.

Barnes, E.M., P.J. Pinter, B.A. Kimball, G.W. Wall, R.L. LaMorte, D.J. Hunsaker, F. Adamsen, S. Leavitt, T. Thompson, and J. Mathius. 2000. Precision irrigation management using modeling and remote sensing approaches. In *Proc. 4th Decennial Natl. Irrigation Symp.,* R.G. Evans, B.L. Benham, and T. Trooien, Eds., November 14–16, Phoenix, AZ, St. Joseph, MI, 332–337.

Boote, K.J., J.W. Jones, and G. Hoogenboom. 1998a. Simulation of crop growth: CROPGRO. Chapter 18. In *Agricultural Systems Modeling and Simulation,* R.M. Peart and R.B. Curry, Eds., Marcel Dekker, New York, 651–692.

Boote, K.J., J.W. Jones, G. Hoogenboom, and N.B. Pickering. 1998b. The CROPGRO for grain legumes. In *Understanding Options for Agricultural Production,* G.Y. Tsuji, G. Hoogenboom, and P.K. Thornton, Eds., Kluwer Academic Publishers, Boston, 99–128.

Engel, T., J.W. Jones, and G. Hoogenboom. 1995. AEGIS/WIN — a Windows interface combining GIS and crop simulation models, Paper No. 95-3244, ASAE, St. Joseph, MI.

Fallick, J.B., W.D. Batchelor, G. Tylka and T. Niblack. 2001. Coupling soybean cyst nematode damage to CROPGRO-Soybean, *Trans. ASAE,* (accepted).

Goffe, W.L., G.D. Ferrier, and J. Rogers. 1994. Global optimization of statistical functions with simulated annealing, *J. Econometrics,* 60:65–99.

Guérif, M. and C.L. Duke. 2000. Adjustment procedures of a crop model to the site specific characteristics of soil and crop using remote sensing data assimilation, *Agric., Ecosystems Environ.,* 81(1):57–69.

Han, S., R.G. Evans, T. Hodges, and S.L. Rawlins. 1995. Linking a geographic information system with a potato simulation model for site-specific crop management, *J. Environ. Qual.,* 24:772–777.

Hoogenboom, G., J.W. Jones, P.W. Wilkens, W.D. Batchelor, W.T. Bowen, L.A. Hunt, N.B. Pickering, U. Singh, D.C. Godwin, B. Baer, K.J. Boote, J.T. Ritchie, and J.W. White. 1994. Crop models. In *DSSAT V3,* vol. 2-2, G.Y. Tsuji, G. Uehara, and S. Balas, Eds., University of Hawaii, Honolulu, HI, 95–244.

Irmak, A., J.W. Jones, W.D. Batchelor and J.O. Paz. 2001. Estimating spatially variable soil properties for application of crop models in precision agriculture, *Trans. ASAE,* 44(5):1343–1353.

Jones, D.D. and E.M. Barnes. 2000. Fuzzy composite programming to combine remote sensing and crop models for decision support in precision crop management, *Agric. Syst.,* 65:137–158.

Jones, C.A. and J.R. Kiniry. 1986. *CERES-Maize: A Simulation Model of Maize Growth and Development,* Texas A&M University Press, College Station, TX.

Kaspar, T.C., T.S. Colvin, D.B. Jaynes, D.L. Karlen, D.E. James, and D.W. Meek. 2001. Estimating corn yield using six years of field data and terrain attributes, *J. Precision Agric.* (submitted).

Kimball, B.A., R.L. LaMorte, P.J. Pinter Jr., G.W. Wall, D.J. Hunsaker, F.J. Adamsen, S.W. Leavitt, T.L. Thompson, A.D. Matthias and T.J Brooks. 1999. Free-air CO_2 enrichment (FACE) and soil nitrogen effects on energy balance and evapotranspiration of wheat, *Water Resour. Res.,* 35(4):1179–1190.

Maas, S.J. 1988. Using satellite data to improve model estimates of crop yield, *Agron. J.,* 80(4):655–662.

Maas, S.J. 1993. Parameterized model of Gramineous crop growth: II. Within-season simulation calibration, *Agron. J.,* 85:354–358.

Monsi, M. and T. Saeki. 1953. Uber den Litchfaktor in den Pflanzengesellschaften und seine Bedeutung für die Stoffproduktion, *Japanese J. Bot.,* 1953:1422–1452.

Moore, I.D., P.E. Gessler, G.A. Nielsen, and G.A. Peterson. 1993. Soil attribute prediction using terrain analysis, *Soil Sci. Soc. Am. J.,* 57:443–452.

Moran, M.S., S.J. Maas, and P.J. Pinter, Jr. 1995. Combining remote sensing and modeling for estimating surface evaporation and biomass production, *Remote Sensing Rev.,* 12:335–353.

Moulin, S., A. Dondeau, and R. Delecolle. 1998. Combining agricultural crop models and satellite observations from field to regional scales, *Int. J. Remote Sensing,* 19(6):1021–1036.

Nelder, J.A. and R. Mead. 1965. A simplex method for function minimization, *Computer J.,* 7:308–313.

Ortlip, E.W. 1986. Method and apparatus for spreading fertilizer, U.S. Patent No. 4,630,773, U.S. Patent Office, Washington, D.C.

Papajorgji, P., J.W. Jones, J-P. Calixte, F.H. Beinroth, and G. Hoogenboom. 1993. A generic geographic decision support system for estimating crop performance. In *Integrated Resource Management and Landscape Modification for Environmental Protection. Proc. Intl. Symp.,* Chicago, IL, December 13–14, Am. Soc. of Agricultural Engineers, St. Joseph, MI.

Parton, W.J., D.S. Ojima, D.S. Schimel, and T.G.F. Kittel. 1992. Development of simplified ecosystem models for applications in Earth system studies: the CENTURY experience. In *Earth System Modeling, Proc. 1990 Global Change Institute on Earth System Modeling,* D.S. Ojima, Ed., Snowmass, CO, 281–302.

Paz, J.O., W.D. Batchelor, G.L. Tylka and R.G. Hartzler. 2001. A modeling approach to quantifying the effects of spatial soybean yield limiting factors, *Trans. ASAE,* 44(5):1329–1334.

Press, W.H., S.A. Teukolsky, and W.T. Vetterling. 1992. *Numerical Recipes in FORTRAN: The Art of Scientific Computing,* 2nd ed., Cambridge University Press, Cambridge, England.

Robert, P.C. 1996. Appendix I — summary of workshops: precision agriculture research and development needs. In *Precision Agric., Proc. 3rd Int. Conf.,* P.C. Robert, R.H. Rust, and W.E. Larson, Eds., Minneapolis, MN, June 23–26, American Society of Agronomy, Crop Science Society of america, Soil Science Society of America, Madison, WI, 1193–1194.

Sadler, E.J. and G. Russell. 1997. Modeling crop yield for site-specific management, In *The State of Site-Specific Management for Agriculture,* F.J. Pierce and E.J. Sadler, Eds., American Society of Agronomy, Crop Science Society of america, Soil Science Society of America, Madison, WI, 69–79.

Sadler, E.J., W.J. Busscher, K.C. Stone, P.J. Bauer, D E. Evans and J.A. Millen 1998. Site-specific modeling of corn yield in the SE Coastal Plain. In *Proc. 1st Int. Conf. Geospatial Inf. in Agric. and For.,* June 1–3, Lake Buena Vista, FL, I214–I221.

Sadler, E.J., B. K. Gerwig, D.E. Evans, J.A. Millen, P.J. Bauer, and W.J. Busscher. 1999. Site-specificity of CERES-Maize parameters: a case study in the southeastern U.S. Coastal Plain. In *Precision Agric., '99, 2nd Eur. Conf. on Precision Agric.,* J.V. Stafford, Ed., July 11–15, Odense, Denmark, 551–560.

Sadler, E.J., P.J. Bauer, W.J. Busscher, and J.A. Millen. 2000a. Site-specific analysis of a droughted corn crop: II. Water use and stress, *Agron. J.,* 92(3):403–410.

Sadler, E.J., B.K. Gerwig, D.E. Evans, W.J. Busscher and P.J. Bauer. 2000b. Site-specific modeling of corn yield in the southeast Coastal Plain, *Agric. Syst.,* 64(3):189–207.

Schueller, J. 1993. Working group report, section III — engineering technology. In *Soil Specific Crop Management,* P.C. Robert, R.H. Rust, and W.E. Larson, Eds., American Society of Agronomy, Crop Science Society of america, Soil Science Society of America, Madison, WI, 181–196.

Shen, J., W.D. Batchelor, R. Kanwar, J.T. Ritchie and J.W. Jones. 1998. Validation of the water balance model in CROPGRO-Soybean, *Trans. ASAE,* 41(5):1305–1313.

Simmons, F.W., D.K. Cassel, and R.B. Daniels. 1989. Landscape and soil property effects on corn grain yield response to tillage, *Soil Sci. Soc. Am. J.,* 53:534–539.

Sudduth, K.A., J.W. Hummel and S.J. Birrell. 1997. Sensors for site-specific management. In *The State of Site-Specific Management for Agriculture,* F.J. Pierce and E.J. Sadler, Eds., American Society of Agronomy, Crop Science Society of america, Soil Science Society of America, Madison, WI, 183–210.

USDA-SCS. 1972. *SCS National Engineering Handbook, Section 4: Hydrology,* U.S. Department of Agriculture, Washington, D.C. 548 pp.

Vansichen, R. and J. De Baerdemaeker. 1991. Continuous wheat yield measurements on a combine. In *Automated Agriculture for the 21st Century,* Publication No. 1191, Am. Soc. of Agricultural Engineers, St. Joseph, MI, 346–355.

Weigand, C.L., A.J. Richardson, and E.T. Kanemasu. 1979. Leaf area index estimates for wheat from LANDSAT and their implications for evapotranspiration and crop modeling, *Agron. J.,* 71:336–342.

Weiss, M., D. Troufleau, F. Baret, H. Chauki, L. Prevot, A. Olioso, N. Bruguier and N. Brisson. 2001. Coupling canopy functioning and radiative transfer models for remote sensing data assimilation, *Agricultural For. Meteorol.,* 108(2):113–128.

CHAPTER 13

Topographic Analysis, Scaling, and Models to Evaluate Spatial/Temporal Variability of Landscape Processes and Management

Lajpat R. Ahuja, Timothy R. Green, Robert H. Erskine, Liwang Ma, James C. Ascough II, Gale H. Dunn, Marvin J. Shaffer, and Ana Martinez

CONTENTS

INTRODUCTION

Topography is an important factor to consider in land management systems. Through its effects on long-term soil formation and short-term seasonal effects, the topography greatly influences the spatial variability of soil properties, soil processes, and their interaction with variable weather conditions. With the availability of high precision global positioning systems, a landscape's topography can be determined to within a few centimeters. This chapter explores the hypothesis that accurate topographic analysis, available soil map data, and process-level agricultural ecosystem models can be combined to improve spatial characterization of soils and soil processes on a landscape, and this characterization can then be used to evaluate and guide spatial and temporal land management. The authors also review the literature on the relationship of certain topographic attributes (elevation, slope, aspect, contributing or catchment area, and their combinations) with the variability of soil properties (e.g., texture, depth, organic matter content), soil processes (soil water and related variables), and crop growth. Contributing area, slope, and curvature may be used to delineate subunits of land for differential management, and then appropriate models can be used to evaluate their responses to alternate management practices or land uses under varying weather conditions during a season and from year to year.

The soil properties and their hydrologic responses to weather conditions, such as infiltration, runoff, and soil water storage, vary spatially and temporally over a landscape, even at a field scale. As a result, crop production and environmental impacts, such as soil erosion, nutrients and pesticides in runoff and groundwaters, vary spatially and temporally even under uniform management. The environmental consciousness, global competition, and advances in global positioning systems (GPS) and computer technologies have generated an interest in precision agriculture worldwide. Precision agriculture aims to vary management (e.g., fertilizer, seeding rates, and irrigation) spatially within a field and between fields with varying soil water and other conditions, so as to optimize production while enhancing water quality. For this purpose, researchers need to develop a sound scientific basis to delineate spatial subunits based on topographic and soil variabilities that determine their hydrologic status and production potential in response to variable weather conditions. In this chapter, a hypothesis is posed that a combination of topographic analysis, scaling of soil properties, and models will provide such a scientific basis to evaluate spatial and temporal variability of landscape processes and spatial management. This framework will also accomplish important related objectives — to transfer improved management results from plots to large variable fields and farms by aggregating or scaling up through distributed modeling. This transfer will also be essential for managing off-site effects, including the TMDL (total maximum daily load) of pollutants in surface waters.

Many attempts have been made in the last three decades to characterize the spatial variability of soil properties and crop yield using geostatistics (e.g., Webster, 1985; Warrick et al., 1986; Jaynes and Colvin, 1997). These attempts have enabled the determination of statistical patterns of variabilities and unbiased interpolations of values rather than linking of these patterns to causative factors and processes. The use of topographic attributes may help address these factors and processes.

This chapter is divided into three sections:

1. Topography in relation to soil variability
2. Topography in relation to crop yield variability
3. A framework for scaling and modeling landscape and climate variability to evaluate and enhance site-specific agricultural management

TOPOGRAPHY IN RELATION TO SOIL VARIABILITY

Jenny (1941) presented a system of pedology by which the soil formation is a function of five factors:

$$\text{Soil Formation} = f\left(\text{climate, parent material, topography, biota, time}\right) \qquad (13.1)$$

Therefore, on field or farm scales and assuming generally uniform climate, parent material and biota, geomorphic stability, and minimal and uniform human disturbance:

$$\text{Soil Formation} \sim f\left(\text{topography}\right) \qquad (13.2)$$

Milne (1936) explained this concept using a soil catena on a hillslope of uniform parent material. The topography determines the flow paths of water over the surface that causes a soil texture variation over the landscape, and determines the amount of water available for infiltration at different points that cause variability in soil profile development.

The topography is generally quantified in terms of a number of attributes. The most commonly used topographic or terrain attributes are: elevation, slope, aspect, curvature (plan and profile), specific catchment area, and wetness index (ln[specific catchment area/slope] following Beven,

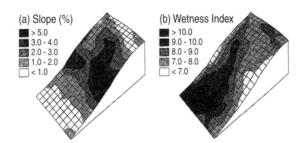

Figure 13.1 Topographic slope and wetness index computed from 15.24-m grid digital elevation data for the Sterling, CO, site. (From Moore et al., 1993. With permission.)

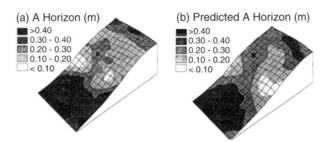

Figure 13.2 Measured A-Horizon depths and A-Horizon depths predicted from the regression relations with topographic attributes at Sterling, CO, site. (From Moore et al., 1993. With permission.)

1986). For describing spatial variability of soils, soil processes, and plant growth on an agricultural landscape (probably at scales greater than 5 or 10 m), topographic attributes provide a continuous and quantitative approach. These attributes can now be economically and accurately measured with GPS technology. The topographic analysis should be combined with available soil survey data and measured on-site soils data. It can also be combined with kriging, fractal, or other types of geostatistical spatial analysis (Odeh et al., 1994). Some typical examples from the literature, presented below, provide the current state-of-science on relating topographic attributes to soil variability at field and farm scales.

Moore et al. (1993) presented a study on soil attribute prediction using topographic analysis on one of our cooperative research sites near Sterling, CO. This 5.4-ha area is a sloping landscape with a catena of three soil types along the slope as mapped by soil survey. The topographic attributes were computed from 15.24-m grid-based elevation data. The distributions of slope and wetness index on this site are shown in Figure 13.1. The figure shows how the slope convergence in the middle part of the terrain results in a tongue of high values of the wetness index. Step-wise multiple regression relations of topsoil properties with topographic attributes showed that the A-horizon depth was significantly ($P < 0.01$) related to the slope and wetness index; organic matter to wetness index, stream power index, and aspect; and sand and silt contents to slope, wetness index, and profile curvature. The coefficients of determination (R^2) varied from 0.48 to 0.64. Measured and predicted A-horizon depth distributions are presented in Figure 13.2. The A-horizon depth generally increases down slope. Based on the literature, the authors felt that it may be possible to improve the predictions if a 5-m grid digital elevation data were used instead of 15.24-m grid; however, the R^2 will probably always be <0.70 (Walker et al., 1968).

Moore et al. (1993) also presented a method of enhancing conventional soil survey maps using topographic attributes in a simple way, without the availability of the regression relations. As a first approximation, soil properties are assumed linearly related to selected terrain attributes, and the maximum range of variation of the soil properties are assumed known. The terrain attribute distributions on the landscape are then used to scale the known range of soil properties to obtain spatially distributed estimates of soil properties within a field or map unit.

Odeh et al. (1994) also reported highly significant multilinear regression relations between several soil variables and certain topographic attributes, with R^2 values that range from 0.45 to 0.75. They also showed that there was a clear advantage for combining the geostatistical kriging technique with the regression method for predicting soil attributes that had a small correlation with topographic attributes. The kriging methods proposed by Delhomme (1978, 1979) and Ahmed and DeMarsily (1987) — kriging with uncertainty due to regression (Model A) and kriging with a guess field (Model B) — were used for this purpose. A comparison of prediction root mean square error (RMSE) from a multilinear regression and several methods of kriging are presented in Table 13.1. In the regression-kriging Model A, the regressed value of the soil property was kriged. In Model B, ordinary kriging was done for both this regressed value and the regression error at points where both are measured. The sum of these two kriged values gave the final estimate. The results in Table 13.2 show that except for topsoil gravel, the soil properties were predicted best by Model B, followed by Model A. In a subsequent paper, Odeh et al. (1995) showed another model, Model C, was even better overall than Model B. In Model C, regression relations were used to predict soil attributes at a finer grid at which the landscape attributes were determined (this grid being finer than the grid at which soil attributes were measured for use in regression). Then, an ordinary kriging of regression errors was conducted from available data to predict errors at the finer grid. The sum of the predicted soil property and regression error at this finer grid gave the final estimate.

Zhu et al. (1997) presented a fuzzy inference scheme for estimating spatial distributions of soil properties in a landscape. This scheme was based on the Soil–Land Inference Model (SOLIM) given earlier (Zhu et al., 1996). The word "fuzzy" implies that this scheme is based on expert knowledge of the factors of soil formation in a landscape. It is assumed that in a given landscape, there are a number of soil series present, where each series is an ideal concept, and for each ideal concept there is an associated set of environmental factors which are clearly defined by an expert. At any given point on the landscape (i,j) the soil may not be the same as any of the ideal series, but may bear some resemblance to each of these series. These partial resemblances are expressed as a soil similarity vector at point (i,j, which represents a point). The values of this vector are determined by comparing the similarity of environmental factors of soil formation (mainly topographic attributes) at point (i,j) with these factors associated with each of the ideal series concepts. The minimum value of similarity among these factors gives the similarity vector value for that series. Zhu et al. (1997) showed that A-horizon depths and transmissivities inferred from this model on two different catchments were closer to measured values than the mapped values based on a soil survey. Lark (1999) has presented a conceptually similar approach, called the continuum classification or the fuzzy c-mean classification, for estimating soil–landscape relationships.

Several other investigators have tried to relate soil water storage directly with topographic attributes, with generally good results (Tomer and Anderson, 1995; Zheng et al., 1996; Western et al., 1999). Zaslavsky and Sinai (1981) measured soil water content and elevation on a grid over a hillslope area (70×70 m) at two depths (20 or 40 cm) and found that the topographic curvature explained much of the variability ($R^2 = 0.81$). Tomer and Anderson (1995) reported high R^2 values (0.52 to 0.77) between soil water storage and topographic attributes in a sand plain hillslope. Zheng et al. (1996) reported that the R^2 value increases if the data were aggregated over larger areas. Western et al. (1999) showed that, under dry conditions, much of the variation in soil moisture in the surface 30-cm layer was random. Spatial patterns of soil moisture were apparent under wet conditions, with the wettest areas in areas of high local convergence of flow paths. The log-transformed specific contributing area (ln a) showed the highest correlation with soil moisture under wet conditions, whereas a combination of ln a and a potential radiation index was best over all seasons.

Table 13.1   Prediction Root Mean Square Error for Predicting Certain Soil Properties from a Multilinear Regression with Topographic Attributes and from a Variety of Kriging Methods

Soil Variable	Prediction Method							
	Multi-linear Regres-sion	Ordinary Kriging	Univer-sal Kriging	Isotopic Co-Kriging	Hetero-topic Co-Kriging	Regres-sion-Kriging Model A	Regres-sion-Kriging Model B	Regres-sion-Kriging Model C
Depth of solum (cm)	11.92	15.76	13.89	15.70	21.74	8.45	11.20	10.01
Depth to bedrock (cm)	21.04	26.71	26.43	24.86	22.45	20.22	19.89	16.51
Topsoil gravel (%)	4.97	12.82	10.31	8.98	3.72	9.65	4.54	5.01
Subsoil clay (%)	10.22	15.20	14.63	10.24	5.89	9.11	9.26	8.04

Source: From Odeh et al., 1995. Models A, B and C use combinations of regression and Kriging.

Based on this review, an appropriate combination of topographic indices may explain between 50 to 65% of the variability in soil attributes at scales greater than 10 m (or perhaps 5 m). Smaller scale variability, caused by microtopography, plants, management, and measurement errors or other factors, is not as predictable. Large human-induced changes will also be difficult to predict.

TOPOGRAPHY IN RELATION TO CROP YIELD

A number of investigators have related crop yield variability within a field to topographic attributes. For example, Simmons et al. (1989) found that corn yield of three different tillage treatments was correlated most strongly with in-row curvature. The second significant attribute was the slope in two cases and specific contributing area in the third case. With two attributes, the R^2 ranged from 0.57 to 0.68. Kaspar et al. (2000) found that the correlation of average corn yield for 6 years with topographic attributes was poor; however, the correlation was high when dry years were separated from wet years (which caused water logging), with $R^2 = 0.66$ with a pothole area included and $R^2 = 0.78$ without the pothole area. The significant attributes were elevation, slope, plan curvature, and profile curvature. Specific catchment area and wetness index were not included in their analyses.

Our own results for a field in Sterling, CO, are shown in Figure 13.3. Winter wheat yield variability seems to be significantly related to soil wetness index based on visual similarities in thespatial patterns, but univariate point-to-point linear regressions leave much uncertainty ($R^2 = 0.37$). Detailed multivariable nonlinear regression analysis is still in progress. Differences in measurement scale and accuracy of measurement influence such results.

A FRAMEWORK FOR SCALING AND MODELING LANDSCAPE AND CLIMATE VARIABILITY TO EVALUATE AND ENHANCE SITE-SPECIFIC MANAGEMENT

A computer simulation framework is needed to integrate the type of spatial prediction based on landscape topography, outlined above, with spatial scaling and distributed modeling of the variables of interest. A program to develop such a framework is being initiated by the authors. The framework would have broad applicability, but is initially targeted at agricultural management. Integration of geospatial data and computer technology is an important part of the framework development. Gaps also exist in scientific knowledge of property and process scaling, parameter estimation and field-scale distributed modeling that need to be addressed in this context.

Four major steps in the scaling and modeling are:

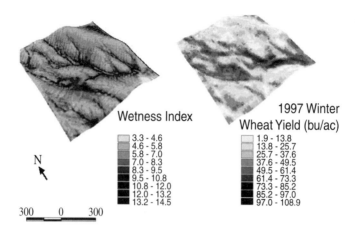

Wetness Index

1997 Winter
Wheat Yield (bu/ac)

N

3.3 - 4.6	1.9 - 13.8
4.6 - 5.8	13.8 - 25.7
5.8 - 7.0	25.7 - 37.6
7.0 - 8.3	37.6 - 49.5
8.3 - 9.5	49.5 - 61.4
9.5 - 10.8	61.4 - 73.3
10.8 - 12.0	73.3 - 85.2
12.0 - 13.2	85.2 - 97.0
13.2 - 14.5	97.0 - 108.9

300   0   300

Figure 13.3 Topographic patterns of the (a) computed wetness index and (b) measured winter wheat yield for 1997 in bushels per acre, at Lindstrom Farm near Sterling, CO. Topographic attributes were computed from a 10-m grid digital elevation data obtained using a survey-grade global positioning system.

Step 1: Subdivide a landscape into modeling units based on topographic attributes and available soil information. A detailed elevation survey will be conducted for the landscape at 5m grid using a sensitive, survey-grade GPS unit. These data will be used to compute selected topographic attributes, such as the slope, specific contributing area, and profile curvature. Using maps of these attributes, the landscape will subdivided into several units. Within each unit, the soil profile will be sampled on a coarse grid for soil horizons, their textural class, and soil moisture by quick field methods. For soil moisture, it may be best to do the sampling 2 to 3 days after a rainfall.

Step 2: Determine the model-input soil properties within each unit using simple scaling approaches. The most important soil properties are the soil bulk density, texture, organic matter content, and hydraulic properties of each horizon. A first estimate of these properties will be obtained from the soil textural class information and soil survey data. The first estimates may be refined by measurements made on a few places over the landscape. For soil hydraulic properties, measurements of bulk density and field capacity (100-kPa, 333-kPa, or field value) greatly improve the estimation of water retention curves (Ahuja and Williams, 1991; Williams and Ahuja, 1992) and saturated hydraulic conductivity (Ahuja et al. 1984, 1989, 1993).

Step 3: Model soil–water–nutrient–plant processes accounting for surface and subsurface water flow processes. In each spatial unit, a model such as the Root Zone Water Quality Model (RZWQM) (Ahuja et al., 2000), enhanced by linking the best available crop growth modules to it, will be used to model soil–water–nutrient–plant growth processes under a given set of management practices and climate data. Overland flow can be routed from one land area (simulation unit) to another using a simple cascade approach.

Step 4: Evaluate and identify the best management practices. For each unit, a variety of management alternatives will be evaluated using Step 3 for a number of climate years. A set of best management practices will be identified for average climatic conditions and for 25% and 75% probability levels of rainfall.

SUMMARY

Based on the literature and on-going research, landscape topography can explain much of the spatial variability in soil properties and crop yield, if topographic attributes are analyzed and combined appropriately. The authors believe the proposed combination of topographic analysis, scaling, and models can be used to evaluate the spatial/temporal variability of landscape processes and help identify an appropriate spatial management scheme. This scheme will recognize the variabilities and achieve management goals of crop production and environmental quality. This combination framework should help scale the management effects from small plots to field and farm scales, as well as improve spatial predictions at different management scales.

REFERENCES

Ahmed, S. and G. DeMarsily. 1987. Comparison of geostatistical methods for estimating transmissivity using data on transmissivity and specific capacity. *Water Resour. Res.*, 23:1717–1737.

Ahuja, L.R. and R.D. Williams. 1991. Scaling of water characteristic and hydraulic conductivity based on Gregson–Hector–McGowan approach, *Soil Soc. Am. J.,* 55:308–319.

Ahuja, L.R., D.K. Cassel, R.R. Bruce, and B.B. Barnes. 1989. Evaluation of spatial distribution of hydraulic conductivity using effective porosity data, *Soil Sci.,* 148:404–411.

Ahuja, L.R., J.W. Naney, R.F. Green, and D.R. Nielsen. 1984. Macroporosity to characterize spatial variability of hydraulic conductivity and effects of land management, *Soil Sci. Soc. Am. J.,* 48:699–702.

Ahuja, L.R., O. Wendrath, and D.R. Nielsen. 1993. Relationship between initial surface drainage and soil-profile average saturated hydraulic conductivity, *Soil Sci. Soc. Am. J.,* 57:19–23.

Ahuja, L.R., K.W. Rojas, J.D. Hanson, J.J. Shaffer and L. Ma, Eds. 2000 *Root Zone Water Quality Model*, Water Resources Publication, Englewood, CO. 1–360.

Beven, K. 1986. Toward a new paradigm in hydrology. In *Water for the Future: Hydrology in Perspective.* IAHS Pub. No. 164, 393–403.

Delhomme, J.P. 1978. Kriging in the hydrosciences. *Adv. in Water Res.* 1:251–266.

Delhomme, J.P. 1979. Spatial variability and uncertainty in groundwater flow parameters; a geostatistical approach, *Water Resour. Res.* 15:269–280.

Jenny, H. 1941. *Factors of Soil Formation: A System of Quantitative Pedology,* McGraw-Hill, New York.

Jaynes, D.B, and T.S. Colvin. 1997. Spatiotemporal variability of corn and soybean yield, *Agronomy J.,* 89(1):30–37.

Kaspar, T.C., T.S. Colvin, D.B. Jaynes, D.L. Karlen, D.E. James, and D.W. Meek. 2000. Estimating corn yield using 6 years of field data and terrain attributes, (submitted).

Lark, R.M. 1999. Soil-landform relationships at within-field scales: an investigation using continuous classification, *Geoderma,* 92:141–165.

Milne, G. 1936. Normal erosion as a factor in soil profilce develoopment. *Nature (London),* 138:548–549.

Moore, I.D., P.E. Gessler, G.A. Nielsen, and G.A. Peterson. 1993. Soil attribute prediction using terrain analysis, *Soil Sci. Soc. Am. J.,* 57:443–452.

Odeh, I.O.A., A.B. McBratney, and D.J. Chittleborough. 1994. Spatial prediction of soil properties from landform attributes derived from a digital elevation model, *Geoderma,* 63:197–214.

Odeh, I.O.A., A.B. McBrateney, and D.J. Chittleborough. 1995. Further results on predcitio of soil properties from terrain attributes: heterotopic cokriging and regression-kriging, *Geoderma,* 67:215–226.

Simmons, F.W., D.K. Cassel, and R.B. Daniels. 1989. Landscape and soil property effects on corn grain yield response to tillage, *Soil Sci. Soc. Am. J.,* 53:534–539.

Tomer, M.D., and J.L. Anderson. 1995. Variation of soil water storage across a sand plain hillslope, *Soil Sci. Soc. Am. J.,* 59:1091–1100.

Walker, P.H., G.F. Hall, and R. Protz. 1968. Relation between landform parameters and soil properties, *Soil Sci. Soc. Am. Proc.,* 32:101–104.

Warrick, A.W., D.E. Myers, and D.R. Nielsen. 1986. Geostatistical methods applied to soil science. In *Methods of Soil Analysis,* part 1. 2nd Ed., A. Klute, Ed., Agronomy 9:53–82.

Webster, R. 1985. Quantitative Spatial Analysis of soil in the field, *Adv. Soil Sci.,* vol 3, Springer-Verlag, New York.

Western, A.W., R.B. Grayson, G. Bloschl, G.R. Willgoose, and T.A. McMahon. 1999. Observed spatial organization of soil moisture and its relation to terrain indices, *Water Resour. Res.,* 35:797–810.

Williams, R.D. and L.R. Ahuja. 1992. Evaluation of similar-media scaling and a one-parameter model for estimating the soil water characteristic, *J. Soil Sci.,* 153:172–184.

Zaslavsky, D. and G. Sinai. 1981. Surface hydrology: I. Explanation of phenomena, *J. Hydraulic Div., ASCE,* 107:1–16.

Zheng, D., E.R. Hunt, and S.W. Running. 1996. Comparison of available soil water capacity estimated from topography and soil series information, *Landscape Ecol.,* 11:3–14.

Zhu, A.-X, L.L. Band, B. Dutton, and T.J. Nimlos. 1996. Automated soil inference under fuzzy logic, *Ecol. Modell.,* 90:123–145.

Zhu, A.-X, L.E. Band, R. Vertessy, and B.Button. 1997. Derivation of soil properties using a soil land inference model (SOLIM), *Soil Sci. Soc. Am. J.,* 61:523–533.

CHAPTER **14**

Parameterization of Agricultural System Models: Current Approaches and Future Needs

Lajpat R. Ahuja and Liwang Ma

CONTENTS

1-56670-0563/02/$0.00+$1.50
© 2002 by CRC Press LLC

INTRODUCTION

Determining the correct parameters for different components of a system model is a major and difficult task that currently restricts the widespread application of models in field research and technology transfer. Several reasons account for these difficulties. For theoretical process-level modules, most of the required parameters are specific to soil type or to a particular plant variety and species. Measured values of these parameters are generally limited or unavailable for different soils and plants, because measurements are expensive and time consuming. Examples include soil physical and hydrologic properties, evapotranspiration parameters and crop growth parameters. Values of these basic parameters may depend on environmental factors, which require additional sub-parameters for further characterization. For example, soil hydrologic properties vary with soil temperature and crop growth parameters can vary with temperature and with stresses due to water or nutrient deficiencies.

On the other hand, some modules in a system model are currently a simplified version of reality, either because of limited theoretical understanding or because of the high level complexity that is difficult to execute. In such cases, the model parameters reflect a simple chosen concept and thus they are difficult to measure. Examples of such simplified, conceptual models are the partitioning of soil organic matter and crop residues into a number of pools (e.g., slow, medium and fast) and then simulating the mineralization and immobilization of carbon and nitrogen in each pool, while doing some accounting for interpool transfers. In reality, all the soil organic components comprise a single continuum. Even in such simplified conceptual modules, the parameters may strongly depend upon environmental factors, which need to be characterized with additional but immeasurable sub-parameters.

For both the previous cases, parameterization is further complicated by the spatial and temporal variability of the land area simulated. For application at the field scale, spatial variability of the soil within the field causes variability in soil physical, chemical and biological parameters that can result in nonuniform plant growth or development. Depending upon the purposes of simulation, one may attempt to find average values of these parameters in modeling (lumped parameters), use probability distributions of the parameters in modeling (stochastic simulation), or run the model for several parts of the field separately (distributed simulation) with spatial and temporal integration. This latter distributed modeling option may be the only choice if the purpose is to evaluate site-specific management. For this option, water and other mass transfers between different parts of the field have to be a part of the simulation.

Temporal variability of soil parameters is caused primarily by land management practices, such as tillage and subsequent reconsolidation, no-tillage and plant-residue cover changes, macropore dynamics, drying and cracking of clayey soils, implement wheel compaction during various farm operations and compaction or disturbances from grazing livestock. These practices change the physical and energy transfer properties at the soil surface and physical, chemical and biological properties of the upper horizons. Weather conditions may enhance or moderate soil surface changes. For example, high intensity rainfall causes surface crusts to form on bare soil surfaces, whereas freezing and thawing tend to relieve surface compaction as well as drying and cracking on heavier texture soils. Model users have to define how these management practices change the soil's physical, chemical and biological parameters initially and how subsequent changes occur temporally.

Current approaches to parameterization include:

1. Using measured values where available
2. Estimating values from available simple properties based on established empirical relationships or by assuming simple conceptual or heuristic relationships
3. Creating databases of measured or estimated parameter values from the literature for various soils, plants and conditions and using values selected from the databases
4. Calibrating or refining the initial selection of parameters by comparing the model results with a set of observed data

The major problems associated with the fourth approach to parameterization, model calibration, are that:

1. Models are mostly partially calibrated due to lack of experimental data for all the system components.
2. No standards (e.g., procedure and statistics) are established for model calibration.
3. Calibration depends on the purpose and experience of model users and it is by trial and error.
4. Parameters are time- and location-dependent.
5. Not all the calibrated parameters are reported in the literature and, therefore, results are generally not reproducible by other users.
6. Model users are generally not sure whether the calibrated parameters are the true representation of their experimental results.
7. Validation of the calibrated parameters are not extensive.
8. No applicable range and sensitivity of the calibrated parameters are given.

This chapter presents current approaches and future research needs to improve these approaches for the selected, most commonly needed set of required parameters. Not all agricultural system models have incorporated detailed processes as described in this chapter, nor do they use the same approaches in simulating various components of agricultural systems. The presentation of this chapter is organized into seven sections:

1. Soil physical parameters
2. Soil hydrologic parameters
3. Evapotranspiration parameters
4. Soil carbon/nitrogen dynamics parameters
5. Crop growth parameters
6. Soil-pesticide parameters
7. Statistical methods for model calibration

Finally, two examples on how to parameterize agricultural system models are presented. Other processes and parameters not covered in this chapter are overland flow, soil erosion, chemical (N and P) movement, parameters related to rangeland and grazing processes, dairy models, insect damage, weed damage and effects of natural disasters. Interested readers may refer to other system models in the literature, such as GLEAMS (Leonard et al., 1987), Opus (Smith, 1990), WEPP (Flanagan and Nearing, 1995), EPIC (Sharpley and Williams, 1993), CENTURY (Parton et al., 1994), SPUR2 (Foy et al., 1999), APSIM (McCown et al., 1996) and GRAZPLAN (Donnelly et al., 2002).

SOIL PHYSICAL PARAMETERS

The main soil physical parameters of interest are: soil horizon delineation by depth; soil texture as sand, silt, clay and organic matter contents; soil bulk density; and soil porosity. Information on these properties, as well as soil structure, pH and cation exchange capacity (CEC) for major soil series is available in the soil characterization pedon database of the Natural Resources Conservation Service (NRCS). The common 1:200,000 soil survey reports provide general ranges in values of these properties for the soil series and, as a result, are less definitive. Even in the pedon database, the information is generally not site-specific for a given farm or field. Hence, for field research applications the user may want to take a few soil core samples of the soil profile in the experimental field to obtain or verify this information. Recommended methods of determining these parameters are given in the monograph on *Methods of Soil Analysis, Part 1*, published by the Soil Science Society of America (Klute, 1986). Large spatial variability of these soil parameters may require division of the field into a few subunits. Soil core sampling will also identify the long-term effects of tillage practices on the changes in the surface soil horizons, such as the existence of a plow pan and in texture and density of the modified horizons.

The total soil porosity, generally denoted by ϕ, is related to soil bulk density as:

$$\phi = 1 - \rho/\rho_p \tag{14.1}$$

where ρ is the soil bulk density and ρ_p is the soil particle density (g cm^{-3} or Mg m^{-3}). The ρ_p varies somewhat with soil type and should be determined in the laboratory by standard methods (Klute, 1986); however, a commonly accepted average value of ρ_p is 2.65 g cm^{-3}.

For field conditions, it is becoming important to separate the total porosity, ϕ, into soil matrix porosity, ϕ_s and soil macroporosity, ϕ_m. In this context, the macropores are defined as large voids in a soil, such as decayed root channels, worm holes and structural cleavages or cracks with radii ≥ 0.25 mm. Under surface ponded conditions, the continuous macropores allow a rapid downward movement of water that bypasses the soil matrix. This bypass or preferential flow is not considered in classical soil physical models of water flow in the soil matrix. Continuous macroporosity in the field is very difficult to determine (Ma and Selim, 1997). The best way might be to estimate it indirectly from infiltration measurements, which will be discussed in the section on Soil Hydrologic Parameters.

Temporal Changes Due to Tillage and Other Management Practices

Tillage decreases soil bulk density of the tilled zone (one or two thin horizons), which later gradually reverts back to the original state due to reconsolidation by natural forces. These changes depend upon soil type and implements used for tillage. No information exists in the literature on quantifying these changes. An approximate empirical equation used in the EPIC model (Williams et al., 1984) is:

$$\rho_t = \rho_{t-1} - \left[\left(\rho_{t-1} - 0.667\rho_c \right) I_i \right] \tag{14.2}$$

where ρ_t is the bulk density after tillage, ρ_{t-1} the bulk density before tillage (g cm^{-3}), ρ_c the consolidation bulk density at 33 kPa pressure (g cm^{-3}) and I_i is the tillage intensity (a 0-1 factor) that depends on the implement used and crop residue type on the soil surface. Williams et al. (1984) provide values for the tillage intensity for 29 different tillage implements for corn and soybean residues (Table 14.1). The ρ_c may be set equal to the bulk density after complete natural reconsolidation.

The bulk density after tillage increases due to natural reconsolidation during cycles of wetting and drying (Cassel, 1983; Onstad et al., 1984; Rousseva et al., 1988) in an asymptotic manner. Onstad et al., (1984) gave the following empirical equation to describe changes in bulk density of a tilled soil over time:

$$\rho_{at} = \rho_t + a \left[P/(1+P) \right] \tag{14.3}$$

where ρ_t is the bulk density just after tillage, ρ_{at} is the bulk density over time, P is the cumulative rainfall or applied water (cm) and a is an empirical constant. Thus, time is expressed in terms of

Table 14.1   Tillage Implements and Parameters

	Implement Name	Tillage Intensity (I_i) Corn	Tillage Intensity (I_i) Soybeans	Effective Depth (cm)
1	Moldboard plow	0.93	0.96	15
2	Chisel plow, straight	0.25	0.45	12.5
3	Chisel plow, twisted	0.45	0.65	12.5
4	Field cultivator	0.25	0.35	10
5	Tandem disk	0.50	0.65	10
6	Offset disk	0.55	0.70	10
7	One-way disk	0.40	0.50	10
8	Paraplow	0.20	0.25	15
9	Spike tooth harrow	0.20	0.25	2.5
10	Spring tooth harrow	0.30	0.45	5
11	Rotary hoe	0.10	0.15	2.5
12	Bedder ridge	0.75	0.80	15
13	V-blade sweep	0.10	0.15	7.5
14	Subsoiler	0.20	0.30	35
15	Rototiller	0.55	0.70	7.5
16	Roller package	0.10	0.10	0
17	Row planter w/smooth coulter	0.08	0.11	0
18	Row planter w/fluted coulter	0.15	0.18	0
19	Row planter w/sweeps	0.20	0.30	0
20	Lister planter	0.40	0.50	0
21	Drill	0.15	0.15	0
22	Drill w/chain drag	0.15	0.15	0
23	Row cultivator w/sweeps	0.25	0.30	0
24	Row cultivator w/spider wheels	0.25	0.30	0
25	Rod weeder	0.15	0.20	0
26	Rolling cultivator	0.50	0.55	0
27	NH_3 applicator	0.15	0.20	0
28	Ridge-till cultivator	0.60	0.75	0
29	Ridge-till planter	0.50	0.70	0

From Williams, J.R. et al., *Trans. ASAE,* 27(1):129–142, 1984.

cumulative rainfall. Comparison of Eq. (14.3) with data showed that the increase in bulk density reached a near maximum value at about 10 cm of simulated rainfall application. Results of Rousseva et al., (1988) showed that bulk density did not reach a plateau even at much higher amounts of rainfall. Equation (14.3) can be modified to allow this behavior by replacing the term $(1 + P)$ by $(b + P)$ where b is another constant. Linden and Van Doren (1987) gave another algorithm for change in total porosity ϕ (directly related to bulk density) of a tilled soil as a double exponential function of cumulative rainfall energy, E and cumulative rainfall amount, P:

$$\phi = \phi_I - \left(\left(\phi_I - \phi_c \right) \left(2 - e^{-0.01E} \right) - e^{-0.25P} \right) \qquad (14.4)$$

where ϕ_I is the initial porosity just after tillage and ϕ_c is the final stable porosity. This equation has given reasonable results in the Root Zone Water Quality Model (RZWQM) applications (Ahuja et al., 2000).

SOIL HYDROLOGIC PARAMETERS

Application of soil water flow theory to describe water flow, soil water storage and plant uptake in the soil matrix requires knowledge of two basic soil hydrologic relationships (Ahuja et al., 2000):

1. Volumetric soil water content (commonly denoted as θ) as a function of soil-matric pressure (h) or soil water suction (τ); i.e., $\theta(h)$ or $\theta(\tau)$, with $\tau = -h$. The units of h or τ are kPa or meters. The $\theta(h)$ relationship is commonly called the soil water retention curve.
2. Soil hydraulic conductivity (K) as a function of soil water content, matric pressure, or suction; i.e., $K(\theta)$, $K(h)$ or $K(\tau)$. The hydraulic conductivity when the soil is saturated ($\theta = \theta_s = \phi$) is called the saturated hydraulic conductivity, K_{sat}. The commonly used units are cm hr^{-1}, that correspond to dimensionless units of gradients of pressure head in Darcy's equation.

Standard methods for measuring these basic soil hydrologic properties in the laboratory and field are detailed in the Klute (1986). Knowledge of these parameters at matric pressures between 0 and about -100 kPa is very important because significant water movement occurs only in this range. Perhaps the most reliable method for determining hydraulic conductivities in this pressure range, for field conditions, is the Darcian analysis of *in situ* tensiometric measurements made during infiltration and the subsequent drainage, using the water content matric pressure relationships (Richards et al., 1956; Ogata and Richards, 1957; Nielsen et al., 1964; Rose et al., 1965; Watson 1966; van Bavel et al., 1968; Flühler et al., 1976). This method is the instantaneous profile method (Green et al., 1986) in which a water content/matric pressure relationship can be obtained by periodic measurement of soil water content during the drainage phase by gravimetric, neutron thermalization, or time domain reflectrometry (TDR) techniques with the soil surface covered to minimize evaporation.

More commonly, however, the relationship between θ and h is measured in the laboratory on undisturbed soil cores for hydrostatic conditions. In general, methods for determining soil hydraulic and water storage properties are time-consuming, tedious and expensive, especially because a large number of measurements are required to characterize the combined effects of inherent and management-induced spatial variability of these properties in a field. Innovative approaches that require less time and effort are needed to increase applications of the theory.

Ahuja et al. (1999) presented a summary of recently proposed methodologies for determining soil hydrologic properties using simpler measurements and less data. For $\theta(h)$, the methods included estimation from soil composition and bulk density using regression equations and more recent scaling approaches using soil bulk density and -33 kPa (one-third bar) or -10 kPa soil water content. For the important property of saturated hydraulic conductivity (K_{sat}), a major new development involved estimation from effective porosity, obtained from bulk density and -33 kPa soil water content. This method, as well as the estimation of K_{sat} from the pore-size distribution based on further development of the Marshall (1958) approach, were described. For unsaturated hydraulic conductivity ($K(h)$ for $h < 0$), the methods included simplified field measurement methods, such as the unit-hydraulic gradient approach and a simplified functions technique involving only field tensiometric data, as well as the estimation of $K(h)$ from $\theta(h)$ and K_{sat}. Temporal changes in $\theta(h)$ and $K(h)$ brought about by tillage, residue management and cropping practices can be important, but were only briefly addressed because of limited knowledge of these changes available at present.

For use in models, each of the basic hydrologic relationships is fitted with a suitable mathematical equation or function. Three commonly used mathematical representations of $\theta(h)$ relationship proposed by Brooks and Corey (1964), Campbell (1974) and van Genuchten (1980), respectively, are shown in Table 14.2, along with the definition of their parameters and the correspondence among them. Each of these models are based upon some simplifying assumption, such as isothermal conditions and no drying–wetting hysteresis. Only the van Genuchten model prescribes a smooth curve for the entire range of $\theta(h)$, assuming the saturated water content, θ_s, occurs at $h = 0$. The other two models assume θ_s occurs at the air-entry pressure, h_b and that water content is equal to θ_s between $h = h_b$ and $h = 0$. The corresponding forms of $K(h)$ functions are presented in Table 14.3. The parameterization problem then is to determine or estimate the independent parameters of a set of water retention and hydraulic conductivity functions in Table 14.2 and 14.3.

Table 14.2   Commonly Used Soil Water Retention Models

Soil Water Retention	Parameters	Parameter Correspondence with Brooks–Corey Model
	Brooks–Corey (1964)	
$$\frac{\theta - \theta_r}{\theta_s - \theta_r} = \left(\frac{h_b}{h}\right)^{\lambda}$$	θ = volumetric water content θ_s = saturated water content λ = pore size distribution index h = soil water pressure head h_b = bubbling pressure or air entry pressure head θ_r = residual water content ϕ = total porosity	$\lambda = \lambda$ $h_b = h_b$ $\theta_r = \theta_r$ $\theta_s = \phi$
	Campbell (1974)	
$$\frac{\theta}{\theta_s} = \left(\frac{H_b}{h}\right)^{1/b}$$	θ_s = saturated water content H_b = bubbling pressure or air entry pressure head b = constant	$\theta_s = \phi$ $H_b = h_b$ $b = \dfrac{1}{\lambda}$
	van Genuchten (1980)	
$$\frac{\theta - \theta_r}{\theta_s - \theta_r} = \left(\frac{1}{1 + (-\alpha h)^n}\right)^m$$	θ_s = saturated water content θ_r = residual water content α = constant n = constant m = constant	$\theta = \phi$ $\theta_r = \theta_r$ $\alpha = (h_b)^{-1}$ $n = \lambda + 1$ $m = 1 - \dfrac{1}{\lambda + 1}$

From Rawls, W.J. and D.L. Brakensiek, Prediction of soil water properties for hydrologic modeling, in *Proc. Watershed Manage. in the Eighties,* ASCE, New York, 1985.

Determining Parameters of θ(h) Curve

Perhaps one of the best ways to determine these parameters will be to fit Brooks–Corey or one of the other two equations in Table 14.2 to the measured θ(h) data; however, the measurements are expensive and time-consuming. In the absence of measured data, some simpler methods of estimating these parameters are given in the following subsection, in a hierarchical order of increasing accuracy.

From Textural Class

Rawls et al. (1982) collected measured data for 500 soils with a total of about 2453 horizons from 18 states. From these data, they calculated mean Brooks–Corey parameters and related information for 11 USDA textural classes. These average values by textural class are presented in Table 14.4. Clapp and Hornberger (1978) and DeJong (1982) reported similar sets of textural class mean values for the Campbell water retention function (Table 14.2) based on smaller datasets. Unfortunately, there was a considerable variation of parameters from the three sources (Ahuja et al., 1999). Nonetheless, in the absence of any other relevant information, the textural-class mean values in Table 14.4 provide the simplest estimates of the required θ(h) parameters.

Table 14.3   Commonly Used Hydraulic Conductivity Functions, K(θ) and K(h)

K(θ)	K(h)	Comments

Brooks–Corey (1964)

$$K(\theta) = K_{sat}\left(\frac{\theta - \theta_r}{\theta_s - \theta_r}\right)^{\frac{2+3\lambda}{\lambda}} \qquad K(h) = K_{sat}\left(\frac{h_b}{h}\right)^{2+3\lambda}$$

Campbell (1974)

$$K(\theta) = K_{sat}\left(\frac{\theta}{\theta_s}\right)^{2b+3} \qquad K(h) = K_{sat}\left(\frac{h_b}{h}\right)^{2+3/b}$$

van Genuchten (1980)

$$K(\theta) = K_{sat}\left(\frac{\theta - \theta_r}{\theta_s - \theta_r}\right)^{0.5}\left[1 - \left(1 - \left(\frac{\theta - \theta_r}{\theta_s - \theta_r}\right)^{1/mn}\right)^m\right]^2$$

$$K(h) = \frac{\left\{1 - (\alpha h)^{n-1}\left[1 + (\alpha h)^n\right]^{-m}\right\}^2}{\left[1 + (\alpha h)^n\right]^{m/2}}$$

Note: Parameters are the same as those in Table 14.2.

From Soil Texture, Organic Matter and Bulk Density

Rawls et al. (1982) used their national database to develop 12 separate multiple linear regression equations to relate percent of silt, clay, organic matter and bulk density to soil water retention at 12 matric pressures within the range from –4 to –1500 kPa. Two further regression equations were provided for 10 pressure heads which included either one or two measured soil water content values, θ at –1500 kPa pressure or θ at –33 and –1500 kPa pressures as additional variables. Later, Rawls and Brakensiek (1985) developed regression equations for Brooks–Corey parameters applicable over the entire range of matric pressures, in place of one equation for each matric pressure. These regression equations are presented in Table 14.5, along with the regression equations for the van Genuchten (1980) model parameters derived by Vereecken (1988) based on a more limited dataset.

There is no independent validation of the regression equations in Table 14.5 available and no information exists on the uncertainty associated with the equations introduced; however, the original 12 sets of regressions equations, one for each matric pressure, were tested against experimental data for silty clay loam soils (Ahuja et al., 1985) and for sandy to sandy clay type soils (Williams et al., 1992). Because of the many more overall parameters and degrees of freedom, the resulting 12 sets of equations should be much more accurate than the regression equations in Table 14.5. Even so, the validation tests described previously showed that the regression equations for individual matric pressures based only on soil texture, organic matter and bulk density generally overpredicted soil water contents at various pressures. The mean relative error ranged from 8 to 29%, with standard deviation of errors ranging from 17 to 36%. The equations that incorporated the measured –1500 kPa water content value, as an additional variable, did not improve the results, but the equations that also incorporated –33 kPa values improved the estimates considerably.

Table 14.4 Hydrologic Soil Properties Classified by Soil Texture (Modified from Rawls et al., 1982.)

Texture Class	Sample Size	Total Porosity (θ_s) cm³/cm³	Residual Saturation (θ_r) cm³/cm³	Bubbling Pressure (h_b) Arithmetic (cm)	Geometric (cm)	Pore Size Distribution Index (λ) Arithmetic	Geometric	Saturated Hydraulic Conductivity (K_s) (cm/h)
Sand	762	0.437[a] (0.374–0.500)[a]	0.020 (0.001–0.039)	15.98 (0.24–31.72)	7.26 (1.36–38.74)	0.694 (0.298–1.090)	0.592 (0.334–1.051)	21.00
Loamy sand	338	0.437 (0.368–0.506)	0.035 (0.003–0.067)	20.58 (0.0–45.20)	8.69 (1.80–41.85)	0.553 (0.234–0.872)	0.474 (0.271–0.827)	6.11
Sandy loam	666	0.453 (0.351–0.555)	0.041 (0.0–0.106)	30.20 (0.0–64.01)	14.66 (3.45–62.24)	0.378 (0.140–0.616)	0.322 (0.186–0.558)	2.59
Loam	383	0.463 (0.375–0.551)	0.027 (0.0–0.074)	40.12 (0.0–100.3)	11.15 (1.63–76.40)	0.252 (0.086–0.418)	0.220 (0.137–0.355)	1.32
Silt loam	1206	0.501 (0.420–0.582)	0.015 (0.0–0.058)	50.87 (0.0–109.4)	20.76 (3.58–120.4)	0.234 (0.105)–0.363)	0.211 (0.136–0.326)	0.68
Sandy clay loam	498	0.398 (0.332–0.464)	0.068 (0.0–0.137)	59.41 (0.0–123.4)	28.08 (5.57–141.5)	0.319 (0.079–0.559)	0.250 (0.125–0.502)	0.43
Clay loam	366	0.464 (0.409–0.519)	0.075 (0.0–0.174)	56.43 (0.0–124.3)	25.89 (5.80–115.7)	0.242 (0.070–0.414)	0.194 (0.100–0.377)	0.23
Silty clay loam	689	0.471 (0.418–0.524)	0.040 (0.0–0.118)	70.33 (0.0–143.9)	32.56 (6.68–158.7)	0.177 (0.039–0.315)	0.151 (0.090–0.253)	0.15
Sandy clay	45	0.430 (0.370–0.490)	0.109 (0.0–0.205)	79.48 (0.0–179.1)	29.17 (4.96–171.6)	0.223 (0.048–0.398)	0.168 (0.078–0.364)	0.12
Silty clay	127	0.479 (0.425–0.533)	0.056 (0.0–0.136)	76.54 (0.0–159.6)	34.19 (7.04–166.2)	0.150 (0.040–0.260)	0.127 (0.074–0.219)	0.09
Clay	291	0.475 (0.427–0.523)	0.090 (0.0–0.195)	85.60 (0.0–176.1)	37.30 (7.43–187.2)	0.165 (0.037–0.293)	0.131 (0.068–0.253)	0.06

[a] First line is the mean value.
[b] Second line is ± one standard deviation about the mean.
From Rawls, W.J. et al., *Trans. ASAE*, 25:1316–1320, 1328, 1982.

Table 14.5   Equations for Estimation of Water Retention Model Parameters

Brooks–Corey Model Parameters (Rawls and Brakensiek, 1985)

h_b = Brooks–Corey bubbling pressure head (cm)[a]

$h_b = \exp[5.340 + 0.185(C) - 2.484(\phi) - 0.002(C^2) - 0.044(S)(\phi) - 0.617(C)(\phi) + 0.001(S^2)(\phi^2) - 0.009(C^2)(\phi^2) - 0.000\,01(S^2)(C) + 0.0009(C^2)(\phi) - 0.0007(S^2)(\phi) + 0.000\,00(C^2)(S) + 0.500(\phi^2)C]$

λ = Brooks–Corey pore size distribution index

$\lambda = \exp[-0.784 + 0.018(S) - 1.062(\phi) - 0.000\,05(S^2) - 0.003(C^2) + 1.111(\phi^2) - 0.031(S)(\phi) + 0.000\,3(S^2)(\phi^2) - 0.006(C^2)(\phi^2) - 0.000002(S^2)(C) + 0.008(C^2)(\phi) - 0.007(\phi^2)(C)]$

θ_r = Brooks–Corey residual water content (m³m⁻³)

$\theta_r = -0.018 + 0.0009(S) + 0.005(C) + 0.029(\phi) - 0.000\,2(C^2) - 0.001(S)(\phi) - 0.0002(C^2)(\phi^2) + 0.0003(C^2)(\phi) - 0.002(\phi^2)(C)$

van Genuchten Model Parameters (Vereecken, 1988)

$\theta_r = 0.015 + 0.005(C) + 0.014(Ca)$

$\alpha = 10^{[-2.486 + 0.025(S) - 0.351(Ca) - 2.617(bd) - 0.023(C)]}$

$n = 10^{[0.053 - 0.009(S) - 0.013(C) + 0.00015(S^2)]}$

$m = 1 - 1/n$

[a] C = % clay, S = % sand, Ca = % carbon, ϕ = porosity (m³m⁻³), bd = bulk density.

From Ahuja, L.R. et al., Determining soil hydraulic properties and their spatial variability from simpler measurements, in *Agricultural Drainage, Agronomy Monograph No. 38,* ASA, Madison, WI, 1999.

Table 14.6   Groups of Soils from the U.S. and Their Average p and q Values[a]

Group	Soils	Textural Groups	Average p Value ln(kPa)	Average q Value ln(m³m⁻³)
1	Oxisols, Kirkland Renfrow, Pima	Loam-silty clay; loam-clay loam	1.415	0.839
2	Norfolk, Teller, Bernow (45–90 cm)	Sandy loam; sandy clay loam	0.343	1.072
3	Lakeland, Bernow (0–45 cm)	Sand	0.541	1.469
—	Australian and British soils	A mixture of textures	–0.982	0.585

[a] Average p and q values for several Australian and British soils pooled together, reported by Gregson et al. (1987), are given in the same units for comparison.

From Ahuja, L.R. and R.D. Williams, *Soil Sci. Soc. Am. J.,* 55:308–319, 1991.

From a One-Parameter Model

A one-parameter model is based on a strong linear correlation ($R^2 > 0.95$) observed between the slope b and intercept a of a log–log linear form of the Brooks–Corey or Campbell water retention equation (Gregson et al., 1987; Ahuja and Williams, 1991). Either of these equations may be written in the form:

$$\ln|h| = a + b\ln(\theta - \theta_r) \tag{14.5}$$

$$a = \ln|h_b| - b\ln(\theta_s - \theta_r) \tag{14.6}$$

where a is a lumped constant and b = $-1/\lambda$ (Table 14.2). For the Campbell equation, $\theta_r = 0$. Furthermore, a vs. b linear relationships for a group of soils merged very nicely into one common relationship (Gregson et al., 1987; Ahuja and Williams, 1991):

$$a = p + qb \tag{14.7}$$

where p and q are constants for all soils in a group. Substituting Eq. (14.7) into Eq. (14.5) yields a one-parameter model, provided that θ_s (porosity) or bulk density and an approximate value of θ_r for the soil type under consideration are known or can be estimated from other information in the literature:

$$\ln|h| = p + b\left[\ln\left(\theta - \theta_r\right) + q\right] \tag{14.8}$$

This equation can be used to estimate the parameter b of the $h(\theta)$ or $\theta(h)$ function simply from one measured value of the $h(\theta)$ curve and given p and q values for the soil group. Then, from Eq. (14.6) the parameter h_b can be found because b, θ_s and θ_r are now known. The average values of p and q for soil groups based on the experimental data analyzed by Ahuja and Williams (1991) and Gregson et al. (1987) are given in Table 14.6 for general application. In comparison with other methods, this one-parameter model gave better results for several soils (Williams and Ahuja, 1992).

Determining Parameters of K(θ) or K(h) Curve

Again, the best way to determine the parameters of $K(\theta)$ or $K(h)$ function, such as the forms given in Table 14.3, will be to fit these functional forms to field measured data on these properties; however, field measurement of this function is extremely tedious and time consuming. Accepting the premise of the commonly used $K(\theta)$ or $K(h)$ functions given in Table 14.3, it turns out that all the parameters are the same as for the corresponding $\theta(h)$ functions, except for K_{sat}. Hence, the issue addressed here is the estimation of K_{sat}.

Estimating K_{sat}

Saturated hydraulic conductivity (K_{sat}) is probably the most critical soil hydraulic parameter in soils, although K(h) under unsaturated condition may yield much more realistic results (Ehlers, 1977). This parameter is difficult to obtain since it is highly dependent on soil conditions, such as compaction, soil aggregates, macropores, sample size, temperature and entrapped air and thus is highly variable. In a field, K_{sat} may vary between one to two orders of magnitude. Because of this, it is much more meaningful to try to determine reasonably accurate spatial distribution of K_{sat} (e.g., mean and standard deviation) in a field, rather than highly accurate point values. Using textural-class mean values, such as given in Table 14.4, gives a general idea of the magnitudes. The simplified methods based on soil texture and bulk density have failed to yield more accurate results.

Recent studies have shown that K_{sat} is strongly related to an effective soil porosity, ϕ_e (Ahuja et al., 1984). In these studies, ϕ_e of a soil was defined as total porosity minus the volumetric soil water content at 33 kPa suction and was related to K_{sat} by a generalized Kozeny–Carman equation:

$$K_{sat} = B\phi_e^n \text{ and } \phi_e = \theta_s - \theta_{33kPa} \tag{14.9}$$

where B and n are constants. Figure 14.1 shows a combined relationship for nine soils combined (473 data points). These soils came from diverse locations in Hawaii, Arizona, Oklahoma and several states of the southeastern U.S. The strength of this relationship in Figure 14.1 is that it was as good for all soils as for any one soil individually, which indicates that Eq. (14.9) is applicable across soil types. Realizing that the measurement of K_{sat} can be subject to an error as large as one to two orders of magnitude, due to unknown effects of entrapped air and macropores, the empirical equation given in Figure 14.1 is useful in estimating K_{sat} from the simpler measurements of effective porosity for a soil. Tests of this equation against carefully measured field data for several Indiana soils and core data for several Korean soils, have shown good results (Franzmeier, 1991; Ahuja et al., 1989). Equation (14.9) also can be used to estimate the spatial distributions (cumulative

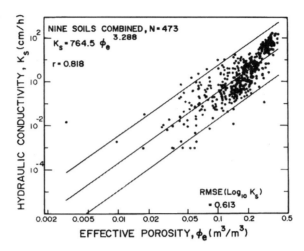

Figure 14.1   Saturated hydraulic conductivity as a function of effective pore porosity for nine soils. The middle line is the best-fit line and the other two lines are ±1.0 RMSE around the fitted line. (From Ahuja, L.R. et al., *Soil Sci.,* 148:404–411, 1989.)

frequency distributions) of K_{sat} from spatial distributions of effective porosity (Ahuja et al., 1984, 1989). Rawls et al. (1998) found that $n = 3 - \lambda$ for their textural class mean values, where λ is the Brooks–Corey pore size distribution index.

Determining Flow through Soil Macropores

As noted earlier, macropores can cause tremendous amounts of preferential water movement during ponded infiltration conditions in the field. Connectivity and continuity of macropores with depth are critical factors for this flow. No simple ways are available to estimate continuous macroporosity and the magnitude of flow through them. Some specialized measurements of infiltration in the field are the only method available at this time.

Watson and Luxmoore (1986) measured infiltration under ponded-water conditions and under conditions where water was applied at a small negative pressure (–3 cm for macropore radius ≥0.5 mm) to separate the macropores and soil matrix flow. The latter measurements have to be made within a small ring, 10 to 20 cm in diameter. The technique works well, but the small area of measurement may not meet the minimum representative volume requirement for all soil-macropore situations. To allow for a large sample volume, Timlin et al. (1993) have shown that the contribution of continuous macropores to field K_{sat} can be obtained from measurements of ponded-water infiltration rate in a large ring and tensiometric data during redistribution of water immediately following infiltration. The soil-matrix K_{sat} at a given depth is obtained from redistribution data obtained near saturation. The difference between this K_{sat} and the steady-state infiltration rate gives the contribution of macropores at this depth.

Changes in Soil Hydraulic Parameters Due to Tillage Practices

The tillage and subsequent reconsolidation by natural forces cause temporal changes in soil hydrologic properties. Wheel track compaction enhances the temporal changes in the tilled zone, but it may also cause permanent changes in properties of the subsoil beneath the tracks.

Tillage Effects on the θ(h) curve

Very limited quantitative data are available in the literature on this subject. Based on limited data, Ahuja et al. (1998) have presented two semiempirical algorithms for determining changes in

θ(h) parameters caused by tillage and subsequent natural reconsolidation, for the Brooks–Corey form of θ(h) curve. The simpler of the two algorithms is summarized below:

1. The change in soil bulk density and hence soil porosity, ϕ or θ_s, due to tillage are assumed known from Eq. (14.2) presented earlier.
2. The residual water content, θ_r and the bubbling pressure head parameter h_b of the soil on tillage stay the same as ones before tillage or at full natural reconsolidation.
3. The parameter λ increases on tillage in the wet range only, between $h = h_b$ and $h = 10\,h_b$. In this range of h, the tilled soil value, $\lambda_{till,}$ is computed from tilled soil saturated water content, θ_{still}:

$$\lambda_{till} = \frac{\log(\theta_{still} - \theta_r) - \log[\theta(10\,h_b) - \theta_r]}{\log(h_b) - \log(10\,h_b)}\qquad(14.10)$$

4. Below the h range, i.e., for h values >10 h_b, the λ value does not change.

The effectiveness of the previously outlined algorithms is shown for four cases in Ahuja et al., (1998). These equations can also be used to estimate curves for partially reconsolidated soil conditions, with the corresponding transient soil bulk density calculated from Eqs. (14.3) or (14.4).

Effects on K_{sat} Parameter

Once the θ(h) curve of the tilled soil is obtained from the previous equations, the new θ_s and θ_{33kPa} are used to obtain the new ϕ_e in Eq. (14.9). Substituting the ϕ_e in Eq. (14.9) then provides the new $K_{sat.}$ The other parameters of the K(h) curve of the tilled soil are determined from the parameters of the new θ(h) curve of the tilled soil, from relationships shown in Tables 14.2 and 14.3.

Surface Sealing–Crusting

Tillage tends to destroy the soil surface cover. Rainfall on a freshly tilled soil generally results in formation of a dense surface seal. On drying, this seal becomes a mechanical crust. Saturated hydraulic conductivity for a 2- to 5-mm thick seal-crust can be 5- to 100-fold smaller than that of the original soil (McIntyre, 1958; Tackett and Pearson, 1965; Sharma et al., 1981); the most common range is 5- to 20-fold. A thin seal-crust gets quickly saturated at the start of rainfall; therefore, its unsaturated hydraulic conductivity and θ(h) can be neglected and only the K_{sat} needs to be determined. Some measurements for a given soil will be the best way to determine this K_{sat}. A first estimate may be 1/10th of the K_{sat} of the underlying soil. Details of the temporal changes in seal-crust K_{sat} are given in Ahuja et al. (1999).

Often, tillage does not bury all the crop residues that are on the soil surface. The residue cover then protects a part of soil from sealing–crusting. Recent studies indicate that for practical purposes, hydraulic conductivity of a partially formed crust layer can be estimated as an area-weighted mean K_{sat} of crusted and uncrusted areas of the soil surface (Ruan et al., 2001). The burial of residues is generally assumed proportional to the tillage intensity I_i in Eq. (14.2) (Wagner and Nelson, 1995).

EVAPOTRANSPIRATION PARAMETERS

Modeling of evapotranspiration (ET) in agricultural systems is generally divided into two parts:

1. Modeling of potential evapotranspiration (PET), defined as "the amount of water transpired in a unit time by a short, green crop, completely shading the ground, of uniform height and never short of water" (Penman, 1948)
2. Modeling of actual ET (AET) when the above conditions are not satisfied, such as with incomplete cover and limiting soil moisture

Modeling Potential ET

Methods of modeling the potential ET (PET) available in the literature range from simple empirical methods, that require fewer parameters, to more mechanistic methods based on energy balance and aerodynamics of vapor and heat transfer, requiring more parameters. Thornthwaite (1948) related monthly PET to mean monthly air temperature for a large number of locations in the east-central U.S. and provided fixed empirical values of coefficients or parameters in the equation. The value of PET is modified by an additional factor of mean monthly percentage of annual daytime hours. Blaney and Criddle (1962) refined the above method for Western U.S. conditions. Jensen and Haise (1963) derived an equation based on air temperature and mean daily solar radiation and provided empirical methods of determining the parameters from local meteorological data. As discussed by Monteith (1998), the equations based on mean monthly temperature do not give correct estimates of daily PET values.

Other empirical methods may provide good daily values only if local data are used to calibrate parameters (Hatfield, 1990). Because most agricultural system models operate on a daily level, they generally utilize the more mechanistic methods to estimate PET. These mechanistic models, of course, also use some empirical methods to obtain driving variables and parameters. The commonly used mechanistic-empirical models of PET are Penman (Penman, 1948), Penman–Monteith (Monteith, 1965), Priestley–Taylor (Priestley and Taylor, 1972) and Shuttleworth–Wallace, an expansion of the Penman–Monteith model (Shuttleworth and Wallace, 1985). Hatfield (1990) and Monteith (1998) describe these methods in detail.

In the Penman (1948) method, PET is obtained from net radiation at the evaporating surface, mean relative humidity deficit of air, known thermodynamic constants and an empirical wind function to represent vapor and heat transfer coefficients. Heat stored in the soil or vegetation is neglected. Penman also provided empirical equations to calculate net radiation from its components: (1) short-wave solar radiation estimated from hours of sunshine; and (2) net long-wave radiation estimated from air temperature and vapor pressure. Besides the meteorological data, the Penman method requires seven empirical parameters in all; however, all these parameters are fixed and given, except a parameter to estimate PET for a soil or vegetation from the PET for open water. The user may calibrate this parameter for specific soil–crop conditions.

In the Penman–Monteith equation, the empirical wind function used in Penman's equation was replaced by the reciprocal of an aerodynamic resistance and a surface or stomatal resistance. Monteith and Unsworth (1990) also made some improvements in calibrating net radiation. For simplified, ideal conditions of soil or crop canopies in the field, Monteith (1998) described methods to calibrate aerodynamic resistance from wind speed profile above the canopy and crop height or leaf area index, along with some suggested values of unknown constants. For best results, however, the user has to calibrate the values of both aerodynamic and surface or stomatal resistances for the site-specific conditions. The latter resistance is also a function of weather conditions and soil water contents.

Priestley and Taylor (1972) simplified the Penman equation for PET for conditions where the saturated deficit of air above the wet surface is constant with height or the air above the wet surface is saturated. The PET then becomes independent of the wind speed. The previously described conditions are, however, seldom met and the user has to be careful in using this approach. This approach is generally considered applicable in humid climate zones only.

Shuttleworth and Wallace (1985) removed the constraint in the Penman–Monteith model that the vegetation completely covers the ground. They constructed a two-dimensional network of resistances to account for separate contributions from soil and foliage. The various resistances are: a soil surface resistance and an aerodynamic resistance for transfer between soil and mean canopy levels; an aerodynamic resistance between mean canopy level and screen (measurement) height; and an aerodynamic and a surface resistance within the canopy. The authors provide equations or values for these resistances. The user is advised to calibrate some of these values.

The Shuttleworth–Wallace model has been extended to a layer of residue cover above the surface, with the resistances then operating between the soil surface and residue plane and between the residue plane and canopy (Farahani and Ahuja, 1996). The values of additional resistance used in RZWQM model are provided by Farahani and DeCoursey (2000).

Modeling Actual ET

The PET models of Penman–Monteith and Shuttleworth–Wallace allow modeling of actual ET (AET) through the adjustment of soil surface resistance, residue surface resistance and stomatal or canopy resistances. In this approach, the root zone of the soil is treated as one uniform layer. Root distribution with depth and water movement within the soil are neglected. Most agricultural system models, therefore, use the above models to compute only PET and link this to detailed models of root growth, root water uptake and soil water movement. For incomplete canopy cover conditions, the PET is divided into potential transpiration (PT) and potential residue and soil evaporation (PE), based on an empirical function of leaf area index (LAI) and residue cover. The PT then serves as an upper boundary condition for a root water uptake model, such as Nimah and Hanks (1973) and PE from soil as the upper condition for solution of the Richards' equation for evaporation flux at the soil surface.

Surface Residue Cover

Crop residues on the soil surface play an important role in protecting the surface against sealing–crusting by raindrop impact; hence, maintaining high infiltration rates into the soil. Between rains, they also influence the rate of soil evaporation. The main residue parameters that affect these processes are residue mass per unit area and percent cover. Each of these parameters may be further apportioned into flat and standing residues. These parameters generally have to be estimated based on some measurements. For a given type of crop residues, percent cover and mass may be related. For example, Gregory (1982) derived the following empirical relation between percent cover (C_r) and mass (M_r) for residues:

$$C_r = 100\left(1 - e^{-F M r}\right) \tag{14.11}$$

where F is a constant. The standing residue cover may be estimated based on stubble population per unit area and average basal area per stubble. For soil evaporation modeling, the surface residue mass, cover and thickness are utilized empirically to estimate the solar radiation absorbed by the residues and aerodynamic resistance of the residue layer.

SOIL CARBON/NITROGEN DYNAMICS PARAMETERS

In all major system models, such as Century (Parton et al., 1983); Phoenix (Juma and McGill, 1986); and RZWQM (Shaffer et al., 2000), the soil organic matter transformations, i.e., mineralization and immobilization, are simulated by dividing the soil crop residues and humus organic matter into a number of pools. The pools are commonly based on carbon/nitrogen (C/N) ratios of the components. Soil microbial population is also divided into pools based on their functional activities, such as aerobic heterotrophs, autotrophs and facultative heterotrophs. Each organic matter pool is assumed to undergo a first-order decay, facilitated by aerobic heterotrophs. Mass transfers between the pools, including the active microbial pool. The primary products of decay (e.g., NH_4) are subject to further nitrification by nitrifier microbes under aerobic conditions, or to denitrification under anaerobic conditions. An example schematic diagram of the soil organic matter pools, primary transformations and interpool transfers is pictured in Figure 14.2. It is generally recognized that

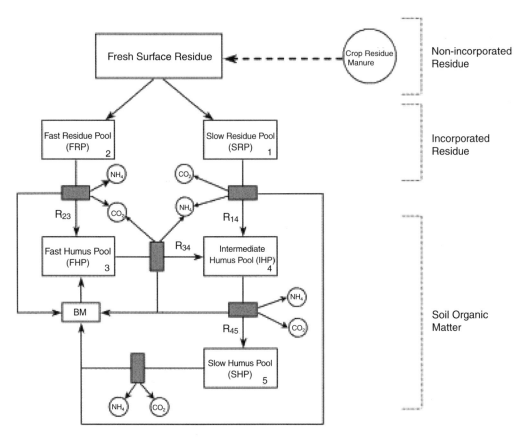

Figure 14.2   A schematic diagram of residue and soil organic matter pools, transformation and interpool transfer. R_{14}, R_{23}, R_{34} and R_{45} are interpool mass transfer coefficients. BM is microbial biomass. (From Ma, L. et al., *Soil Sci. Soc. Am. J.*, 62:1006–1017, 1998a. With permission.)

the organic matter in soil is, in fact, a continuum and its division into pools for modeling is an approximate arbitrary approach.

Based on Figure 14.2, the following sets of parameters are required to model both primary and secondary transformations of soil organic matter:

1. Soil surface crop residue decomposition parameters
2. Pool sizes for incorporated residues, humus organic matter; and microbial pools
3. Transformation rate coefficients and rate constants for primary transformations and coefficients for their strong dependence upon soil temperature, moisture content, O_2 concentration, ion strength, pH and microbial populations
4. Interpool transfer coefficients
5. Rate coefficients and rate constants for nitrification and denitrification and their dependence upon environmental factors

Decomposition of Surface Residues with Time

Ma et al., (1999) reviewed and evaluated the models that have been used to describe the decomposition of surface crop residues. Douglas and Rickman (1992) assumed a first-order decay with respect to daily growing degree-days. In this model, the decomposition coefficient is a multiple of three factors — f_N, f_w and k accounting for initial crop residues N, soil moisture effects and a rate constant, respectively. Douglas and Rickman (1992) provide fixed equations or values for these factors, which are assumed to apply universally. Gregory et al. (1985) used a second-order decay equation, but the parameters can still be determined from fixed equations from weather data and initial C/N ratio of the residues. Other models used in the literature have similar parameters.

Determining Soil Organic Matter Pool Sizes

When there are only two pools, such as for the incorporated crop residues in Figure 14.2, the pool sizes can be obtained approximately if the average C/N ratio of the total material in both pools is known, as well as the assumed average C/N ratio within each pool (Ma et al., 1998a). When they are more than two pools, however, such as for the humus (OM) pools in Figure 14.2, determining pool sizes is a very difficult problem. To our knowledge, no laboratory procedures have been developed to determine the sizes of these pools. The potentially mineralizable N frequently mentioned in the literature may cover more than one pool, such as both fast and medium OM pools (Landa et al., 1999). The pool sizes are greatly influenced by the past management practices. Based on our experience with RZWQM (Ma et al., 1998a), our recommendation is to start with a first guess for the three humus OM pool sizes, such as 5, 10 and 85%, respectively, for fast, medium and slow pools and set the microbial pools at a minimum level of 50,000, 500 and 5,000 organisms/g soil respectively, for aerobic heterotrophs, autotrophs and facultative heterorophs. Then run the model for 10 to 15 years with past management and climate conditions to obtain stabilized size for all pools and their distribution with depth in the soil. Measured total organic matter of the soil serves as a check on the derived stable values. Other suggested methods of partitioning organic carbon pools are the lignin content (CENTURY model) and years of cultivation of the soil (EPIC model) (Ma and Shaffer, 2001).

Application and incorporation of animal manures, compost and other organic waste changes the soil residue pools. Solid components of these materials are generally divided into organic waste and bedding, based on their C/N ratios. Each of these is then partitioned into the existing soil residue pools of prescribed C/N ratios (Rojas and Ahuja, 2000). Incorporation of surface crop residues by tillage is treated in the same way as manures. Earthworms and other soil fauna also bring about some incorporation of decomposed surface residue and manures to soil residue pools.

Determining Rate Coefficients and Interpool Transfer Coefficients

A limited amount of laboratory-measured data is available for mineralization and nitrification rates with time for certain soils, management and environmental conditions. From these data, initial guesses of the rate coefficients for available soil and management conditions and their dependencies may be made. These guesses are then refined by calibration of model results against measured field data. With time, we hope to have a database of these parameters for different soils, climates and management practices assembled for future model users.

For interpool transfer coefficients, the only method appears to be to calibrate these parameters. Ma et al. (1998a) parameterized the interpool transfer coefficients through calibrating corn yield and crop N uptake in a manure study in Colorado. Jaynes and Miller (1999), on the other hand, adjusted these interpool coefficients by assuming that total soil organic matter content remains stable and should be equal to the measured organic matter content in soils under consistent long-term management.

Most agricultural system models simulate mineralization/immobilization, nitrification, denitrification, urea hydrolysis and ammonia volatilization. Ma and Shaffer (2001) summarized these processes in nine major U.S. models and McGechan and Wu (2001) reviewed the same processes for several European models. Each process may be simulated as zero-order, half-order, first-order, Michaelis–Menten, or Monod kinetics (Ma and Shaffer, 2001) and the rate coefficients depend on the type of kinetics. Rate coefficients may be derived from experimental data, literature values, or by calibration. For simple models, such as NLEAP and NTRM, most of the parameters are from some experimental results and fixed (Shaffer et al., 1991). For more complex models, such as RZWQM and ecosys, parameters have to be derived from literature and refined by calibration (Ma et al., 2001, Grant, 2001).

Rate coefficients depend on how the organics are partitioned and whether soil microbial activities are included in model simulation (Ma and Shaffer, 2001; McGechan and Wu, 2001). Rate coefficients for mineralization also depend on how environmental factors (e.g., temperature and soil moisture) are considered in the process. In the various models reviewed by Ma and Shaffer (2001) and McGechan and Wu (2001), temperature and water effects are treated very much differently and, therefore, the rate coefficients can be considerably different from each other. As a result, it is meaningless to compare rate coefficients from different models and to use rate coefficients from another model without knowing the exact definition of the rate coefficients. For example, RZWQM includes five soil organic C pools — two for surface residue (fast and slow) and three for soil humus pools (fast, intermediate and slow). Decomposition of soil organic C is simulated individually for each pool with a first-order kinetics (Ma et al., 2001).

$$r_i = -k_i C_i \qquad (14.12)$$

where r_i is the decay rate of ith pool (μg C/g/d) [i = 1 for slow surface residue pool; i = 2 for fast surface residue pool; i = 3 for fast humus pool; i = 4 for intermediate humus pool; i = 5 for slow humus pool]. C_i is carbon concentration (μg C/g soil) and k_i is a first-order rate coefficient (1/d) and is calculated from:

$$k_i = f_{aer} \left(\frac{k_b T}{h_p} \right) A_i e^{-\frac{E_a}{R_g T}} \frac{[O_2]}{\left[H^{kh} \gamma_1^{kh} \right]} P_{het} \qquad (14.13)$$

where A_i is the rate constant for pool i, $[O_2]$ is O_2 concentration in the soil water with assumption that oxygen in soil air is not limited (moles O_2/liter pore water), H is the hydrogen ion concentration (moles H/liter pore water), γ_1 is the activity coefficient for monovalent ions ($1/\gamma_1^{kh}$ = 3.1573 × 10^3 if pH > 7.0 and $1/\gamma_1^{kh}$ = 1.0 if pH ≤ 7.0), kh is hydrogen ion exponent for decay of organic matter (= 0.167 for pH ≤ 7.0 and = –0.333 for pH > 7.0), P_{het} is the population of aerobic heterotrophic microbes [no. of organisms/g soil, minimum 50,000], k_b is the Boltzman constant (1.383 × 10^{-23} J/K), T is soil temperature (K), h_p is the Planck constant (6.63 × 10^{-34} J.s), R_g is the universal gas constant (1.99 × 10^{-3} kcal/mole/K), E_a [= 15.1 + 12.3 U; U is ionic strength (mole)] is the apparent activation energy (kcal/mole) and f_{aer} is a soil aeration factor and is estimated from Linn and Doran (1984).

Based on Eqs. (14.12) and (14.13) and assigned value for each variable in the equations, the model developers derived a set of rate constants (A_i). A_1 and A_2 are derived from measured residue decomposition data and A_3, A_4 and A_5 are derived by assuming a turnover time of 5, 20 and 2000 years for the fast, intermediate and slow humus pools (Ma and Shaffer, 2001). The derived A_i values with a derived unit of s/d (second/day) are A_1 = 1.67 × 10^{-7}, A_2 = 8.14 × 10^{-6}, A_3 = 2.5 × 10^{-7}, A_4 = 5.0 × 10^{-8} and A_5 = 4.5 × 10^{-10}. These values depend on the previous equation and have no meaning to other models of a different structure. The total amount of decayed organic C/N is either transformed into another C pool, released as inorganic CO_2 and NH_4, or assimilated into microbial biomass (immobilization) based on calibrated partitioning fractions.

Similarly, rate constants for nitrification, denitrification, ammonia volatilization and urea hydrolysis are model dependent (Ma and Shaffer, 2001; McGechan and Wu, 2001). McGechan and Wu (2001) compiled a list of model parameters for C/N dynamics and found that their values varied from model to model and from author to author. Obtained values from one model may be used as reference only in the context of model structure. For example, in RZWQM, a zero-order kinetics is used for nitrification and a first-order kinetics for denitrification. The corresponding zero- and first-order rate coefficients are:

$$k_0 = f_{aer}\left(\frac{k_bT}{h_p}\right)A_{nit}e^{-\frac{E_{an}}{R_gT}}\frac{[O_2]^{1/2}}{[H^{kh}\gamma_1^{kh}]}P_{aut}$$ (14.14)

and

$$k_1 = f_{anaer}\left(\frac{k_bT}{h_p}\right)A_{den}e^{-\frac{E_{den}}{R_gT}}\frac{C_s}{[H^{kh}\gamma_1^{kh}]}P_{ana}$$ (14.15)

where P_{aut} is the autotrophic microbial population (nitrifiers) [number of organisms/g soil, minimum 500]; P_{ana} is the population of anaerobic microbes for denitrification [number of organisms/g soil, minimum 5000]; E_{an} and E_{den} are the apparent activation energy for nitrification and denitrification processes; C_s is a weighted soil organic C in the soil; and f_{aer} and f_{anaer} are the water effects (Ma et al., 2001). The rate constants, A_{nit} and A_{den}, have calibrated values of 1.0×10^{-9} and 1.0×10^{-13} second/day/organism. Again, these A values are dependent on Eqs. (14.14) and (14.15) and they are only approximated within an order of magnitude.

Urea hydrolysis is not a major concern in most models (Ma and Shaffer, 2001). For those models that simulate urea hydrolysis such as CERES and RZWQM, urea hydrolysis is simulated as a first-order kinetics. The first-order rate coefficient for urea hydrolysis in RZWQM is written as:

$$k_{urea} = f_{aer}\left(\frac{k_bT}{h_p}\right)A_{urea}e^{-\frac{E_u}{R_gT}}$$ (14.16)

where E_u is the activation energy for urea hydrolysis (12.6 kcal/mole). A_{urea} is rate constant for urea hydrolysis (2.5×10^{-4} second/day).

Ammonia volatilization is usually simulated as first-order kinetics and its rate depends on wind speed, soil CEC, soil pH and air temperature (Ma and Shaffer, 2001). In RZWQM, ammonia volatilization is simulated based on partial pressure gradient of NH_3 in the soil (P_{NH_3}, atm) and air (P'_{NH_3}, 2.45×10^{-8} atm):

$$VOL = -K_vT_f\left(P_{NH_3} - P'_{NH_3}\right)C_{NH_4}$$ (14.17)

where C_{NH_4} is the concentration of NH_4 in the soil (mole N/liter of pore water). K_v is a volatilization rate coefficient affected by wind speed (W, km/d) and soil depth (z, cm):

$$K_v = 4.0 \times 10^3 \ln(W)e^{-0.25Z}$$ (14.18)

The temperature factor, T_f is calculated from:

$$T_f = 2.9447 \times 10^4 e^{\frac{-6.0}{1.99 \times 10^{-3}(T+273)}}$$ (14.19)

where T is soil temperature (°C). P_{NH_3} is calculated from equilibrium chemistry between soil NH_4 and NH_3.

Table 14.7   Regression Variables and Constants for Various N Processes in NRTM

Urea Hydrolysis Equation[a]

Urea hydrolysis rate (ppm/day) = Con + b_1 $\log_{10}(T_s)$ + b_2 $\log_{10}$ (Urea-N)

$$Con = 4.13 \ 10^2 \qquad b_1 = -1.56 \ 10^2 \qquad b_2 = -1.53 \ 10^2$$

Minearlization–Immobilization Equation

Mineralization–immobilization rate (ppm/day) = Con + b_1 T_s + b_2 (Organic-N) + b_3 $\log_{10}$ (NH_4-N)

$$Con = 8.92 \ 10^{-1} \qquad b_1 = 2.16 \ 10^{-3} \qquad b_2 = 2.70 \ 10^{-2} \qquad b_3 = 3.92 \ 10^{-1}$$

Nitrification Equation

Nitrification rate (ppm/day) = Con + b_1 T_s (NH_4-N) + b_2 $\log_{10}(NH_4$-N) + b_3 $\log_{10}(NO_3$-N)

$$Con = 4.64 \qquad b_1 = 1.62 \ 10^{-3} \qquad b_2 = 2.38 \ 10^{-1} \qquad b_3 = -2.51$$

Nitrate-N Immobilization Equation

Nitrate-N immobilization rate (ppm/day) = Con + b_1 T_s/(organic-N)^2 + b_2 exp (T_s) + b_3 (T_s (organic-N) – (NO_3-N))/(organic-N)

$$Con = 0.0 \qquad b_1 = 1.52 \qquad b_2 = 3.23 \ 10^{-15} \qquad b_3 = -4.90 \ 10^{-3}$$

[a] T_s is soil temperature.
From Dutt, G.R. et al., *Tech. Bull.*, 196, Arizona Agricultural Experiment Station, Tucson, 1972.

Some simpler models do not simulate all the soil C and N processes and thus require fewer model parameters. For example, the CENTURY model does not simulate nitrification process and nitrate is only inorganic N form. Similarly, NTRM does not simulate denitrification and ammonia volatilization processes. NLEAP and CENTURY models do not simulate urea hydrolysis. In case of urea application, the models assume equivalent NH_4 application. Although ammonia volatilization was simulated as first-order kinetics in most models, volatilization loss of N is assumed to be 5% of total N mineralized in the CENTURY model. Some simpler models, such as NTRM, use regression equations to calculate all the processes (Table 14.7). Although these equations may be experiment dependent and may not be applicable to other conditions, this type of model does release the burden of obtaining model parameters from model users.

Environmental Stress Factors

As shown previously, the C and N soil processes are affected by a variety of environmental factors, such as soil water, soil temperature, soil pH and soil aeration. The Arrhenius equation was used to describe various C and N dynamics in RZWQM as described in Eqs. (14.13) to (14.16); however, these factors and their interactions are simulated differently in different models (Ma and Shaffer, 2001; McGechan and Wu, 2001). Most models use a 0-1 factor to reduce an "ideal" rate constant under no stress conditions (e.g., CERES and EPIC). These factors may affect a process independently (multiply the factors), such as in EPIC (Williams, 1995), or dependently (the most stressful factor prevails), such as in CERES (Godwin and Jones, 1991; Godwin and Singh, 1998). Parameters associated with each stress factor are generally calibrated and experiment dependent (McGechan and Wu, 2001).

CROP GROWTH PARAMETERS

Process-level plant growth models have a large number of parameters that control plant growth and development processes, such as photosynthesis, respiration, carbon allocation, phenology, development and yield, as well as their dependence on temperature, water and nutrient stresses. These parameters are generally crop species and variety specific. For simple models developed for environmental quality purposes, such as NLEAP, a plant growth curve based on yield goal and

Table 14.8   Crop Growth-Related Model Parameters for Users to Calibrate RZWQM

Parameter Name	Definition
CNUP1	Maximum daily N uptake (g/plant/day) used in the Michaelis–Menten equation — Increasing this parameter causes an increase in active uptake, which will result in an increase in yield.
ALPHA	Proportion of photosynthate used for maintenance respiration — Increasing this parameter results in a decrease in biomass.
CONVLA	Conversion factor from biomass to leaf area index (g/LAI) — Increasing this parameter causes a decrease in total plant production.
CLBASE	Plant density on which CONVLA is based (no. of plants/ha)
SLA3	Factor to reduce photosynthetic rate at propagule stage (0-1) — This parameter along with SLA4 will adjust harvest index.
SLA4	Factor to reduce photosynthetic rate at seed production stage (0-1) — It is the same as SL3, which is used to adjust harvest index.
RDX	Maximum rooting depth under optimal conditions (cm) — Increasing rooting depth will promote early penetration of roots and generally increase total plant production.
RST	Minimum leaf stomatal resistance (s/m)
SUFNDX	Nitrogen sufficiency index threshold below which automatic fertilization is triggered (0-1)
EFFLUX	Plant luxurious nitrogen uptake efficiency factor (0-1)

Table 14.9   Parameter Estimates for Field Corn and Summary Statistics for the Five Regional Parameters of the RZWQM Generic Crop Growth Component for the Five Primary MSEA Sites and Colorado

Parameter	IA	MN	MI	NE	OH	CO[d]	CO[i]	Mean	SD[a]	CV[b] (%)
CNUP1	1.5	1.5	2.0	1.5	2.5	1.5	1.5	1.71	0.39	23.0
ALPHA	0.15	0.25	0.008	0.19	0.29	0.3	0.28	0.21	0.10	49.9
CONVLA	10.0	10.0	9.5	9.5	11.50	24.0	12.0	10.42	1.07	10.3
SLA3	0.90	0.90	0.97	0.85	0.85	0.92	0.78	0.88	0.06	6.9
SLA4	0.85	0.85	0.95	0.64	0.70	0.94	0.78	0.82	0.12	14.2

Note: CO[d] is dryland agriculture in Colorado and CO[i] is irrigated agriculture in Colorado.

[a] Standard deviation of the mean.
[b] Coefficient of variation.

From Hanson, J.D. et al., *Agron. J.,* 91:171–177, 1999.

growth duration is used only for water and nitrogen uptake purposes (Shaffer et al., 2001). Another simple plant growth model is used in the EPIC model (Williams, 1995), where phenological development is solely based on heat units and potential growth is calculated from canopy-intercepted solar radiation. The EPIC model has simple ways to calculate plant biomass, leaf area index (LAI) and root growth. Crop yield is estimated from the harvest index (Williams, 1995). This type of model does not require extensive model parameterization.

More complex models, such as the CERES family model (Ritchie et al., 1998), CROPGRO (Boote et al., 1998), RZWQM (Hanson, 2000) and *ecosys* (Grant, 2001), have more detailed process-based crop growth components and, thus, require more parameters. Model developers, however, have made extraordinary efforts to reduce the number of model parameters that require calibration by the user. For example, RZWQM includes ten crop related parameters for users to calibrate plant growth (Table 14.8) and the rest are either hard coded in the model or in a text file called PLGEN.DAT. Table 14.9 gives calibrated values for the five most sensitive parameters for corn in the Midwestern states of the U.S. Experienced users may modify sections in the PLGEN.DAT file for further model calibration (Hanson et al., 1999; Hanson, 2000).

The CERES-Maize model includes six crop parameters for users to calibrate (Table 14.10). These six parameters are cultivar related and the species-specific parameters are available in a

Table 14.10   Crop Growth-Related Model Parameters for Users to Calibrate CERES-Maize

Parameter Name	Definition
P1	Thermal time from seedling emergence to the end of the juvenile phase (expressed in degree days above a base temperature of 8°C), during which the plant is not responsive to changes in photoperiod
P2	Extent to which development (expressed as days) is delayed for each hour increase in photoperiod above the longest photoperiod at which development proceeds at a maximum rate (which is considered to be 12.5 hours)
P5	Thermal time from silking to physiological maturity (expressed in degree days above a base temperature of 8°C)
G2	Maximum possible number of kernels per plant
G3	Kernel filling rate during the linear grain filling stage and under optimum conditions (mg/day)
PHINT	Phylochron interval — the interval in thermal time (degree days) between successive leaf tip appearances

Table 14.11   Default Parameters for Maize in CERES-Maize Model

Variety	P1	P2	P5	G2	G3	PHINT
Long season	320.0	0.520	940.0	620.0	6.00	38.90
Medium season	200.0	0.300	800.0	700.0	8.50	38.90
Short season	110.0	0.300	680.0	820.4	6.60	38.90
Very short season	5.0	0.300	680.0	820.4	6.60	38.90

Note: See Table 14.10 for definitions of parameters.

species file with extension of SPE (Tsuji et al., 1994). Table 14.11 gives calibrated values of the six parameters for typical corn cultivars. The model also provides a database of the six parameters for all the cultivars tested.

The CROPGRO-Soybean model has 15 crop parameters that can be modified by users for each cultivar (Table 14.12) and the model also provides a database for tested cultivars. In addition, one file contains parameters specific to soybean (with extension SPE) and another file contains parameters for each ecotype (maturity groups) with an extension of ECO (Tsuji et al., 1994). Table 14.13 gives the 15 calibrated parameters by maturity groups. Experienced users may modify these two files to achieve better model calibration. Model developers are trying to keep these minimum sets of parameters for any future release of CERES-Maize and CROPGRO models.

Environmental Stress Factors

One of the most difficult areas of agricultural system modeling is how to quantify environmental stresses and their effects on various biological processes. Various biological processes are affected by water, nitrogen and temperature stresses. Examples of these processes are photosynthesis, leaf expansion, root growth, carbon allocation/partitioning, tillering, internode elongation, phenology development, grain filling, seed growth rate, pod/grain number and senescence (Hanson, 2000; Ritchie et al., 1998; Boote et al., 1998). The responses of individual processes to these stress factors may be different; for example, in CROPGRO and CERES models, two water stress factors are used — one for leaf expansion, tillering and internode elongation and one for photosynthesis and transpiration (Ritchie, 1998). Similarly, RZWQM includes two N stress factors — one for photosynthesis and vernalization and one for carbon allocation and plant development (Hanson, 2000). In addition, different stress factors are used for above- and belowground processes. For less process-

Table 14.12   Crop Growth-Related Model Parameters for Users to Calibrate CROPGRO-Soybean

Parameter Name	Definition
CSDVAR/CSDL	Critical daylength below which reproductive development proceeds unaffected by daylength and above which development rate is reduced in proportion to hours above CSDVAR (hr)
PPSEN	Slope of relative rate of development for daylength above CSDVAR or sensitivity to photoperiod (1/hr)
PH2T5/EM-FL	The time from end of juvenile phase to first flower in photothermal days under optimal conditions (photothermal days)
PHTHRS(6)/FL-SH	The time from first flower to first pod greater than 0.5 cm in photothermal days, under optimal conditions (photothermal days)
PHTHRS(8)/FL-SD	The time from first flower to first seed in photothermal days under optimal conditions (photothermal days)
PHTHRS(10)/SD-PM	The time from first seed to physiological maturity in photothermal days, under optimal conditions (photothermal days)
PHTHRS(13)/FL-LF	The time from first flower to end of leaf growth in photothermal days, under optimal conditions (photothermal days)
LFMAX	Maximum leaf photosynthesis rate at saturated light level, optimal temperature and CO_2
SLAVAR	Specific leaf area (SLA) for new leaves during peak vegetative growth
SIZLF	Maximum size of fully expanded leaf on the plant under optimal conditions (cm^2)
XFRUIT	Maximum fraction of daily available gross photosynthesis (PG) that is allowed to go to seeds plus shells
WTPSD	Maximum weight per seed under non-limiting substrate (g)
SFDUR	Seed filling duration for a cohort of seed (photothermal days)
SDPDVR/SDPDV	Average seed per pod under standard growing conditions
PODUR	Photothermal days for cultivar to add full pod load under optimal conditions (used to compute rate of pod and flower addition)

based models where crop yield is converted from aboveground biomass using a harvest index, the environmental effect may be added directly onto the harvest index (Williams, 1995).

The way in which stress factors are quantified and applied in various models significantly affects the values and applicability of model parameters. The following subsection describes how these stress factors are calculated and used in RZWQM as a demonstration. Further research is needed to improve the stress responses of plants and their parameterization.

Belowground Processes

Root growth is an important aspect of soil–plant interactions. The amount of photosynthate partitioned into root depends on environmental stress factors (water, temperature, light, nitrogen, etc.). In RZWQM, daily photosynthate is partitioned between root and shoot based on a root/shoot ratio (RATRS):

$$RATRS = \frac{RATRS*}{max(0.5, min(EWP, PNS))} \qquad (14.20)$$

where RATRS and RATRS* are the root/shoot ratio with and without water stress. PNS is whole plant nitrogen stress factor and EWP is water stress factor, which are defined by:

$$PNS = 1.0 - \frac{N_{dmd} - N}{N_{dmd} - N_{min}} \qquad (14.21)$$

$$EWP = 0.15 + 0.85 \times \frac{AT}{PT} \qquad (14.22)$$

Table 14.13   Default Model Parameters for Soybean Provided in the CROPGRO Model

Maturity Group	CSDL	PPSEN	EM-FL	FL-SH	FL-SD	SD-PM	FL-LF	LFMAX	SLAVR	SIZLF	XFRT	WTPSD	SFDUR	SDPDV	PODUR
0	14.10	0.171	16.8	6	13.0	31.00	26	1.030	375	180	1	0.19	23	2.20	10
1	13.84	0.203	17.0	6	13.0	32.00	26	1.030	375	180	1	0.19	23	2.20	10
2	13.59	0.249	17.4	6	13.5	33.00	26	1.030	375	180	1	0.19	23	2.20	10
3	13.40	0.285	19.0	6	14.0	34.00	26	1.030	375	180	1	0.19	23	2.20	10
4	13.09	0.294	19.4	7	15.0	34.50	26	1.030	375	180	1	0.19	23	2.20	10
5	12.83	0.303	19.8	8	15.5	35.00	18	1.030	375	180	1	0.18	23	2.05	10
6	12.58	0.311	20.2	9	16.0	35.50	18	1.030	375	180	1	0.18	23	2.05	10
7	12.33	0.320	20.8	10	16.0	36.00	18	1.030	375	180	1	0.18	23	2.05	10
8	12.07	0.330	21.5	10	16.0	36.00	18	1.030	375	180	1	0.18	23	2.05	10
9	11.88	0.340	23.0	10	16.0	36.50	18	1.030	375	180	1	0.18	23	2.05	10
10	11.78	0.349	23.5	10	16.0	37.00	18	1.030	375	180	1	0.18	23	2.05	10

See Table 14.12 for definitions of parameters.

where AT is actual transpiration and PT is potential transpiration, N_{dmd} is plant nitrogen demand, N_{min} is the minimum N concentration below which the plant will not grow and N is current nitrogen concentration in the plant. PNS and EWP are fitness indices, where 1.0 means optimal growth condition and 0.0 reflects maximum stresses.

In RZWQM, the effect of environmental stresses on root growth per se is adapted from the CERES-Maize model (Jones et al., 1991). In the model, calcium and aluminum stresses are considered also, which will not be discussed here. Interested readers may refer to Jones et al. (1991) or Hanson (2000) for further discussion. Other than the effects on carbohydrate supply (Eq. 14.20), the soil water content influences root growth in two ways: soil strength and soil aeration. The first effect is incorporated into the layer strength stress factor (SST):

$$SST = \sqrt{\frac{SBD \times (SWC - LL)}{(UL - LL)}} \qquad (14.23)$$

where SBD is the soil bulk density, SWC is current soil water content in a soil layer and UL and LL are the upper and lower drain limits and are set to 1/3 bar and wilting point water contents, respectively, in RZWQM. Another soil water related factor is the soil aeration index (SAI):

$$SAI = \frac{1.0 - (1.0 - GMN) \times (WFP - CWP)}{1 - CWP} \qquad (14.24)$$

where GMN is fraction of normal root growth at saturation, WFP is water filled pore space in a soil layer and CWP is critical water filled pore space at which root growth stops. CWP is estimated from the clay content of a soil layer (CLA) as:

$$CWP = 0.7 - 0.002 \times CLA \qquad (14.25)$$

The SST and SAI factors affect rooting depth but also root length/weight ratio, root weight accumulation and root senescence (Jones et al., 1991). Daily rooting depth (DRD) increment is calculated by:

$$DRD = RDX \times \frac{GS - GSY}{GSR} \times min\left(STP, \sqrt{Min(SST, SAI)}\right) \qquad (14.26)$$

where STP is soil temperature stress:

$$STP = sin\left(\frac{\frac{\pi}{2} \times (T_s - TBS)}{TOP - TBS}\right) \qquad (14.27)$$

where T_s is soil temperature in a soil layer, TOP and TBS are the optimum and base temperature for root growth, RDX is maximum rooting depth, GSR is the growth stage at which maximum rooting depth reaches and GS and GSY are growth stages of current and previous days, repectively.

Root length/weight ratio (LWA) is calculated from:

$$LWA = \frac{LWN}{1.0 + 3.0 \times (1.0 - SST)} \qquad (14.28)$$

where LWN is the root length/weight ratio under normal conditions as determined by growth stage.

Potential root growth in a soil layer (DPL) is a function of soil plasticity, soil thickness, amount of biomass partitioned to root system daily and environmental stress factors.

$$DPL = WCP \times DMDRT \times THF \times min(SST, STP, SAI) \qquad (14.29)$$

where WCP is a weighting coefficient for plasticity, DMDRT is the amount of biomass partitioned to root system daily and THF is a coefficient to adjust soil layer thickness.

Daily root death in a soil layer is calculated from existing root biomass and daily biomass increment.

$$DWL = \left(RWL \times RTMNT + RGA \times (1.0 - ETP)\right) \times \left(1.0 - \sqrt{min(SST, SAI)}\right) \qquad (14.30)$$

where DWL is daily root death rate, RWL is root weight in the layer, RTMNT is root maintenance requirement, RGA is actual root growth in the layer and ETP is whole plant temperature stress factor:

$$ETP = \left(a_1 \times a_2^{a_3}\right)^z \qquad (14.31)$$

where z is a constant and a value of 1.328 is used in RZWQM and a_1, a_2 and a_3 are functions of soil temperature (T). Also,

$$a_1 = \frac{T_{max} - T}{T_{max} - T_{opt}}, \, a_2 = \frac{T - T_{min}}{T_{opt} - T_{min}}, \, a_3 = \frac{T_{opt} - T_{min}}{T_{max} - T_{opt}} \qquad (14.32)$$

where T is the soil temperature. T_{max}, T_{opt} and T_{min} are the corresponding maximum, optimum and minimum temperatures for growth. Parameters associated with root growth are in the PLGEN.DAT file.

Aboveground Processes

Environmental stress (temperature, water and nitrogen) effects on various aboveground processes can vary considerably from model to model. In simple models such as EPIC, stresses mainly affect biomass, leaf area index, plant height and harvest index (Williams, 1995). In more complex models such as CERES and CROPGRO, stresses can affect other processes such as internode enlongation, pod addition and grain filling (Ritchie et al., 1998; Boote et al., 1998). Parameterization of stress effects on each individual process is more difficult for detailed modeling approaches, since all measurable experimental observations result from multiple processes. Default values are usually provided in a database distributed with the model (Tsuji et al., 1994).

RZWQM has a generic crop growth model that has a complexity inbetween EPIC and CROPGRO. It simulates biomass, leaf area index, plant height, yield and plant population development (Hanson, 2000). Plant growth and development are based on calendar days and the temperature stress factor (ETP) is used to incorporate the heat effect. For example, the photosynthesis rate is calculated by:

$$PLPROD = PLPROD* \times ETP \times min(ENP, EWP) \qquad (14.33)$$

where PLPROD and PLPROD* are daily photosynthesis rates with and without stresses, respectively. ENP is another N stress factor used in RZWQM:

$$ENP = \begin{cases} 0.25 + 0.75 \times \left(1.0 - e^{-EFFN \times SPCTN}\right), \; SPECTN > 0 \\ 0.25, \; SPECTN \leq 0 \end{cases} \qquad (14.34)$$

where EFFN is nitrogen use efficiency coefficient and SPECTN is nitrogen concentration in leaf and stem above minimum value required for growth.

In RZWQM, shoot death may also be initiated at water stress:

$$DS1 = SDWMAX \times BIOSHT \times (1.0 - EWP) \qquad (14.35)$$

where DS1 is daily shoot death due to water stress, SDWMAX is maximum death in percentage of shoot biomass and BIOSHT is biomass of shoot.

In RZWQM, not all the plants are advancing at the same speed throughout the growing cycle. The model divides a plant cycle into seven phenology stages:

1. Dormant
2. Germinating
3. Emerged
4. Established
5. Vegetative
6. Reproductive
7. Senescent

Plants in one stage may stay in that stage or advance to the next stage. A probability function P(i,j) is used to represent the percentage of plants in stage j advanced to stage i. When i = j, plants are staying at stage j. Probability P(i,j) may be modified based on environmental fitness. P(2,1) is based on germination rate after the seeds encounter with certain ten-day average soil temperature and five-day average soil moisture content. For j = 2, 3, 4, 5, or 6, P(j+1, j) is modified as:

$$P(j+1, j) = P^*(j+1, j) \times max(0.9, ETP \times min(EWP, PNS)) \qquad (14.36)$$

where P* is the probability without environmental stresses. When vernalization is required, P(4,3) is calculated as:

$$P(4,3) = FV \times P^*(4,3) \times ETP \times min(EWP, ENP) \qquad (14.37)$$

where FV is a vernalization factor. Parameters required to quantify the stress effects in RZWQM are provided in the PLGEN.DAT file.

PESTICIDE PARAMETERS

Simulation models for pesticide processes in soils have been developed for more than two decades because of its importance in environmental protection. It could be one of the few areas where simulation models have extensive impact on policy makers. The U.S. Environmental Protection Agency (EPA) has officially used PRZM to register pesticide usage in agriculture (Carsel et al., 1985). The common parameterization needs in pesticide chemistry are for adsorption–desorption, degradation, volatilization, leaching and losses to runoff water. Depending on the time scale of intended use of a model, these processes may be simplified and combined.

Extensive studies have been conducted on adsorption equilibrium, kinetics and hysteresis (Ma and Selim, 1994; 1996; Ma et al., 1993; Cheng, 1990; Linn et al., 1993). For most agricultural field scale models, an equilibrium adsorption of pesticide onto soil is assumed in the form of the Freundlich or Langmuir equation (Ma and Selim, 1996). The Freundlich equation can be reduced to a linear adsorption isotherm when the exponential constant (n) equals unity:

$$S = K_f C^n \tag{14.38}$$

or

$$S = K_d C \quad \text{when } n = 1 \tag{14.39}$$

where K_f and n are Freundlich constants and K_d is the distribution coefficient. These parameters are usually obtained through batch experiments, although column studies have also been used to derive these parameters (Ma and Selim, 1996; Lion et al., 1990; Green and Karickhoff, 1990). Many studies have been conducted to relate K_d to soil properties, such as soil organic carbon, soil clay content and soil specific surface (Ma and Selim, 1996; Green and Karickhoff, 1990) and found that K_d is highly correlated to soil organic carbon. Therefore, a K_{oc} [= K_d/(% of soil organic carbon)] value is defined to normalize pesticide adsorption among different soils.

For soils where no experimental data are available, K_{oc} values may be estimated from pesticide physical properties. These properties include octanol-water partition coefficient (K_{ow}), aqueous solubility, molecular surface area, melting point and pure pesticide molar volume (Green and Karickhoff, 1990; Gerstl, 1990). Table 14.14 shows some of the semiempirical equations from the literature compiled by Green and Karickhoff (1990). K_{oc} values for commercial pesticides are listed by Wauchope et al. (1992) and Hornsby et al. (1996). It should be noted that K_{oc} for a pesticide is not a constant for a given soil, because factors other than soil organic carbon, such as clay content and soil specific surface, may also be important in pesticide adsorption (Ma and Selim, 1996; Green and Karickhoff, 1990). It is always recommended to measure K_d for each pesticide–soil combination.

Table 14.14 Semiempirical Equations for Estimating K_{oc} from Solute Physical Properties

	Calibration Compounds[a]	K_{oc} Range	No. of Compounds	r^2	Reference
Log K_{oc} = −0.68 log $S(\mu g/ml)$ + 4.273	pn, ha, aa, ch	10–10^6	23	0.930	Hassett et al., 1980
Log K_{oc} = −0.557 log $S(\mu mol/L)$ + 4.277	ch	10–10^5	15	0.990	Chiou et al., 1979
Log K_{oc} = −0.83 log S(mole fraction)[b] −0.01 (mp − 25°C)[c] − 0.93	pn, ch, ct, cb, op	10^2–10^6	47	0.930	Karickhoff, 1984
Log K_{oc} = −0.729 log S(molar) −0.0073 (mp − 25°C) + 0.24	pn, ch	10–10^5	12	0.996	Chiou et al., 1983
Log K_{oc} = −0.808 log $[vS(molar)]$[d] −0.0081 (mp − 25°C) − 0.74	pn, ch	10–10^5	12	0.997	Chiou et al., 1983
Log K_{oc} = 0.904 log K_{ow} − 0.539	pn, ch	10–10^5	12	0.989	Chiou et al., 1983
Log K_{oc} = 1.029 log K_{ow} − 0.18	ch, ct, cb, op, ur, pa	10–10^5	13	0.910	Rao & Davidson, 1982
Log K_{oc} = log K_{ow} − 0.317	pn, ha, aa, ch	10–10^6	23	0.980	Hassett et al., 1980
Log K_{oc} = 0.72 log K_{ow} + 0.49	cb, mb	10^2–10^4	13	0.950	Schwartzenbach and Westall, 1981

[a] pn = polynuclear aromatic hydrocarbons; ha = heteronuclear aromatic hydrocarbons; aa = aromatic amines; ch = chlorinated hydrocarbons; ct = chloro-s-triazines; cb = carbamates; op = organophosphates; ur = uracils; pa = phenoxy acids; mb = methylated benzenes.
[b] For hydrophobic compounds, the solubility, S(mole fraction) ≈ S(molar) × 18/1000.
[c] mp = melting point in degrees (Celsius; reference temperature 25°C; for liquids mp set at 25°C, crystal term vanishes).
[d] v = pure solute molar volume (L/mol).

From Green, R.E. and S.W. Karickhoff, Sorption estimates for modeling, in *Pesticides in the Soil Environment: Processes, Impacts and Modeling*, SSSA, Madison, WI, 79–101, 1990.

For a more mechanistic pesticide model, adsorption kinetics may be needed. The most commonly used assumption is that two types of adsorption sites are located in the soil — one has a fast adsorption rate and the other has a relative slow rate. The fast adsorption site is assumed to be in equilibrium with pesticide in the soil solution and the slow site is kinetic in nature (Ma and Selim, 1994, 1996). In formulating reaction rate equations for the two sites, one can assume that the reaction rates are only functions of pesticide concentration in the solution phase, or functions of both pesticide concentrations in the liquid phase and adsorption sites on the soil. Ma and Selim (1998) compared these two approaches and found that the latter is better in predicting pesticide retention in soils. When multiple reaction site models are used for pesticide adsorption/desorption, one needs to partition the total adsorption sites between the two sites. Ma and Selim (1994) proposed a new model formulation that does not need such a partitioning parameter. Selim et al. (1999) further compared the new formulation with the ones in the literature with different recommended partitioning parameter values and found that the new model provided better pesticide retention and transport in soils. Parameters for kinetic reactions are largely obtained from individual experiments (batch or column) by fitting the model to measured pesticide in soil solution with time in batch slurry or column effluent (Ma and Selim, 1994, 1996), although efforts have been seen in the literature to use directly measured pesticide adsorbed on the soil solid phases (Gaston and Locke, 1994).

Pesticide degradation in soils may be from the solution phase or the adsorbed phase depending on the mechanism (chemical or biological) (Ma and Selim, 1994; Xue et al., 1995). The pesticide degradation constant is expressed as a half-life ($t_{1/2}$) based on pesticide disappearance from the soil system. Apparent half-lives in the field for most pesticides are documented in Wauchope et al. (1992) and Hornsby et al. (1996). When a first-order kinetics is used, one can estimate a disappearance rate constant (k_t) from the half-life:

$$k_t = \frac{0.6932}{t_{1/2}} \tag{14.40}$$

Pesticide volatilization process and associated parameters have been reviewed by Taylor and Spencer (1990). Pesticide water solubility and vapor pressure are the determining factors for pesticide volatilization, as described by Henry's law. Databases for pesticide solubility and vapor pressure are available in Taylor and Spencer (1990), Wauchope et al. (1992) and Hornsby et al. (1996). As with the degradation process, volatilization is usually treated as first-order kinetics. Field measured disappearance rates (k_t) are usually the sum of degradation and volatilization rates (Taylor and Spencer, 1990).

Wauchope et al. (2000) presented the most comprehensive discussion of modeling pesticide processes to date, which has incorporated into RZWQM. This pesticide model takes into account pesticide application efficiency, washoff from foliage and crop residue, volatilization, degradation, multiple-site pesticide adsorption, pesticide plant uptake, multiple pesticide species (neutral, cationic and anionic) and pesticide extraction into runoff water. In Table 14.15, we present a list of all possible parameters that could be used in their pesticide model, but all are not required to run the model.

The so-called "Molecular Parameters" listed in Table 14.15 are available from a handbook of chemistry and are hard-coded into the model including their associated temperature or other dependencies. Most other parameters listed vary with the soil type, climate and cropping systems and they have to be measured for each specific location. For the parameters listed in boldface, several authors have assembled available values in pesticide databases (Wauchope et al., 1992; Hornsby et al., 1996; Kellog et al., 1994). The databases provide the range of values for any given parameters reported in the literature. They are not likely to be accurate for any specific location, but serve as a starting point for further refinement by calibration. For other parameters not available

Table 14.15 Pesticide Parameters Used in R2WQM

Identity	Soil "Surface" Parameters[d]
Name	Runoff mixing constant
Ancestry Code	Volatilization half-life[e]
Transformation Process Code	Photodegradation half-life
Molecular Paramenters	**Soil Subsurface Parameters[d]**
Molecular weight (g mol^{-1})	Soil organic carbon sorption constant (ml g^{-1})
Aqueous solubility (mg L^{-1})	**pH of source soil for organic carbon sorption constant**
Aqueous diffusion coefficient (cm^2/h)	Freundlich organic carbon sorption constant
Vapor pressure (mm Hg)	Freundlich isotherm exponent
Henry's Law Constant[a]	Anion organic carbon sorption constant[e]
Octanol/water partition coefficient	Cation organic carbon sorption constant[e]
Acid dissociation constant[b] (mg L^{-1})	Fraction of kinetic sorption sites
Base dissociation constant[b] (m L^{-1})	Soil desorption half-life
	Soil binding half-life
Application Parameters	Optimal parameters:
Application method	Aerobic biodegradation references half-life
Application rate (kg ha^{-1})	Reference moisture content for reference aerobic half-life
Application deposition efficiency (%)	Reference temperature of reference aerobic half-life
Controlled-release rate (kg ha^{-1} day^{-1})	Activation energy for aerobic half-life (KJ/mol°C)
	Activation energy for anaerobic half-life (KJ/mol°C)
Foliage Parameters	Walker constant for soil moisture effects on aerobic degradation
Half-life (d)	Maximum factor for half-life change with depth
Washoff fraction (%)	Anaerobic soil degradation half-life
Washoff power term (mm^{-1})	Abiotic or user-defined degradation half-life
Photodegradation half-life	Daughter or granddaughter transformation process code
Abiotic or user-defined degradation half-life	Daughter or granddaugher formation yield fraction
Plant Residue Parameters[c]	
Half-life	
Washoff fraction (%)	
Washoff power term (mm^{-1})	
Photodegradation half-life	
Abiotic or user-defined degradation half-life	

Note: Most may be replaced or adjusted by the user. Not all are required to run the model. Parameters that are in the database are indicated by boldface. All half-lives are in days.

[a] Calculated from pesticide vapor pressure, molecular weight and aqueous solubility.
[b] Value in database only if applicable
[c] Set by default to foliar values
[d] All subsurface parameters except anaerobic half-life also apply to soil surface layer.
[e] Calculated by a standard equation or user input.

From Wauchope, R.D. et al., Pesticide processes, in *Root Zone Water Quality Model,* Water Resources Publications, Highland Ranch, CO, 163–244, 2000.

in the databases but essential to run the model, Wauchope et al. (2000) gave default values. Some of the parameters are not essential but are listed in Table 14.15 only for advanced users who may want to try advanced algorithms for certain processes.

STATISTICAL METHODS OF MODEL CALIBRATION

The goodness of model parameterization by calibration is largely based on data availability and tolerance of model users. It depends on how representative measurements are in space and time against measurements in which the model is calibrated. Optimization algorithms have been developed for models where only one or two parameters need to be calibrated (Toride et al., 1995). Problems in using an optimization scheme for agricultural system models include:

1. Too many model parameters/processes and difficulty to deciding which one to optimize
2. Too few complete experimental data sets that can be used to parameterize all the components of a system model
3. Too much uncertainty in field measurements
4. Measured heterogeneous data range from physical, chemical, to biological attributes
5. Ill-Defined initial and boundary conditions and difficulty differentiating simulation errors due to model parameters from those due to initial and boundary conditions

As a result, agricultural system models are usually parameterized by trial-and-error or iterative processes.

In most model parameterization processes, researchers try to match simulated and experimental results as closely as possible, regardless of experimental error, by adjusting a few critical model parameters. They may also play with the initial conditions within a reasonable range when initial conditions are not defined or measured. The goodness of model parameterization is quantified by some statistical indices and judged by model prediction for other experimental conditions. One of the most commonly used statistical indices is the root mean square error (RMSE). For a data set with M measured points, RMSE is defined as:

$$\text{RMSE} = \sqrt{\frac{\sum_{i=1}^{M} w_i \left(P_i - O_i\right)^2}{M}} \tag{14.41}$$

where w_i is the weight factor often set equal to 1.0 and P_i and O_i are the model predicted and experimental measured points, respectively. The M observed data points may be from one treatment or multiple treatments (Ma and Selim, 1994). The RMSE reflects a magnitude of the mean difference between experimental and simulation results. A normalized objective function (NOF) may be calculated from RMSE as (Ma et al., 1998b):

$$\text{NOF} = \frac{\text{RMSE}}{O_{avg}} \tag{14.42}$$

where O_{avg} is the averaged observed value. NOF = 0 indicates a perfect match between experimental and modeling results. NOF < 1 may be interpreted as simulation error of less than one standard deviation around the experimental mean. Another index is the mean bias error (ME) (Shen et al., 1998):

$$\text{ME} = \frac{\sum_{i=1}^{M} \left(P_i - O_i\right)}{M} \tag{14.43}$$

The value of ME indicates whether there is a systematic bias in the prediction. A positive value means an over prediction and a negative value indicates an overall under prediction. Another commonly used approach is to conduct regression analysis between measured and predicted outputs. A coefficient of determination (r^2) is then calculated as:

$$r^2 = \frac{\left[\sum_{i=1}^{M} \left(O_i - O_{avg}\right)\left(P_i - P_{avg}\right)\right]^2}{\sum_{i=1}^{M} \left(O_i - O_{avg}\right)^2 \sum_{i=1}^{M} \left(P_i - P_{avg}\right)^2} \tag{14.44}$$

The r^2 value ranges from 0 to 1. $r^2 = 1$ indicates a perfect correlation between experimental and simulation results and $r^2 = 0$ means no correlation between the two results. The r^2 approach alone can be misleading as it does account for a systematic bias. Some model users simply use the percentage difference between simulated and measured results as a criterion for goodness of model parameterization (Hanson et al., 1999). Wu et al. (1996, 1999) used three other statistical indices to compare RZWQM simulation results with measured ones. They are the maximum difference (MD), the modeling efficiency (EF) and the D-index, which are defined by:

$$MD = \max \left| P_i - O_i \right|_{i=1}^{M} \tag{14.45}$$

$$EF = 1 - \frac{\sum_{i=1}^{M} \left(P_i - O_i \right)^2}{\sum_{i=1}^{M} \left(O_i - O_{avg} \right)^2} \tag{14.46}$$

$$D = 1 - \frac{\sum_{i=1}^{M} \left(P_i - O_i \right)^2}{\sum_{i=1}^{M} \left(\left| P_i - O_{avg} \right| + \left| O_i - O_{avg} \right| \right)^2} \tag{14.47}$$

EF is a measure of the deviation between model predictions and measurements in relationship to the scattering of the observed data. EF = 1 indicates a perfect match between simulation and observed results. The D-index is similar to EF but more sensitive to systematic model bias. It also has values ranging from 0 to 1, where D = 1 means perfect simulation. A slight different version of MD is to add all the absolute differences between the simulated and observed results (Johnsen et al., 1995). Loague and Green (1991) used a coefficient of residual mass (CRM):

$$CRM = \frac{\left(\sum_{i=1}^{M} O_i - \sum_{i=1}^{M} P_i \right)}{\sum_{i=1}^{M} O_i} \tag{14.48}$$

The coefficient of residual mass (CRM) tests more of integrated values. It may be used for chemical load to groundwater.

Paired t-test has also been used to assess model accuracy in the literature. Ma et al. (1998c) applied the paired t-test to evaluate RZWQM simulations of surface runoff. The t-statistic was calculated by:

$$t_{M-1} = \frac{D_m - D_e}{SE_D} \tag{14.49}$$

where M is the number of measurements, t_{M-1} is the calculated t-statistic with M-1 degrees of freedom, D_m is the mean of the differences between measured and predicted results, D_e is the expected difference (zero) and SE_D is the standard error of the paired difference.

Comparison of multiple models for their accuracy in simulating a common data set is rather difficult at this time. One method would be to conduct a paired t-test for each individual model and find out which one provides better simulation results compared with observed data (Ma et al., 1998d). When the models are only different in the number of parameters, an F test may be applicable to test significant improvement by adding more model parameters (Amacher et al., 1998).

$$F(p_2 - p_1, M - p_2) = \frac{(rss_1 - rss_2)/(p_2 - p_1)}{rss_2/(M - p_2)} \quad (14.50)$$

where p is the number of model parameters and rss is residual sum of squares. The subscripts 1 and 2 represent the two models.

Besides these statistical indices, simulation results may be compared to observed values in graphical displays (Loague and Green, 1991) or simple tables (Singh and Kanwar, 1995). Graphic displays can intuitively show trends, types of errors and distribution patterns. Graphics may be plotted as linear scale, semi-log scale, or log–log scale depending on how the authors want to present the data (Ma et al., 1995). A table presentation may be good for data with only a few measurements. It is also worthwhile to mention that observed data can be represented in many ways. For example, for a data set with pesticide concentration measurements in a soil profile, one can compare pesticide concentrations in the soil profile, center of mass of pesticide movement, total pesticide leached, maximum depth of pesticide leaching, total pesticide amount in the soil profile, maximum pesticide concentration in the soil profile and average pesticide concentration in the soil profile (Loague and Green, 1991; Ma et al., 2000). Statistical methods may vary with type of observed data.

EXAMPLES OF MODEL CALIBRATION

Efforts needed to parameterize a model largely depend upon the complexity of the model. Some models have detailed simulation of all the processes in an agricultural system, e.g., RZWQM (Ahuja et al., 2000) and *ecosys* (Grant, 2001), while others have only simplified approaches to some processes. For example, the CERES and CROPGRO family of crop growth models have been used in many parts of the world for their detailed plant growth components, but they use a simple soil water balance routine, which requires only saturated soil water content and the drained upper and lower limits of soil water contents (Ritchie, 1998). Thus, it is relatively easy to parameterize the soil water component of CERES and CROPGRO. Also, because many of the processes are related, calibration of one component will affect others in the agricultural system. Therefore, model calibration is usually an iterative process.

The Root Zone Water Quality Model (RZWQM)

Parameterization of RZWQM has been outlined by Hanson et al. (1999) and Rojas et al. (2000). This model provides default values for most of the model parameters (e.g., soil hydraulic properties for each soil class, plant growth, C/N soil dynamics, etc.), although it does require minimum data, which users must supply (Table 14.16). It is recommended to calibrate soil water and chemical transport first, then soil C/N dynamics and finally plant growth (Hanson et al., 1999; Rojas et al., 2000). The process is then iterated a couple of times to obtain final calibration. The model has different debug flags allowing users to check on detailed water, nitrogen and pesticide information (Rojas et al., 2000).

Hanson et al. (1999) has laid out the detailed calibration procedures for RZWQM and potential effects of calibrated model parameters. Table 14.17 shows how to calibrate the water balance component of RZWQM and the processes that need to be calibrated for selected soil properties. In most cases, calibrating the 1/3 bar soil water content has a major effect on soil water balance. Figure 14.3 shows the calibration procedure for soil C/N dynamics submodel. Hanson et al. (1999) also suggested adjusting five plant parameters to match measured plant growth data (Tables 14.8 and 14.9). Pesticide parameters are calibrated after soil water balance. Model users need to iterate

Table 14.16 Minimum Data Required to Run RZWQM

Data Type	Minimum Data Required
Break-point rainfall	Breakpoint rainfall data with a minimum of two pairs of rainfall amounts and times
Daily meteorology	Daily meteorology data (minimum and maximum air temperature, wind run, solar radiation and relative humidity)
Site description	Soil horizon delineation by depth
	Soil horizon physical properties — bulk density, particle size fractions for each horizon
	Optional — soil horizon hydraulic properties: 330- or 100-cm suction water content and saturated hydraulic conductivity if available for each horizon
	Estimate of dry mass and age of residue on the surface
	General pesticide data such as common name, half-life, K, dissipation pathway (this information can be found in the ARS pesticide database)
	Specifying a crop from supplied database with regional parameters
	Management selections and additions as needed
Initial state	Initial soil moisture contents
	Initial soil temperatures
	Initial soil pH, CEC (cation exchange capacity) values
	Initial nutrient model inputs (soil residue, humus, microbial populations, mineral NO_3-N, NH_4-N; use RZWQM98 wizards to determine)

From Ahuja, L.R. et al., *Root Zone Water Quality Model,* Water Resources Publications, Englewood, CO, 1–360, 2000.

Table 14.17 Helpful Steps in Calibrating Soil Water Balance Components of RZWQM

Water Balance Component	Adjustment
Precipitation	No adjustment-driving variable.
Runoff	Add a crust, if there was an evidence for it.
	Change crust conductivity until the system responds as expected.
	Add macropores, if there was an evidence for them.
	If macropores are already present, change the geometry of the radial pores to change the maximum flux of the pores.
	Change the field saturation fraction to match the soil water content during infiltration.
	Change the soil albedo to modify the loss of water due to evaporation and match antecedent water contents.
Evapotranspiration	Change the field saturation fraction to match soil water contents.
	Change the soil albedo to modify the loss of water due to evaporation and match antecedent water contents.
	Verify that the rooting depth of the crop is correct so that water extraction from soil layers is correct.
	Modify the bottom boundary condition to allow the water content in the soil change based on the nature of drainage.
Seepage	Add macropores, so that any generated runoff water would be diverted down the pores.
	Modify the bottom boundary condition to allow the water content in the soil change based on drainage.
	Generally, decrease or increase the effect from runoff and evapotranspiration to make more or less water available in the soil profile to seep.
Storage	Adjust runoff, evapotranspiration and seepage to achieve a balance of inputs and losses and the storage will fall in line.

From Hanson, J.D. et al., *Agron. J.,* 91:171–177, 1999.

the procedure a few times to make sure all the components are balanced since these components are related. For example, nitrate leaching is affected by water movement, N soil dynamics and plant N uptake. Similarly, pesticide fate is affected not only by pesticide parameters (e.g., equilibrium and kinetic rate constants, degradation, etc.), but also by soil water movement, soil organic

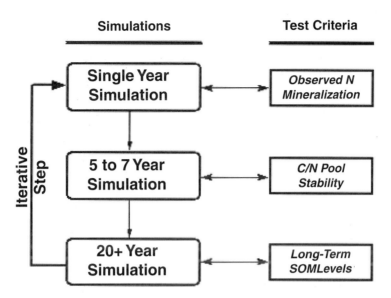

Figure 14.3 Procedures for calibrating the nitrogen submodel of RZWQM. "SOM" is soil organic matter. (From Hanson, J.D. et al., *Agron. J.,* 91:171–177, 1999. With permission.)

carbon content and macropore flow. Also, crop yield depends on correct soil water prediction and soil nitrate content in addition to crop growth parameters, Therefore, it is essential to follow the iterative calibration steps for a model such as RZWQM.

The DSSAT Family Models (CERES and CROPGRO)

Model calibration is facilitated through a simple disc operating system (DOS) interface in the DSSAT (Decision Support System for Agrotechnology Transfer) family of models. It has several good features that help users to calibrate the DSSAT models, such as graphing simulated against experimental data; crop genetic coefficient calculator; and sensitivity analysis (Hoogenboom et al., 1994; Hunt and Pararajasingham, 1994; Boote, 1999). As in other models, soil and weather data are required to run the DSSAT models. The DOS interface will guide users step-by-step on model input, but users have to provide their best estimates for required soil, weather and management information (Hunt et al., 1994). Once users go through all the steps, they can run the model and compare simulation results with experimental data in table or graphic forms. If discrepancies exist, users may conduct a sensitivity analysis for some model parameters (Hoogenboom et al., 1994). When good experimental data are available, the users may also use the Genotype Coefficient Calculator to optimize selected plant parameters (Hunt and Pararajasingham, 1994). Table 14.18 shows the data needed to calculate genotype coefficients.

Calibrating DSSAT family of models is also an iterative process. Because DSSAT put main emphasis on crop production with less consideration of environmental quality, soil water balance and crop production are the two major goals in model calibration. Boote (1999) has outlined ten steps in calibrating CROPGRO and corresponding parameters for each step (Table 14.19). Along with the ten steps to parameterize plant growth, model users should also constantly check soil water and nitrogen balances by matching soil N concentration, plant N uptake, crop water stress and soil water profile (Boote, 1999). The calibration process is not completed until all the observed values are predicted to the user's satisfaction.

Table 14.18   Data Required to Calculate Crop Genotype Coefficients

Crop	Required Data	Crop	Required Data
Peanut	Flowering date	Soybean	Flowering date
	Maturity date		Maturity date
	Seed yield		Grain yield
	Biomass at maturity		Biomass at maturity (no leaves)
	Seed number/m^2		Grain number/m^2
	Seed number per pod		Grain number per pod
	Seed dry weight		Grain dry weight
	First pod date	Wheat	Anthesis date
	Full pod date		Maturity date
	Pod yield (dry)		Grain yield (dry)
Maize	Silking date		Biomass at maturity
	Maturity date		Grain number/m^2
	Grain yield (dry)		Grain number per spike
	Biomass at maturity		Grain dry weight
	Grain number/m^2		
	Grain number per ear		
	Grain weight (dry)		

From Hunt, L.A. and S. Pararajasingham, Genotype coefficient calculator, in *1994 DSSAT version 3*, vol. 3, University of Hawaii, Honolulu, HI, 1994.

FUTURE NEEDS FOR PARAMETERIZATION

Parameterization of agricultural system models is a forbidding task for many model users. Model developers are looking for a balance between simple models with less parameters and mechanistic models with more parameters. Although parameters for a more mechanistic model should be less dependent upon experimental conditions, in reality, most calibrated parameters are more or less experimentally dependent. Such dependency is most often due to lack of a complete dataset for calibration; it is also due to an incomplete understanding of all the processes in an agricultural system. Ideally, a system model need not be calibrated except for the inputs to correctly characterize the soil, weather and management practices. For each of the components discussed in this chapter, further research and development are needed in the following areas:

1. Develop new simple methods to determine the required critical parameters and their dependencies. These parameters should be based on knowledge of the process and independent of a particular experimental setting. Also, guidelines and standardized methods of parameterization should be developed for different interacting components of system models. When there is a correlation between model parameters, efforts should be made to reduce unnecessary model parameters.
2. Make parameter determination a regular part of field research. There is generally a mismatch between collected experimental data and data required to run a system model. Model developers generally have to work with available data and conduct an incomplete model calibration due to lack of quality data. On the other hand, experimental scientists complain about the number of parameters needed to run a system model and they may not have heard about some of the parameters at any point in their careers. This miscommunication should be corrected.
3. Expand and improve existing parameter databases. To help users, parameter databases or default values are distributed along with system models; however, these databases contain parameters measured or calibrated from selected experimental sites or conditions. Therefore, they should be used, at most, as guidelines by users. These databases need to be expanded beyond the original sites or conditions and need to be better documented as to how they are derived and how they may be correlated to experimental conditions.
4. Improve the scientific concepts underlying the models, so that the parameters have a clear physical meaning and are measurable. For crop varieties, effort should be made to relate the variety-specific parameters to genetic characteristics and markers used in plant breeding. The most troublesome

Table 14.19 Steps to Parameterize Crop Growth in CROPGRO

Steps	Related Model Parameters
1. Crop life cycle	The first step should be to simulate crop development by initially selecting a maturity group of the cultivar. Then, adjust EM-FL and SD-PM to match flowering and maturity dates. It may be necessary to adjust CSDL and PPSEN. Maturity date is also affected by FL-SD.
2. Dry matter accumulation	This step simulates the rate of dry matter accumulation and also looks at LAI and specific leaf area (SLA). The two primary parameters to adjust are SLPF and LFMAX.
3. LAI and SLA	Parameters to be adjusted at this step are SLAVR, FL-LF, SIZLF, FL-SH and FL-SD. Increase SLAVR to increase SLA if needed. Then, increase FL-LF to delay LAI peak if necessary. The parameter SIZLF can be used to limit leaf area expansion. For dwarf cultivars, it may be necessary to adjust some parameters in the ecotype database (with extension ECO).
4. Recalibrate dry matter accumulation	Use the calibrated SLA and leaf area timing aspects from step 3 and recalibrate dry matter accumulation in step 2.
5. Species parameters	"Species" parameters that affect photosynthesis and dry matter accumulation include temperature and N effects on photosynthesis, rate of canopy expansion, degree of vertical layering of specific leaf weight, partitioning to root, nodule growth and N-fixation. Users should not change these parameters unless solid information is obtained.
6. Initial calibration for seed size, seeds per pods and seed filling duration	Parameters that need to be evaluated are WTPSD, SDPDV and SFDUR. SFDUR should be proportional to, but less than SD-PM.
7. Initial timing and initial rise in pod and seed dry weight	Now adjust FL-SH, FL-SD and PODUR to get the correct timing to the initial rise in pod dry weight and seed dry weight. Move FL-SH and FL-SD in the same direction.
8. Recalibrate time from first seed to maturity	If FL-SD is calibrated, adjust SD-PM to correctly predict the observed date of physical maturity.
9. Seed size, shelling percentage and seed filling duration	Now, go back to fine-tune seed size and final shelling percentage. Increase or decrease WTPSD to match the seed size. Adjust SFDUR to correctly predict shelling percentage.
10. Reevaluate total dry matter accumulation and relative partitioning between vegetative and reproductive stages	Are seed size, shelling percentage, pod addition and phenology timing correct? If so, reevaluate dry matter accumulation in seed mass and total aboveground biomass, as well as the relative partitioning between seed and shoot mass. If both seed and total biomass accumulation are too rapid or too slow, then adjust SLPF or LFMAX to attain the correct slopes of aboveground dry matter and seed dry weight, as well as final seed yield. If the fit is good, the calibration process is completed.

Note: See Table 14.12 for parameter names.

From Boote, K.J., Concepts for calibrating crop growth models, in *DSSAT version 3*, vol. 4, University of Hawaii, Honolulu, HI, 1999.

parameters are those that are immeasurable experimentally. Efforts are needed to better define these parameters and design experimental methods to estimate them independently and directly or indirectly. Because measured crop variables (e.g., yield, plant biomass and phenology) are environmentally dependent, calibrated crop parameters are an expression of a genotype under certain experimental conditions. Extra efforts are needed to differentiate which crop parameters are due to the genotype and which ones are due to changes in experimental conditions.

5. Further research to understand and quantify the effect of water, nitrogen and temperature stresses as well as their interactions on different plant growth processes, in order to derive improved stress factors. Experimental conditions affect genotype expression through a series of stresses. Most of these environmental stresses are simulated empirically and have not incorporated any of the new findings in plant sciences on stresses at the molecular level, such as changes in hormone and enzyme levels during stress, reposition of plant cells and adaptability of plants to stresses.

6. Further research on how management practices cause changes in soil properties and processes in order to define and determine the parameters better. Management practices have profound effects on soil properties and plant growth and are mainly responsible for human-induced temporal

variability in model parameters. Several empirical equations have been used to simulate management effects, but more efforts are needed to improve and integrate these effects into system models.

7. Quantify and employ spatial and temporal variability of parameters. Most models are one-dimensional and do not take into account the change of model parameters in space and time. For model applications under field conditions, system models should accommodate this variability, such as multiple model inputs and multiple model runs across a field.

In conclusion, more efforts have been made on developing system models and less efforts have been made on how to parameterize the models. Thus, a system model can be released and used without rigorous parameterization. Most of the time, a system model is released with one test case for selective model components and further tests are left to model users. To improve model parameterization process and promote better model use in agriculture, model developers should work more closely and actively with field scientists to conduct a more complete model parameterization and define the applicable range of derived model parameters. On the other hand, field scientists should voluntarily collaborate with model developers to collect and document as complete data sets as possible, so that the data can be used to better synthesize and transfer knowledge.

REFERENCES

Ahuja, L.R., D.K. Cassel, R.R. Bruce and B.B. Barnes. 1989. Evaluation of spatial distribution of hydraulic conductivity using effective porosity data, *Soil Sci.,* 148:404–411.

Ahuja, L.R., F. Fiedler. G.H. Dunn. J.G. Benjamin and A. Garrison. 1998. Changes in soil water retention curve due to tillage and natural reconsolidation, *Soil Sci. Soc. Am. J.,* 62:1228–1233.

Ahuja, L.R. et al. 1984. Macroporosity to characterize spatial variability of hydraulic conductivity and effects of land management, *Soil Sci. Soc. Am. J.,* 48:699–702.

Ahuja, L.R., J.W. Naney and R.D. Williams. 1985. Estimating soil water characteristics from simpler properties or limited data, *Soil Sci. Soc. Am. J.,* 49:1100–1105.

Ahuja, L.R., J.W. Naney, R.E. Green and D.R. Nielsen. 1999. Determining soil hydraulic properties and their spatial variability from simpler measurements. In *Agricultural Drainage, Agronomy Monograph No. 38,* R.W. Skaggs and J. van Schilfgaards, Eds., American Society of Agronomy, Madison, WI.

Ahuja, L.R., K.W. Rojas, J.D. Hanson, M.J. Shaffer and L. Ma., Eds. 2000. *Root Zone Water Quality Model,* Water Resources Publications, Englewood, CO, 1–360.

Ahuja, L.R. and R.D. Williams, 1991. Scaling of water characteristic and hydraulic conductivity based on Gregson–Hector–McGowan approach, *Soil Sci. Soc. Am. J.,* 55:308–319.

Amacher, M.C., H.M. Selim and I.K. Iskandar. 1998. Kinetics of Cr and Cd retention in soils — a nonlinear multireaction model, *Soil Sci. Soc. Am. J.,* 52:398–408.

Blaney, H.F. and W.D. Criddle. 1962. Determining consumptive use and irrigation water requirements, *Tech. Bull. 1275,* USDA-ARS, Washington, D.C.

Boote, K.J. 1999. Concepts for calibrating crop growth models. In *DSSAT version 3,* vol. 4, Hoogenboom, G., P.W. Wilkens and G.Y. Tsuji, Eds., 1999. University of Hawaii, Honolulu, HI.

Boote, K.J., J.W. Jones, G. Hoogenboom and N.B. Pickering. 1998. The CROPGRO model for grain legumes. In *Understanding Options for Agricultural Production,* G.Y. Tsuji, G. Hoogenboom, P.K. Thornton, Eds., Kluwer Academic Publishers, 99–128.

Brooks, R.H. and A.T. Corey. 1964. Hydraulic properties of porous media, *Hydrology Paper 3,* Colorado State University, Fort Collins, CO.

Campbell, G.S. 1974. A simple method for determining unsaturated conductivity from moisture retention data, *Soil Sci.,* 117:311–314.

Carsel, R.F., L.A. Mulkey, M N. Lorber and L.B. Baskin. 1985. The pesticide root zone model (PRZM): A procedure for evaluating pesticide leaching threats to groundwater, *Ecol. Modeling,* 30:49–69.

Cassel, D.K. 1983. Spatial and temporal variability of soil physical properties following tillage of Norfolk loamy sand, *Soil Sci. Soc. Am. J.,* 47:196–201.

Cheng, H.H., Ed., 1990. *Pesticides in the Soil Environment: Processes, Impacts and Modeling,* Soil Science Society of America, Inc., Madison, WI.

Chiou, C.T., L.J. Peters and V.H. Freed. 1979. A physical concept of soil-water equilibria for nonionic organic compounds, *Science,* 206:831–832.

Chiou, C.T., P.E. Porter and D.W. Schmedding. 1983. Partition equilibria of nonionic organic compounds between soil organic matter and water, *Environ. Sci. Technol.,* 17-227-231.

Clapp, R.B. and G.M. Hornberger. 1978. Empirical equations for some hydraulic properties, *Water Resour. Res.,* 15:601–604.

De Jong, R. 1982. Assessment of empirical parameters that describe soil water characteristics, *Can. Agric. Eng.,* 24:65–70.

Donnelly, J.R., R.J. Simpson, L. Salmon, A.D. Moore, M. Freer and H. Dove. 2002. Forage-livestock models for the Australian livestock industry. In *Agricultural System Models in Field Research and Technology Transfer,* L.R. Ahuja, L. Ma and T.A. Howell, Eds., CRC Press, Boca Raton, FL, pp. 2–31.

Douglas, C.L., Jr. and R.W. Rickman. 1992. Estimating crop residue decomposition from air temperature, initial nitrogen content and residue placement, *Soil Sci. Soc. Am. J.,* 56:272–278.

Dutt, G.R., M.J. Shaffer and W.J. Moore. 1972. Computer simulation model of dynamic bio-physicochemical processes in soils, *Tech. Bull. 196,* Arizona Agricultural Experiment Station, Tucson, AZ.

Ehlers, W. 1977. Measurement and calculation of hydraulic conductivity in horizons of tilled and untilled loess-derived soil, *Geoderma,* 19:293–306.

Farahani, H.J. and L.R. Ahuja. 1996. Evapotranspiration modeling of partial canopy/residue covered field, *Trans. ASAE,* 39:2051–2064.

Farahani, H.J. and D.G. DeCoursey. 2000. Potential evaporation and transpiration processes. In *Root Zone Water Quality Model,* L.R. Ahuja, K.W. Rojas, J.D. Hanson, M.J. Shaffer and L. Ma, Eds., Water Resources Publications, Highland Ranch, CO, 50–80.

Flanagan, D.C. and M.A. Nearing. 1995. USDA-Water Erosion Prediction Project: Hillslope profile and watershed model documentation, *NSERL Report No. 10,* USDA-ARS National Soil Erosion Research Laboratory, West Lafayette, IN

Flühler, H., M.S. Ardakani and L.H. Stolzy. 1976. Error propagation in determining hydraulic conductivities from successive water content and pressure head profiles, *Soil Sci. Soc. Am. J.,* 40:830–836.

Foy, J.K., W.R. Teague and J.D. Hanson. 1999. Evaluation of the upgraded SPUR model (SPUR2.4), *Ecol. Modeling,* 118:149–165.

Franzmeier, D.P. 1991. Estimation of hydraulic conductivity from effective porosity data from some Indiana soils, *Soil Sci. Soc. Am. J.,* 55:1801–1803.

Gaston, L.A. and M.A. Locke. 1994. Predicting alachlor mobility using batch sorption kinetic data, *Soil Sci.,* 158:345–354.

Gerstl, Z. 1990. Estimation of organic chemical sorption by soils, *J. Contaminant Hydrol.,* 6:357–375.

Godwin, D.C. and C.A. Jones. 1991. Nitrogen dynamics in soil-plant systems. In *Modeling Plant and Soil Systems,* J. Hanks and J.T. Ritchie, Eds., American Society of Agronomy, Madison, WI, 287–321.

Godwin, D.C. and U. Singh. 1998. Nitrogen balance and crop response to nitrogen in upland and lowland cropping systems. In *Understanding Options for Agricultural Production,* G.Y. Tsuji, G. Hoogenboom, P.K. Thornton, Eds., Kluwer Academic Publishers, Boston, 55–77.

Grant, R.F. 2001. A review of the Canadian ecosystem model — *ecosys.* In *Modeling Carbon and Nitrogen Dynamics for Soil Management,* M.J. Shaffer, L. Ma and S. Hansen, Eds., CRC Press, Boca Raton, FL, 173–263.

Green, R.E., L.R. Ahuja and S.K. Chong. 1986. Hydraulic conductivity, diffusivity and sorptivity of unsaturated soils: Field methods. In *Methods of Soil Analysis,* Part 1, 2nd ed., A. Klute, Ed., Agron. Monogr. 9, ASA/SSSA, Madison, WI, 771–798.

Green, R.E. and S.W. Karickhoff. 1990. Sorption estimates for modeling. In *Pesticides in the Soil Environment: Processes, Impacts and Modeling,* H.H. Cheng, Ed., Soil Science Society of America, Inc., Madison, WI, 79–101.

Gregory, J.M. 1982. Soil cover prediction with various amounts and types of crop residue, *Trans. ASAE,* 25:1333–1337.

Gregory, J.M., T.R. McCarty, F. Ghidey and E.E. Alberts. 1985. Derivation and evaluation of a residue decay equation, *Trans. ASAE,* 28:98–101.

Gregson, K., D.J. Hector and M. McGowan. 1987. A one-parameter model for the soil water characteristic, *J. Soil Sci.,* 38:483–486.

Hanson, J.D. 2000. Generic crop production. In *Root Zone Water Quality Model,* L.R. Ahuja et al., Eds., Water Resources Publications, Highland Ranch, CO, 81–118.

Hanson, J.D., K.W. Rojas and M.J. Shaffer. 1999. Calibrating the Root Zone Water Quality Model, *Agron. J.,* 91:171–177.

Hassett, J.J., Means. J.C., Banwart. W.L. and Wood, S.G. 1980. Sorption properties of sediments and energy-related pollutants, *EPA-600/3-80-041,* U.S. Environmental Protection Agency, Washington, D.C.

Hatfield, J.L. 1990. Methods of estimating evapotranspiration. In *Irrigation of Agricultural Crops, Agronomy 30,* B.A. Stewart and D.R. Nielsen, Eds., American Society of Agronomy, Madison, WI, 435–474.

Hornsby, A.G., R.D. Wauchope and A.E. Herner. 1996. Pesticide Properties in the Environment, Springer-Verlag, New York.

Hoogenboom, G., J.W. Jones, P.W. Wilkens, W.D. Batchelor, W.T. Bowen, L.A. Hunt, N.B. Pickering, U. Singh, D.C. Godwin, B. Baer, K.J. Boote, J.T. Ritchie and J. W. White. 1994. Crop models. In *1994 DSSAT version 3,* vol. 2, Tsuji, G.Y., G. Uehara and S. Balas, Eds., University of Hawaii, Honolulu, HI.

Hunt, L.A. and S. Pararajasingham. 1994. Genotype coefficient calculator. In *1994 DSSAT version 3,* Vol. 3, Tsuji, G.Y., G. Uehara and S. Balas, Eds., University of Hawaii, Honolulu, HI.

Hunt, L.A., J.W. Jones, P.K. Thornton, G. Hoogenboom, D.T. Imamura, G.Y. Tsuji and S. Balas. 1994. Accessing data, models and application programs. In *1994 DSSAT version 3,* vol. 1, Tsuji, G.Y., G. Uehara and S. Balas, Eds., University of Hawaii, Honolulu, HI.

Jaynes, D.B. and J.G. Miller. 1999. Evaluation of the Root Zone Water Quality Model using data from the Iowa MSEA, *Agron. J.,* 91:192–200.

Jensen, M.E. and H.R. Haise. 1963. Estimating evaportranspiration from solar radiation, *J. Irrig. Drainage, ASCE,* 89:15–41.

Johnsen, K.E., H.H. Liu, J.H. Dane, L.R. Ahuja and S.R. Workman. 1995. Simulating fluctuating water tables and tile drainage with a modified Root Zone Water Quality Model and a new model, WAFLOWN, *Trans. ASAE,* 38:75–83.

Jones, C.A., W.L. Bland, J.T. Ritchie and J.R. Williams. 1991. Simulation of root growth. In *Modeling Plant and Soil Systems, Monograph 31,* J. Hanks and J.T. Ritchie, Eds., American Society of Agronomy, Madison, WI.

Juma, N.G. and W.B. McGill. 1986. Decomposition and nutrient cycling in agro-ecosystems, In *Microfloral and Faunal Interactions in Natural and Agro-ecosystems,* M.J. Mitchell and J.P. Nakas, Eds., Dordrecht, Boston, 74–136.

Karickhoff, S.W. 1984. Organic pollutant sorption in aquatic systems, *J. Hydraul. Eng.,* 110:707–735.

Kellog, R.L., M.S. Maizel and D.W. Goss. 1994. The potential for leaching of agricultural chemicals used in crop protection: a national perspective, *J. Soil Water Cons.,* 49:294–298.

Klute, A., Ed., 1986. *Methods of Soil Analysis, Part 1,* 2nd ed., Agron. Monogr. 9, ASA/SSSA, Madison, WI.

Landa, F.M., N.R. Fausey, S.E. Nokes and J.D. Hanson. 1999. Evaluation of the Root Zone Water Quality Model (RZWQM3.2) at the Ohio MSEA, *Agron. J.,* 91:220–227.

Leonard, R.A., W.G. Knisel and D.S. Still. 1987. GLEAMS: groundwater loading effects of agricultural management systems, *Trans. ASAE,* 30:1403–1418.

Linden, D.R. and D.M. van Doren, Jr. 1987. Simulation of interception, surface roughness, depression storage and soil settling. In *NTRM, A Soil Crop Simulation Model for Nitrogen, Tillage and Crop-Residue Management,* M.J. Shaffer and W.E. Larson, Eds., USDA-ARS. Conservation Research Rept. 34-1, Washington, D.C., 90–93.

Linn, D.M., T.H. Carski, M.L. Brusseau and F.H. Chang., Eds., 1993. *Sorption and Degradation of Pesticide and Organic Chemicals in Soil,* Soil Science Society of America, Inc., Madison, WI.

Linn, D.M. and J.W. Doran. 1984. Effect of water-filled pore space on carbon dioxide and nitrous oxide production in tilled and non-tilled soils, *Soil Sci. Soc. Am. J.,* 48:1267–1272.

Lion, L.W., T.B. Stauffer and W.G. MacIntyre. 1990. Sorption of hydrophobic compounds on aquifer materials: analysis methods and the effect of organic carbon, *J. Contaminant Hydrol.,* 5:215–234.

Loague, K. and R.E. Green. 1991. Statistical and graphical methods for evaluating solute transport models: overview and application, *J. Contaminant Hydrol.,* 7:51–71.

Ma, L. et al. 1993. Hysteretic characteristics of atrazine adsorption-desorption by a Sharkey soil, *Weed Sci.,* 41:627–633.

Ma, L. and H.M. Selim. 1994. Predicting atrazine adsorption-desorption in soils: A modified second-order kinetic model, *Water Resour. Res.,* 30:447–456.

Ma, L. and H.M. Selim. 1996. Atrazine retention and transport in soils, *Rev. Environ. Contamin. Toxicol.* 145:129–173.

Ma, L. and H.M. Selim. 1997. Physical nonequilibrium modeling approaches to solute transport in soils, *Adv. Agron.,* 58:95–150.

Ma, L. and H.M. Selim. 1998. Coupling of retention approaches to physical nonequilibrium models. In *Physical Nonequilibrium in Soils: Modeling and Application,* Selim, H.M. and L. Ma, Eds., Ann Arbor Press, MI, 83–115.

Ma, L., M.J. Shaffer, J.K. Boyd, R. Waskom, L.R. Ahuja, K.W. Rojas and C. Xu. 1998a. Manure management in an irrigated silage corn field: experiment and modeling, *Soil Sci. Soc. Am. J.,* 62:1006–1017.

Ma, L., H.D. Scott, M.J. Shaffer and L R. Ahuja. 1998b. RZWQM simulations of water and nitrate movement in a manured tall fescue field, *Soil Sci.,* 163:259–270.

Ma, L., G.A. Peterson, L.R. Ahuja, L. Sherrod, M.J. Shaffer and K.W. Rojas. 1999. Decomposition of surface crop residues in long-term studies of dryland agroecosystems, *Agron. J.,* 91:393–401.

Ma, L., L.R. Ahuja, J.C. Ascough II, M.J. Shaffer, K.W. Rojas, R.W. Malone and M.R. Cameira. 2000. Integrating system modeling with field research in agriculture: applications of the Root Zone Water Quality Model (RZWQM), *Adv. Agron.,* 71:233–292.

Ma, L. and M.J. Shaffer. 2001. A review of carbon and nitrogen processes in nine U.S. soil nitrogen dynamics models. In *Modeling Carbon and Nitrogen Dynamics for Soil Management,* M.J. Shaffer, L. Ma and S. Hansen, Eds., CRC Press, Boca Raton, FL, 55–102.

Ma, L., M.J. Shaffer and L.R. Ahuja. 2001. Application of RZWQM for soil nitrogen management. In *Modeling Carbon and Nitrogen Dynamics for Soil Management,* M.J. Shaffer, L. Ma and S. Hansen, Eds., CRC Press, Boca Raton, FL, 265–301.

Ma, Q.L., R.D. Wauchope, J.E. Hook, A.W. Johnson, C.C. Truman, C.C. Dowler, G.J. Gascho, J.G. Davis, H.R. Summer and L.D. Chandler. 1998c. Influence of tractor wheel tracks and crusts/seals on runoff: Observation and simulations with the RZWQM, *Agric. Syst.,.* 57:77–100.

Ma, Q.L., R.D. Wauchope, J.E. Hook, A.W. Johnson, C.C. Truman, C.C. Dowler, G.J. Gascho, J.G. Davis, H.R. Summer and L.D. Chandler. 1998d. GLEAMS, Opus and PRZM-2 model predicted versus measured runoff from a coastal plain loamy sand, *Trans. ASAE,* 41:77–88.

Ma, Q.L., L.R. Ahuja, K.W. Rojas, V.F. Ferreira and D.G. DeCoursey. 1995. Measured and RZWQM predicted atrazine dissipation and movement in a field soil, *Trans. ASAE,* 38:471–479.

Marshall, T.J. 1958. A relationship between permeability and size distribution of pores, *J. Soil Sci.,* 9:1–8.

McCown, R.L, G.L. Hammer, J.N.G. Hargreaves, D.L. Holzworth and D.M. Freebairn. 1996. APSIM: a novel software system for model development, model testing and simulation in agricultural systems research, *Agric. Syst.,* 50: 255–271.

McGechan, M.B. and L. Wu. 2001. A review of carbon and nitrogen processes in European soil nitrogen dynamics models. In *Modeling Carbon and Nitrogen Dynamics for Soil Management,* M.J. Shaffer, L. Ma and S. Hansen, Eds., CRC Press, Boca Raton, FL, 103–171.

McIntyre, D.S. 1958. Permeability measurements of soil crusts formed by raindrop impact, *Soil Sci.,* 85:185–189.

Monteith, J.L. 1965. Evaporation and the environment, *Symp. Soc. Exper. Biol.* 19:205–234.

Monteith, J.L. 1998. Evaporation models. In *Agricultural System Modeling and Simulation,* R.M. Peart and R.B. Curry, Eds., Marcel Dekker, Inc., New York, 197–234.

Monteith, J.L. and M.H. Unsworth. 1990. *Principles of Environmental Physics,* Edward Arnold, London.

Nielsen, D.R., J.M. Davidson, J.W. Biggar and R.J. Miller. 1964. Water movement through Panoche clay loam soil, *Hilgardia,* 35:491–506.

Nimah, M. and R.J. Hanks. 1973. Model for estimating soil-water-plant –atmospheric interrelation: description and sensitivity, *Soil Sci. Soc. Am. Proc.,* 37:522–527.

Ogata, G. and L.A. Richards. 1957. Water content change following irrigation of bare field soil that is protected from evaporation, *Soil Sci. Soc. Am. Proc.,* 21:355–356.

Onstad, C.A., J.L. Wolfe, C.L. Larson and D.C. Slack. 1984. Tilled soil subsidence during repeated wetting, *Trans. ASAE,* 27:733–736.

Parton, W.J., D.W. Anderson, C.V. Cole and J.W.B. Stewart. 1983. Simulation of soil organic matter formations and mineralization in semiarid ecosystems. In *Nutrient Cycling in Agricultural Ecosystems,* Lowrance et al., Eds., Special Publication 23, University of Georgia, College of Agriculture Experiment Station, Athens, GA.

Parton, W.J., D.S. Ojima, C.V. Cole and D.S. Schimel. 1994. A general model for soil organic matter dynamics: sensitivity to litter chemistry, texture and management. In *Quantitative Modeling of Soil Forming Processes SSSA,* R.B. Bryant and R.W. Arnold, Ed., Special Publication No. 39, Madison, WI, 147–167.

Penman, H.L. 1948. Natural evaporation from open water, bare soil and grass, *Proc. R. Soc. London,* A193:120–145.

Priestley, C.H.B. and R.J. Taylor. 1972. On the assessment of surface heat flux and evaporation using large-scale parameters, *Mon. Weather Rev.,* 100:81–92.

Rao, P.S.C. and Davidson, J.M. 1982. Retention and transformation of selected pesticides and phosphorus in soil-water systems: a critical review, *EPA-600/3-82-060,* U.S. Environmental Protection Agency, Washington, D.C.

Rawls, W.J., D.L. Brakensiek and K.E. Saxton. 1982. Estimation of soil water properties, *Trans. ASAE,* 25:1316–1320, 1328.

Rawls, W.J. and D.L. Brakensiek. 1985. Prediction of soil water properties for hydrologic modeling. In *Proc. Watershed Manage, in the Eighties,* Denver, CO, April 30–May 1, ASCE, New York.

Rawls, W.J., D. Gimenez and R. Grossman. 1998. Soil texture, bulk density and slope of retention curves to predict saturated hydraulic conductivity, *Trans. ASAE,* 41:983–988.

Richards, L.A., W.R. Gardner and G. Ogata. 1956. Physical processes determining water loss from soils, *Soil Sci. Soc. Am. Proc.,* 20:310–314.

Ritchie, J.T. 1998. Soil water balance and plant water stress. In *Understanding Options for Agricultural Production,* G.Y. Tsuji, G. Hoogenboom, P.K. Thornton, Eds., Kluwer Academic Publishers, Dordrecht, The Netherlands, 41–54.

Ritchie, J.T., U. Singh, D.C. Godwin and W.T. Bowen. 1998. Cereal growth, development and yield. In *Understanding Options for Agricultural Production,* G.Y. Tsuji, G. Hoogenboom, P.K. Thornton, Eds., Kluwer Academic Publishers, Dordrecht, The Netherlands, 79–98.

Rojas, K.W. and L.R. Ahuja. 2000. Management practices. In *Root Zone Water Quality Model,* L.R. Ahuja, K.W. Rojas, J.D. Hanson, M.J. Shaffer and L. Ma, Eds. Water Resources Publications, Highlands Ranch, CO. 245–280.

Rojas, K.W., L. Ma, J. D. Hanson and L.R. Ahuja. 2000. *RZWQM98 User Guide,* In *Root Zone Water Quality Model,* L.R. Ahuja, K.W. Rojas, J.D. Hanson, M.J. Shaffer and L. Ma, Eds. Water Resources Publications, Highlands Ranch, CO, 327–364.

Rose, C.W., W.R. Stern and J.E. Drummond. 1965. Determination of hydraulic conductivity as a function of depth and water content for soil *in situ, Aust. J. Soil Res.,* 3:1–9.

Rousseva, S.S., L.R. Ahuja and G.C. Heathman. 1988. Use of a surface gamma-neutron gauge for in-situ measurement of changes in bulk density of the tilled zone, *Soil Tillage Res.,* 12:235–251.

Ruan, H., L.R. Ahuja, T.R. Green and J.G. Benjamin. 2001. Residue cover and surface sealing effects on infiltration: Numerical simulations for field applications, *Soil Sci. Soc. Am. J.,* 65:853–861.

Schwartzenbach, R.P. and Westall, J. 1981. Transport of nonpolar organic compounds from surfacewater to groundwater, *Environ. Sci. Tech.,* 15:1360–1367.

Selim, H.M., L. Ma and H. Zhu. 1999. Predicting solute transport in soils: second-order two-site models, *Soil Sci. Soc. Am. J.,* 63:768–777.

Shaffer, M.J., A.D. Halvorson and F.J. Pierce. 1991. Nitrate leaching and economic analysis package (NLEAP): model description and application. In R.F. Follett, D.R. Keeney and R.M. Cruse, Eds., *Managing Nitrogen for Groundwater Quality and Farm Profitability,* Soil Science Society of America, Inc., Madison, WI, 285–322.

Shaffer, M.J., K.W. Rojas, D.G. DeCoursey and C.S. Hebson. 2000. Nutrient Chemistry processes — OMNI. In *Root Zone Water Quality Model Water Resources Publications,* L.R. Ahuja et al., Eds., Highland Ranch, CO, 119–143.

Shaffer, M.J., K. Lasnik, X. Ou and R. Flynn. 2001. NLEAP internet tools for estimating NO3-N leaching and N2O emissions. In *Modeling Carbon and Nitrogen Dynamics for Soil Management,* M.J. Shaffer, L. Ma and S. Hansen, Eds., CRC Press, Boca Raton, FL, 403–426.

Sharma, P.P., C.J. Gantzer and G.R. Blake. 1981. Hydraulic gradients across simulated rain-formed soil surface seals, *Soil Sci. Soc. Am. J.,* 45:1031–1034.

Sharpley, A.N. and J.R. Williams. 1993. EPIC, Erosion/Productivity Impact Calculator. 1. Model documentation, *USDA-ARS, Technical Bull.,* 1768, Washington, D.C.

Shen, J., W.D. Batchelor, J.W. Jones, J.T. Ritchie, R.S. Kanwar and C.W. Mize. 1998. Incorporation of a subsurface tile drainage component into a soybean growth model, *Trans. ASAE,* 41:1305–1313.

Shuttleworth, W.J. and J.S. Wallace. 1985. Evaporation from sparse crops-an energy combination theory, *Quart. J. R. Meteorol. Soc.,* 111:839–855.

Singh, P. and R.S. Kanwar. 1995. Modification of RZWQM for simulating subsurface drainage by adding a tile flow component, *Trans. ASAE,* 38:489–498.

Smith, R.E. 1990. Opus: an integrated simulation model for transport of nonpoint source pollutants at field scale, vol. 1. documentation, *USDA-ARS Rept. 98,* Washington, D.C.

Tackett, J.L. and R.W. Pearson. 1965. Some characteristics of soil crusts formed by simulated rainfall, *Soil Sci.,* 99:407–413.

Taylor, A.W. and W.F. Spencer. 1990. Volatilization and vapor transport processes. In *Pesticides in the Soil Environment: Processes, Impacts and Modeling,* Soil Science Society of America, Inc., Madison, WI, 213–269.

Thornthwaite, C.W. 1948. An approach towards a rational classification of climate, *Geogr. Rev.,* 38:55–94.

Timlin, D.J., L.R. Ahuja and M. Ankeny. 1993. Comparison of three methods to measure maximum macropore flow, *Soil Sci. Soc. Am. J.,* 57:278–284.

Toride, N., F.J. Leij and M.Th. Van Genuchten. 1995. The CXTFIT code for estimating transport parameters from laboratory or field tracer experiments, *Res. Rept. 137,* U.S. Salinity Laboratory, USDA-ARS, Riverside, CA.

Tsuji, G.Y., G. Uehara and S. Balas, Eds., 1994. *DSSAT version 3,* vol. 2, University of Hawaii, Honolulu, HI.

van Bavel, C.H.M., G.B. Stirk and K.J. Brust. 1968. Hydraulic propertites of a clay loam soil and field measurement of water uptake by roots: 1. interpretation of water content and pressure profiles, *Soil Sci. Soc. Am. Proc.,* 32:310–317.

van Genuchten, R. 1980. Predicting the hydraulic conductivity of unsaturated soils, *Soil Sci. Soc. Am. J.,* 44:892–898.

Vereecken, H. 1988. Pedotransfer functions for the generation of hydraulic properties of Belgian soils, Ph.D. dissertation, Katholieke University of Leuven, Leuven, Belgium.

Wagner, L.E. and R.G. Nelson. 1995. Mass reduction of standing and flat crop residues by selected tillage implements, *Trans. ASAE,* 38:419–427.

Watson, K.K. 1966. An instantaneous profile method for determining the hydraulic conductivity of unsaturated porous materials, *Water Resour. Res.,* 2:709–713.

Watson, K.W. and R.J. Luxmoore. 1986. Estimating macroporosity in a watershed by use of a tension infiltration, *Soil Sci. Soc. Am. J.,* 50:578–582.

Wauchope, R.D., T.M. Buttler, A.G. Hornsby, P.W.M. Augustijn-Beckers and J.P. Burt. 1992. The SCS/ARS/CES Pesticide properties database for environmental decision-making, *Rev. Environ. Cont. Toxicol.,* 123:1–164.

Wauchope, R.D, R.G. Nash, K.W. Rojas, L.R. Ahuja, G.M. Willis, Q. Ma, L.L. McDowell and T.B. Moorman. 2000. Pesticide processes. In *Root Zone Water Quality Model,* L.R. Ahuja et al., Eds., Water Resources Publications, Highland Ranch, CO, 163–244.

Williams, J.R. 1995. The EPIC Model. In *Computer Models of Watershed Hydrology,* V.P. Singh, Ed., Water Resources Publications, Highland Ranch, CO, 909–1000.

Williams, J.R., C.A. Jones and P.T. Dyke. 1984. A modeling approach to determining the relationship between erosion and soil productivity, *Trans. ASAE,* 27(1):129–142.

Williams, R.D. and L.R. Ahuja. 1992. Evaluation of similar-media scaling and a one-parameter model for estimating the soil water characteristics, *J. Soil Sci.,* 43:237–248.

Williams.R.D., L.R. Ahuja and J.W. Naney. 1992. Comparison of methods to estimate soil water characteristics from soil texture, bulk density and limited data, *Soil Sci.,* 153:172–184.

Wu, L., R.R. Allmaras, J.B. Lamb and K.E. Johnsen. 1996. Model sensitivity to measured and estimated hydraulic properties of a Zimmerman fine sand, *Soil Sci. Soc. Am. J.,* 60:1283–1290.

Wu, L., W. Chen, J.M. Baker and J.A. Lamb. 1999. Evaluation of the Root Zone Water Quality Model using field-measured data from a sandy soil, *Agron. J.,* 91:177–182.

Xue, S.K., I.K. Iskandar and H.M. Selim. 1995. Adsorption-desorption of 2, 4, 6-Trinitrotoluene and Hexahydro-1,3,5-Trinitro-1,3,5-Triazine in soils, *Soil Sci.,* 160:317–327.

CHAPTER **15**

The Object Modeling System

Olaf David, Steven L. Markstrom, Kenneth W. Rojas, Lajpat R. Ahuja and Ian W. Schneider

CONTENTS

INTRODUCTION

The problems facing both users and developers of natural resource models are becoming much more complex. Understanding human management issues such as farming practices, erosion control, pesticide and fertilizer application, reservoir management and habitat restoration become compounded when viewed within the physical, hydrological, chemical and biological responses of the natural world. Computer simulations for conceptualization, prediction and management of agricultural areas, watersheds, water supply and environmentally sensitive sites are likewise becoming more complex. The interdisciplinary nature of these problems usually requires taking into account a significant number of different models, alternatives, data sources and domain experts.

Much domain-specific disciplinary expertise has been developed and captured in the form of computer simulation models. These models usually represent the efforts of an individual or a small group of scientists and are consequently very focused in their scope. Several of these models must often be applied within the context of a single project application. Although these models may share much in common, there may be conflicting data, scales, methodologies, file formats, or even computer hardware and software requirements for these models. To overcome these limitations they are often rebuilt to meet an appropriate system delineation.

On the other hand, some large, monolithic process models have been designed and implemented to cover a range of simulation alternatives (Abel, 1994). Many of these models are closed, stand-alone systems. They require fixed input/output data format with no capability to interface with other models. Many of these models are still operating in batch mode. Modifications and extensions to a model are often handled as additions made to the main body of code developed years ago. Such a tightly wrapped structure limits this integration with other models and makes their update and maintenance difficult.

An important effort in environmental simulation model design, therefore, is to enhance modularity, reusability and interoperability of both science and auxiliary components. Leavesley et al. (2002) pointed out that models applied for different systems or sub-systems required different levels of detail and comprehensiveness, which are driven by problem objectives, data constrains and spatial and temporal scales of application.

Reusability can be increased by establishing standard simulation module libraries. These libraries are comprised of components of simulation models and provide basic building blocks for a number of similar applications. They are designed to allow interoperability, which is essential for the incorporation of various scientific disciplines. Module libraries have been successfully applied in several domains such as manufacturing systems, transport and other systems (Top et al., 1997; Breunese et al., 1998; Praehofer, 1996). An advanced modular modeling framework based on the creation of a library of science and utility modules offers an exciting possibility for developing customized agricultural system models in the 21st Century. These models will use the best science available for their purpose and will be easy to update and maintain. The library may lead to standardization of science, tools and interfaces and will serve as a reference and coordination mechanism for model developers and future research.

This chapter gives an overview of the efforts made in adopting modular modeling principles for the construction and application of natural systems models through the Object Modeling System (OMS) framework. The chapter presents:

1. OMS project background and objectives
2. Basic OMS principles
3. OMS architecture and implementation
4. Results of a model application example

Background and Objectives of the OMS Project

The OMS Project is an interagency project between the USDA-ARS (Agricultural Research Service), USGS (U.S. Geological Service) and USDA-NRCS (Natural Resource Conservation Service). The past experiences of the agencies indicated that the development of comprehensive, multidisciplinary simulation models was a very expensive process ($15 to $30 million per model). Many development activities were duplicated among different modeling projects; for example, the hydrologic components in crop models, water quality models and erosion models. The duplicated components generally used different levels of detail and time scales, whereby they did not give the same results in hydrology and, thus, other outputs. Furthermore, the maintenance of several large, monolithic models has been a problem. These considerations led to the initiation of the OMS Project.

The overall project goal is the development of OMS, an integrated modeling framework, which allows integration of models, founded on a standard library of components, tools and data. Models developed within OMS will be based on representing different scientific approaches with regard to components that address data constrains and spatial/temporal scale of application. The objectives of the OMS project are to:

1. Identify modeling library parts (modules or components) and glean them from existing nonmodular simulation models.
2. Formalize the linkages between these components to support model building.
3. Develop generic software tools to support models and modeling.
4. Develop the framework which supports these objectives.

To provide comprehensive modeling assistance, the following functional components will be part of the framework:

1. A module-building component that will facilitate the integration of existing (legacy) code into the framework (this adaptation support will simplify the technical procedure for module implementation)
2. A module repository that will contain modules that can be readily utilized to assemble a working model (types of modules in the library will include science-, control-, utility-, assessment-, data access- and system modules)
3. A model builder that will assemble modules from the module library into executable models and verify data connectivity and compatibility in scale and comprehensiveness
4. A dictionary framework that will manage extended modeling data type information and provide extended semantics checking for module connectivity verification
5. An extensible user interface that will facilitate an appropriate user interaction for general model development and application (it will be supported by a number of contributing software packages for database management, visualization and model deployment)

BASIC OMS PRINCIPLES

Modularity

The fundamental and underlying approach of OMS is to apply modular design to simulation models. New scientific results are generally developed piecewise; each step is reviewed and validated by other scientists in the community. A similar approach should be taken to model software development. Unfortunately, when formalizing scientific methods and research into operational software constructs, scientists tend to start from a scratch. Moreover, when it comes to implementation there is usually a lack of understanding of appropriate software design.

It is possible that better understanding of the behavior of complex physical systems could be gained through better software practices. Specifically, dissaggregation of large and complex systems into components may reveal inter- and intra- relationships. This has been the interest of many authors, most notably Zeigler (1990), who called this "modeling in the large," and of Cota and Sargent (1992). This is related to a more general software concept known as "programming in the large." Earlier, DeRemer and Kron (1976) pointed out: "… structuring a large collection of modules to form a system is an essentially distinct and different intellectual activity from that of constructing the individual modules." With programming in the large, the emphasis is on partitioning the work into modules whose interactions are precisely specified. Modularity is the overall key to coping with the complexity inherent in large systems. Disaggregating these systems into smaller subsystems that are easier to understand can be achieved on a "divide and conquer" (Pidd and Castro, 1998) strategy.

Modularization has not been a common technique in practical model development, although the inherent structure of resource simulation models support modularization. Leavesley et al. (2002) pointed out different degrees of modularization: fully process modules and models; fully coupled models; loosely coupled models; and uncoupled models for decision support systems, which are all important conceptualizations of model integration and must be design principles for any modeling framework.

Several different modular approaches have been applied to watershed models (Singh, 1995). One of the earliest modular model development efforts was done for the SHE (European Hydrological System) Model (Abbot et al., 1986; Ulgen et al., 1991). The Precipitation Runoff Modeling System (PRMS) (Leavesley et al., 1983) was modularized from a monolithic version and implemented in the Modular Modeling System (MMS) (Leavesley et al., 1996). MMS was an early attempt by the U.S. Geological Survey to support interactive model construction, requiring a specified module structure. Recently, interest in the standardization of model software design has increased. Projects such as APSIM (McCown, 1995) are based on these principles. Model standardization is also gaining momentum within the environmental research and regulatory organizations and agencies (Whelan et al., 1997).

OMS development focuses on the following points to achieve a maximum benefit from modularization:

- Modularization is the key concept to simulation model development. OMS provides an application programming interface (API) for creating new modules. A master library of modules is also available for simulating a variety of water, energy and biochemical processes.
- OMS allows the interactive construction of complex models. OMS graphical user interface (GUI) components facilitate control of module connection, validity, consistency and completeness.
- An extended interface description for modules needs to be developed to specify data semantics. This is important for interdisciplinary module development.
- OMS supports automated module documentation generation. A module communicates with other modules through its public interface. Documentation generated from the module interface specification is sufficient to judge its suitability in a given simulation context.
- OMS potentially supports model scenario management and model customization through module exchangeability. This requires a set of module alternatives stored in a specific module library for that purpose. The library must organize this information and make the model fragments available for reuse.

The validation of module compatibility requires an advanced semantic representation of data objects (variables, parameter, etc.) used to connect modules. This leads to the concept of data dictionaries covering extended modeling related type information such a units, value ranges and description.

OMS Framework

The OMS framework is a domain-specific, reusable architecture with a set of interdependent classes in an object-oriented language. The primary benefits of application frameworks in general are modularity through well-defined and stable interfaces, reusability by using generic components, extensibility through hook methods and inversion of control through a reactive dispatching mechanism. The different kinds of frameworks are distinguished by the ways of adapting the framework: "black-box" and "white-box" frameworks. The black-box framework contains abstract elements, which needs to be specialized in the application. The white-box framework instead consists of already implemented specific components, which needs to be selected and customized in an application. OMS is fundamentally a white-box system as classified by the abstract hook method (Fayad and Schmidt, 1997).

OMS SYSTEM CHARACTERISTICS

The OMS has the following characteristics:

1. OMS models are treated as hierarchical assembled components representing building blocks. Components are independent and reusable software units implementing processing objects for simulation models. They reside in a model library and are categorized into data access components,

science components, control components, utility components and system components. This is described in the "OMS Modules and API" subsection.

2. OMS is able to integrate legacy code components. Due to automated wrapper generation for legacy code, components written in languages such as Fortran can be embedded into OMS at the function level more easily. This is described in the "Legacy Code Implementation" subsection.

3. The "knowledge"-backbone of OMS is the dictionary framework. It enables OMS to verify state variables and parameters according to scientific nomenclatures during model development and application. Dictionaries are also used to specify parameter sets, model control information and the component connectivity. They are implemented in the Extensible Markup Language (XML). This is described in the "Dictionary Framework" subsection.

4. OMS is extensible. Extension packages exist for different aspects in model development and application. Extension packages are used for visual model assembly, model application, an interface to the dictionary framework and GIS. This is described in the "Modeling System GUI Design" subsection.

5. OMS scales from a full-featured, stand-alone development system with tools for model assembly, visualization and analysis to a runtime Web service environment. This is described in the "OMS Model Views" subsection.

The approach being used for component definition is based on modular, hierarchical system modeling. It is an approach to complex dynamic system modeling where modular building blocks, i.e., system components with a well-defined interface in the form of input and output ports, are coupled in a hierarchical manner to form a complex system (Zeigler, 1990; Praehofer, 1996). This conceptual approach was implemented by using object-oriented programming paradigms. Because objects lack an explicit output interface the concept of input and output ports representing data flow was applied. An application programming interface (API) was designed to implement hierarchical modules.

OMS Modules and API

An OMS module is a piece of software implementing a specific function. This function might be complex or a single equation. The scope and complexity of the module depends on the problem the module will address. An OMS module is implemented as an object-oriented class that defines data objects, which are manipulated by the module's methods. A module has a public interface defining the interaction of the module with its environment. All communications must go through this interface. An OMS module has a single purpose, so it is restricted to a fixed set of interface methods. It consists of the following components:

- The OMSComponent class of the OMS core library provides the API for custom module implementation based on inheritance.
- A module encapsulates modeling objects. These data objects can be input data objects, output data objects and other submodules. The data objects are variables or parameter that are used for intermodule communication. The module can be the creator and owner of the data object or it may obtain a data object from other parts of OMS.
- A module encapsulates modeling logic. OMS provides the tools to implement the code in the Java, Fortran, or C programming language.
- Each OMS module must implement the following interface methods:
  - Init — The init method initializes the modules data objects. It is used to set the initial state of the data objects. This method is called prior to a model simulation run.
  - Run — The run method implements the computational part of a module. It operates on all data objects and components with the results of the module processing. This method is called during the simulation run and implements the process logic.
  - Cleanup — The cleanup method performs finalizing tasks for the module. It will release system resources (i.e., memory and sockets). It can also be used for writing summary reports. This method is called past model simulation run.

This OMS module definition API is implemented in the Java programming language (Arnold et al., 2000).

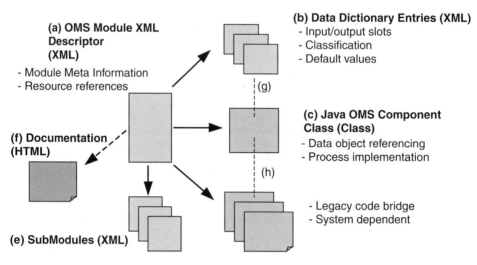

Figure 15.1   Module components and relations in OMS.

From an OMS framework point of view, a module consists of several components important for model development. The two main parts of a module include (see Figure 15.1):

- **Module Interface Specification** — This is represented by an XML descriptor (Figure 15.1a), the public interface is exposed. This includes the input and output data as specified by the data dictionary (Figure 15.1b), a reference to the module implementation component (Figure 15.1c) and sub-module descriptors (Figure 15.1e).
- **Module Implementation** — The implementation of the Java OMSComponent class (Figure 15.1c) and the native libraries (Figure 15.1d) if there is a legacy code attached to the module.

This design has several implications:

1. Module connectivity can be validated by using its interface only. Because modules are referencing data dictionary entries (e.g., parameter or variables), two modules can be connected if both are referencing the same entry. In addition, an extended "type" verification of data objects allows a more semantic type verification of proper data references among modules extending a simple "data type-based" verification (Figure 15.1g).
2. Because the module interface is expressed in XML, modules can be implemented in different programming languages. The design also takes architecture-dependent variations of resources belonging to the model into account. Native libraries are referenced for different operating systems and architectures (Figure 15.1h).
3. OMS uses automated documentation generation for the module interfaces based on XML style sheets into HTML (Figure 15.1f).

In summary, the benefit for an XML/Java combined description of modules results in sophisticated system handling of this module for verification, architecture adaptation and documentation based on a common resource.

Module data objects reside as attributes of each module. A module may encapsulate a number of data objects depending on its complexity. Several factors need to be taken into considerations when adding data objects to a module:

- Data objects may be declared either private or public. Accessibility and visibility is realized by publishing these data objects to the OMS framework.
- Public data objects can be shared among different modules. In this case a module is referencing a previously published data object. The module is responsible for requesting the data object from the modeling framework. If a data object is not shared, it is local only to the module.

Module Declaration	`// Declares basin and HRU physical parameters.`
	`public class basinprms extends OMSComponent {`
	`. . .`

Declared objects.
Initialized with data
generated by
OMS system
calls based on
data dictionary content
	`OMSDouble o_basin_area = getOMSDouble ("basin_are");`
	`OMSDoubleArray o_hru_area = getOMSDoubleArray ("hru_area", "nhru")`
	`OMSDoubleArray o_hru_imperv = getOMSDoubleArray ("hru_imperv", nhru")`
	`OMSDoubleArray o_hru_percent_imperv = getOMSDoubleArray`
	`          ("hru_percent_imperv", "nhru")`
	`OMSDoubleArray o_hru_perv = getOMSDoubleArray ("hru_perv", "nhru");`
	`OMSDimension o_nhru = getOMSDimension ("nhru");`
	`. . .`
	`}`

Figure 15.2   Data object declaration and system initialization.

The declaration of data objects in a module is shown in Figure 15.2. Objects are generated based on OMS API system calls. The identifiers must match the data dictionary entries (see example in Figure 15.4).

Legacy Code Implementation

Wrapping is a technique to embed existing nonobject-oriented software into an object-oriented architecture. Mapping interfaces of conventional systems into an object-oriented syntax enables the access to this legacy code. This mapping procedure can be supported by wrapper generation tools, which generate glue code for both software "worlds." Wrapping provides some general advantages:

- Keeping and reusing the existing software infrastructure as much as possible
- Moving step by step toward object orientation
- Accelerated introduction of object orientation
- "Separation of concerns" (graphical user interface vs. number crunching part, etc.)

A module contains a set of methods. Some of them are inherited from the super class OMSComponent. A module may implement additional methods. A module has to override the methods **init**( ), **run**( ) and **cleanup**( ) for interface functionality. A method for customization is used to provide specific editors for setting up the module data objects. Such editors may be applications implementing a kind of preprocessing prior model run.

OMS uses a sophisticated method to integrate native languages such as Fortran and C. Some OMS framework tools support the integration of native code into modules. The example given in Figure 15.3 shows how module data objects (input/output/local data objects) are mapped to Fortran variables and parameters by mapping Java object properties to Fortran variables..

Dictionary Framework

The OMS dictionary provides and manages dictionary resources for model construction like model data, model parameters, modules and models. In addition, during the model development phase, it provides and manages parameter scenarios, time series data in model application. It also

Intialization routine	
Property map section connects Java data objects to FORTRAN variables	
FORTRAN computation code implements process logic	

```
public native int init ( )   /*@F77
    @propertymap o_basin_area.value          -> basin_area;
    @propertymap o_hru_area.value1D          -> hru_area;
    @propertymap o_hru_imperv.value1D        -> imperv;
    @propertymap o_hru_percent_imerv.value1D -> percent_imperv;
    @propertymap o_hru_perv.value1D          -> hru_perv;
    @propertymap o_nhru_value                -> nhru
    @{
        real *8 totarea
        integer i
        real *8 diff

        totarea = 0.
        do 100 i = 1, nhru
            hru_imperv (i) = hru_percent_imperv (i)  * hru_area (i)
            hru_perv (i) = hru_area (i) - hru_imperv (i)
            totarea = totarea + hru_area (i)
    100 continue
        diff = (totarea - basin_area)/basin_area
        if (abs (diff).ge. .01) then
            omsExeption ('Sum of hru areas is not equal to basin area')
            return
        end if
    @}  init = 0
    */;
```

Figure 15.3   Legacy code integration.

Data dictionary entry ID	
Description	
Version	
Data category, range and default value	
Data type and unit	
Author reference	

```
<coms: data id="hru_elev">
    <oms:descrption>
        <oms:short>Mean elevation</oms:short>
        <oms:full>Mean elevation for each HRU </oms:full>
    </oms:description>
    <oms:version>
        <oms:cdate>2001-08-15 02:24:39</oms:cdate>
        <oms:mdate>2001-08-15 02:24:39</oms:mdate>
        <oms:revision>1.0</oms:revision>
    </oms:version>
    <oms:value cat="par am">
        <oms:min>-300.</oms:min>
        <oms:max>30000</oms:max>
        <oms:default>0.</oms:default>
    </oms:value>
    <oms:unit cat="length">feet</oms:unit>
    <oms:type>float*8</oms:type>
    <oms:author href="rojas" />
<oms:data>
```

Figure 15.4   Data dictionary element.

handles the interaction between several dictionary data sources with database management systems and the modeling system. OMS dictionaries are written in XML.

The OMS dictionary design addresses two major issues:

1. The OMS dictionary XML resource file structure defines a generic and open scheme for describing and processing sets of (heterogeneous) dictionary content.
2. The OMS dictionary GUI tool operates with dictionary XML resources and manages its content in coordination with the other OMS components of the framework to construct and apply a model.

Figure 15.4 shows a dictionary element, which encapsulates all meta-information and information about the parameter "hru_elev." Based on this record the system is able to instantiate objects requested by modules (see Figure 15.1) and initialize them properly.

OMS Dictionary Framework Architecture Design Considerations

A major dictionary design goal was to maximize transparency, openness and extensibility. Format is available for data, module, parameter, or scenario dictionaries. The preferred solution for handling dictionaries is a generic scheme where all of these special dictionaries could be mapped. These issues were taken into account when designing the overall dictionary architecture:

- A dictionary consists of data sets, which may be either homogenous or heterogeneous. Because the dictionaries are written in XML, data sets are not document-oriented and should not limited to a tabular view.
- A permission scheme is required to control the operations associated with the dictionary. Dictionary developers have the responsibility to set up appropriate permissions.
- Dictionaries may contain large sets of XML entries. Well-known approaches such as the XML query language can filter information from an XML data set. A dictionary must deal with this feature. Each dictionary must provide applicable queries to filter specific information. The use of XML as the primary dictionary data format opens a wide range of other data sources. The system should be able to generate XML dictionary documents from relational database management systems. There must also be uniform and transparent support for local, remote (via a URL) and database dictionaries.
- Dictionaries should be self-describing in terms of their data structures. The dictionary data content must drive the behavior of any GUI tools.
- The dictionary GUI tool should be able to operate with other tools in OMS in an easy way. A bi-directional interaction should be supported, so dictionary entries may be "dragged" into other tools to construct or run the model. New entries, such as modeling results, may be "dropped" into the dictionary as modeling results.
- The logical structure of a dictionary must be mapped into a physical tool representation.

Other design considerations including keeping the dictionary file as simple and as transparent as possible, both for humans and for processing in the OMS framework. By using XML as dictionary resource definition format and XPath for accessing elements of a dictionary, a slim architecture for OMS tools is possible.

Modeling System GUI Design

The OMS framework uses a common core user interface (CommonUI). The CommonUI supports different types of OMS extensions. Extensions implement different tools for model development and application. OMS is open and configurable through the extension package. OMS comes with a basic set of extensions, but also offers system developer an API to implement custom extensions.

The OMS CommonUI provides the following functionality for all extension: GUI resource handling of common elements; system logging; and a console interface to interact with the system by using a python shell. The CommonUI enables extensions with slim implementation overhead, common look and feel but also provides maximum flexibility to adapt the overall appearance of the system. One powerful feature of the CommonUI is the python-scripting interface. It allows the dynamic interaction of the CommonUI with extensions, based on shell commands.

OMS extensions implement the toolbox elements that can be used to customize the CommonUI according to modeler needs. OMS comes with a basic set of extensions dealing with various facets of modeling:

- **Development Extension** — allows the interactive assembly of models based on a module library. It provides support for module search, module retrieval and module integration from local and remote sites according model requirements. It allows the validation of module connectivity using input/output data constraints graphically. This extension delivers runable models.
- **Application Extension** — deals with the interactive model application of runable models. It provides automatically generated GUI elements for model parameterization, output customization, graphical components for visualization of variables and parameter scenario management. This extension delivers model results in terms of graphs and numbers.
- **Dictionary Framework Extension** — manages the modeling dictionaries. They are used to handle the following modeling resources: module interfaces; parameters and variables; parameter sets; and meteorological data set descriptors. Other extensions use this extension to validate the proper module connectivity (development extension) or for parameter value assignment (application extension).

Figure 15.5   OMS CommonUI and extensions.

Other types of extensions could be developed for post-run results analysis, GIS connectivity, legacy code integration and model deployment using Web services. Figure 15.5 shows the OMS CommonUI loaded with the extensions for model development, model application and the dictionary framework. Each extension appears as a folder node in the model resource tree on the left side of the window. All extension related resources are sub-nodes of their folder. The common logging panel and the console command line interface appear as tabbed panels on the bottom of the window. The main part of the CommonUI occupies the extension desktop in the center of the window. The desktop is customized according to the extension implementation purpose.

OMS Model Views and Deployment

Enabling simulation models to run under different architectures and computing environments becomes more and more important, especially with the increasing demand for Web-based application of simulation models. OMS is designed to cover a variety of application and deployment and execution paradigms:

- The OMS application gets deployed by Java Web Start Technology over the Internet. This ensures the local client installation, automated update and security of OMS. Models being developed with OMS in such a scenario are running within OMS. This setting is typical for a development application, where the modeler needs flexibility to change the models module structure, parameter sets, as well as input data.
- The application of OMS models as canned models is required when validated OMS models are applied in projects or application scenarios. There is no need to change the structure of the model

or have the flexibility of automatically generated GUI components for parameterization. The usage of an adaptive, model-specific GUI directs the model user.

- OMS models can be executed in a server-centric environment. Handled by a Web server, an OMS runtime environment produces a generic Web interface to enable a Web browser application of models. Preconfigured parameter and data sets can be accessed at the server side, whereas the client site deals only with a Web browser for model input and output.

Within OMS it is possible to transfer an existing model to different execution schemes without changing the model structure. Models being developed with the OMS environment are capable to run under a Web-only environment.

Figure 15.6 Loaded PRMS modules and PRMS generic parameterization GUI.

EXAMPLE APPLICATION OF PRMS

For proof of concept, a prototype model was developed to test component integration and interaction. The hydrological model PRMS, (Leavesley et al., 1983) was selected for integration into OMS. The goal was to transform the model structure into OMS modules and their related XML resource descriptors and compare the model output with the validated version of the model in MMS.

PRMS is a deterministic, distributed, continuous hydrological model (Leavesley et al., 1983). A watershed is divided into homogeneous hydrological response units (HRUs). The modular conceptualization of PRMS is reflected by simulating the hydrological system as a vertical series of interlinked reservoirs. Each hydrological process is represented as a separate module. Variants of PRMS reflect the variable characteristics of different catchments. The basic PRMS model (as tested) consists of 14 process modules (Figure 15.6).

The test application for OMS also contained a meteorological data and a parameter set for the East River basin, a subcatchment of the Gunnison River basin in Colorado. A 20-year set of climate data was used for validation. The East River parameter set was preprocessed and generated by the GIS–Weasel tool (Leavesley et al. 1997). Both, the parameter set and the time series input where in MMS formats.

Figure 15.6 shows OMS with modules forming the PRMS model. The Component Development node in the resource tree contains the PRMS OMS model build upon the 14 sub-modules. The loaded PRMS runtime was configured with the East-River Parameter Set. OMS is automatically generating GUI elements for parameter input such as spreadsheets for one-dimensional arrays.

Although the PRMS source code was already modularized for MMS, some recoding was required for the Java environment. PRMS modules are implemented in Fortran77. Automated code generation could prevent many of the errors that were introduced by the manual procedure. The following steps in code migration were performed:

1. The MMS runtime library was modified for capturing each modules interface calls (reading/writing to variables/parameter/dimensions). XML-intermediate module descriptors for this information were generated.
2. The Java OMSComponent file and the XML component descriptor were generated from the intermediate descriptor. Data objects referenced by MMS system calls in MMS/PRMS modules resulted in OMS/PRMS data objects referencing data dictionary elements. This transformation was done automatically.
3. The Java OMSComponents were extended with Fortran science code, embedded into the initialization, processing and cleanup section of each component. Local variables were identified (compiler test run) and added to the code.
4. OMS tools and compiler were used to generate binary executable code (Java class files, dynamic link libraries).
5. A total of 456 parameters, variables and dimensions were identified and generated. For each of them, an XML data dictionary element was generated and added to a PRMS data dictionary.
6. The East River parameter set was converted into an XML OMS parameter file and added to a parameter dictionary.

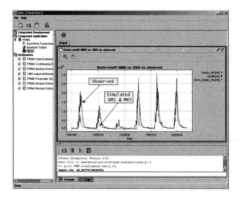

Figure 15.7 PRMS run in OMS — comparison of basin runoff from OMS vs. MMS and observed data.

The modules were loaded into a running OMS framework. The driving variables were traced for each component to verify the input. The output variables were plotted and compared with corresponding MMS/PRMS variables. Finally, the simulated basin runoff for the OMS and the MMS version of PRMS was plotted with the observed runoff. Figure 15.7 shows the result, matching graphs for MMS and OMS values for the basin runoff.

CONCLUSION

The OMS introduced in this chapter is a framework that facilitates the development of customized models from a library of science, data and utility modules, as well as their testing, application and deployment. OMS features component integration techniques, graphical user interface components, graphical visualization features and other utilities supporting model construction and application. The design of OMS was influenced and driven by the needs of agricultural and natural resource agencies to optimize the models, development process and maintenance. Due to the increasing complexity of the simulation problems encountered in natural resource and environmental management model development can only be efficient by using methods and tools such as OMS. Due to its dedication to modularized model development, OMS offers the potential for science building block reuse and easy update and maintenance.

Model management and transfer practices from the research unit development level to the field office application level were considered in OMS design. Hence, the goal of this project is the design and implementation of a modeling system that emphasizes interoperability, connectivity, scalability and reuse of environmental simulation modules. The PRMS model was adapted into OMS, which lead to a significant reduction of module implementation code. Results showed that a major fraction of GUI components for model parameterization could be generated automatically from module data components. Researchers are in the process of disaggregating another large model, the Root Zone Water Quality Model (RZWQM), into OMS modules.

Future Developments

Future OMS development efforts will be leveraged by application and project demands. The integration of GIS capabilities by means of an OMS extension will enable interactive geo-processing in the context of model application. GIS implementation efforts are underway. The overall focus of the GIS extension will be driven by the need of modeling related features. This comprises model parameterization in terms of spatial parameter sets as well as spatial visualization of model progress and results.

Another major OMS effort will focus on a deployment of simulation models using Web services and Universal Description Discovery and Integration (UDDI). This will enable simulation models to be discoverable and directly callable by other simulation environments. Other efforts include post-run analysis extensions that will be integrated with descriptive statistics.

ACKNOWLEDGMENT

The OMS Project is supported and funded by the USDA-ARS, in collaboration with the USDA-NRCS and the U.S. Geological Survey.

REFERENCES

Abbot, M.B., J.C. Bathurst, J.A. Cunge, P.E. O'Connell and J. Rasmussen. 1986. An introduction to the European Hydrological System — Système Hydrologique Européen (SHE), structure of a physically based, distributed modelling system, *J. Hydrology,* 87:61–77

Abel, D.J. and P.J. Kilby. 1994. The systems integration problem, *Int. J. Geographical Inf. Syst.,* 8(1):1–12.

Arnold, K., J. Gosling and D. Holmes. 2000. *The Java™ Programming Language, Java™ Series,* 3rd ed. Addison-Wesley, Reading, MA.

Balmer, D.W. and R.J. Paul. 1990. Integrated support environments for simulation modelling, *Proc. of Winter Simulation Conf.,* December 9–12, New Orleans, Louisiana, 243–249.

Breunese, A.P.J., J.L. Top, J.F. Broenink and J.M. Akkermans. 1998. Library of reusable models: theory and application. Simulation Series, Simulation Councils Inc. 71(1):7–22.

Cota B.A. and Sargent R.G., 1992. A modification of the process interaction world view, *ACM Trans. on Modeling and Computer Simulation,* 2(2):109–129.

DeRemer, F. and H.H. Kron. 1976. Programming-in-the-large vs. programming-in-the-small, *IEEE Trans. Software Eng.,* SE-2(N.2):114–121.

Fayad, M.E. and D. Schmidt. 1997. Object-oriented application frameworks. *Commn. ACM,* 40(10):32–38

Leavesley, G.H., P.J. Restrepo, S.L. Markstrom, M. Dixon and L.G. Stannard. 1996. The modular modeling system (MMS) — user's manual, U.S. Geological Survey Open-File Report 96-151.

Leavesley, G.H., R.J. Viger, S.L. Markstrom and M.S. Brewer. 1997. A modular approach to integrating environmental modeling and GIS, *Proc. 15th IMACS World Congr. on Scientific Computation, Modelling and Applied Mathematics,* August 24–29, Berlin, Germany.

Leavesley, G.H., R.W. Lichty, B.M. Troutman and L.G. Saindon. 1983. Precipitation-runoff modeling system — user's manual, U.S. Geological Survey Water Resources Investigation Report 83-4238.

Leavesley, G.H., S.L. Markstrom, P.J. Restrepo and R.J. Viger. 2002. A modular approach to adressing model design, scale and parameter estimation issues in distributed hydrological modeling, *Hydrological Processes,* 16(2):173–187.

McCown, R.L., G.L. Hammer, J.N.G. Hargraves, D.L. Holzworth and D.M. 1995. APSIM: a novel software system for model development, model testing and simulation in agricultural systems research, *Agricultural Syst.,* 50:255–271.

Panagiotis, K.L., S. Molterer and B. Paech. 1998. Re-Engineering for Reuse: Integrating Reuse Techniques into the Reengineering Process. citeseer.nj.nec.com/linos98reengineering.html.

Pidd M. and R. Bayer Castro. 1998. Hierarchical modular modeling in discrete simulation, *Proc. 1998 Winter Simulation Conference,* D.J. Medeiros et al., Eds., IEEE, Washington, D.C., 383–390.

Praehofer, H. 1996. Object-oriented modeling and configuration of simulation programs. In *Eur. Meet. on Cybernetics and Syst. Res.,* Vienna, Austria, April, 259–264.

Singh, V.P. 1995. *Computer Models in Watershed Hydrology,* Water Resources Publications, Highlands Ranch, CO.

Top, J.L., A.P.J. Breunese, J.F.Broenink and J.M. Akkermans. 1997. Structure and use of a library for physical system models, *Proc. ICBGM'95, 2nd Int. Conf. Bond Graph Modeling and Simulation,* Las Vegas, Nevada, January 15–18, SCS Publishing, San Diego, CA, Simulation Series, 27(1):97–105.

Ulgen, O.M., O. Norm and T. Thomasma. 1991. Reusable models: making your models more user-friendly, *Proc. 1991 Winter Simulation Conf.,* December 8–11, Phoenix, AZ, 148–151.

Whelan, G., K. J. Castleton, J.W. Buck, G.M. Gelston, B.L. Hoopes, M.A. Pelton, D.L. Strenge and R.N. Kickert. 1997. *Concepts of a Framework for Risk Analysis in Multimedia Environment Systems (FRAMES),* PNNL-11748, Pacific Northwest National Laboratory, Richland, WA.

Zeigler, B.P. 1990. *Object-Oriented Simulation with Hierarchical, Modular Models,* New York Academic Press, San Diego.

CHAPTER 16

Future Research to Fill Knowledge Gaps

Jerry L. Hatfield and Bruce A. Kimball

CONTENTS

INTRODUCTION

Simulation models of agricultural systems have been developed over the past 40 to 50 years. These models have ranged in complexity describing physiological, soil, or atmospheric processes within the soil–plant–atmosphere continuum across a range of spatial and temporal scales. Output generated from a wide variety of models has had an impact on our ability to capture the dynamics of processes and to be able to extend this information into other scales. A decade ago, Hanks and Ritchie (1991) assembled the current state of knowledge on models in the plant and soil system. Since that time there has been continual progress on the further refinement and development of models for agricultural systems. There still remains a challenge; however, in determining how the perfect model is to be constructed that will mimic agricultural systems. In this chapter we will explore some of the challenges that future research will face in the refinement of agricultural systems models.

Models are representations of systems. Mathematical representations of physical systems behave with much more predictability than biological systems. For example, we can construct a model to predict the trajectory of an object in space accurately as evidenced by our ability to fire a rocket into space and to intercept the moon. Likewise, a high degree of predictability exists for many chemical reactions because we are able to estimate their yield and products, although our ability to predict biological systems is much less certain. Models themselves are systems and are aggregations of our understanding of biological systems represented by a combination of empirical equations. We learn from the process of assembling models in advancing our understanding of the interactions among parameters that we can easily obtain or infer and those responses which we can also observe. Models have existed throughout the history of modern science. Rossi (1968) describes that Galileo in the 16th century was a proponent of constructing models to describe the behavior of physical systems. Mankind has had a quest to be able to describe all of the phenomena surrounding everyday life.

This chapter represents our examination of the current state of models from the view of looking at models with a goal of providing a challenge. We will admit that is impossible to trace the shortcomings in current models by examining their geneology and philosophy and our goal is to create some imaginative thought in this quest of building better models that describe plant–soil–atmosphere interactions.

APPLICATION OF CURRENT MODELS

One of the long-term goals of agricultural models is to simulate the potential effects of various changes in agricultural systems due to genetics, management, or weather. A number of current agricultural models attempt to describe complex agricultural systems. One major contrast is the acknowledgment of genetic diversity in simulation models. Models such as CERES-Wheat or CERES-Maize incorporate genetic diversity within the physiological functions of the model (Jones and Kiniry, 1986) and are described as genetic variations within phenological and physiological processes. Predicted results from these models provide a glimpse into potential genetic variation on crop growth and yield. In contrast, RZWQM uses a generic crop production model built around a primary production model, constructed from carbon and N budgets, that predicts the responses of plants to environment changes (Hanson, 2000). The purpose of the plant component within RZWQM is to simulate a large number of plant species rather than diversity within a species whereas CERES-Maize accounts for differences within species. Both of these types of models have broad applications to many studies associated with agricultural systems; however, these are not intended to replace one another.

Current agricultural system models are often used to predict the potential impact of future scenarios on crop production. An example of large-scale application of CERES-Wheat (Godwin et al., 1990) and I-Wheat (Meinke et al., 1998) has been to evaluate the potential impact of the El Niño Southern Oscillation (ENSO) on wheat production in the U.S., Canada and Australia (Hill et al., 2000). Rosenthal et al. (1998) has shown that simulation model output from specific sites can be aggregated into an estimate of regional production and this approach was used in the aggregation of crop yield estimated with these models. Another example of aggregation of outputs from individual site simulations to create regional and state level yield estimates has been provided in research by Haskett et al. (1995) who used the GYLSIM model to estimate soybean (*Glycine max* L. Merr.) yields for Iowa.

Models have been evolving over time to replace some of the original interdependence among processes into a modular form. The RZWQM has been developed as a modular system that consists of six submodels. These submodels represent the: physical; soil chemical; nutrient; pesticide; plant growth; and management components within an agricultural system. The input requirements and the output are described in Ahuja et al. (2000b). RZWQM is constructed as a one-dimensional model that places some restriction on large-scale application; however, this was not the intended use of the model in the initial stages of development and implementation. Nevertheless, the model does allow components of agricultural systems to be evaluated.

One important aspect of current models is their range of scales and applications. For example, GLEAMS (Groundwater Loading Effects of Agricultural Management Systems) (Leonard et al., 1987), CERES (Crop-Environment Resource Synthesis) (Hanks and Ritchie, 1991), CENTURY (Parton et al., 1994), DSSAT (Decision Support System for Agrotechnology Transfer) (Tsuji et al., 1998), GPFARM (Great Plains Framework for Agricultural Resource Management) (Ascough et al.,

1998), ECOSYS *ecosys* (Grant, 1995), EPIC (Erosion Productivity Impact Calculator) (Sharpley and Williams, 1993), SPUR (Simulation of Production and Utilization of Rangeland) (Foy et al., 1999), ALMANAC (Agricultural Land Management Alternatives with Numerical Assessment Criteria) (Kiniry et al., 1992), SWAT (Soil Water Assessment Tool) (Arnold et al., 1998), HUMUS (Hydrologic Unit Model for the U.S.) (Srinivasan et al., 1993), AGNPS (Agricultural Nonpoint Pollution Sources) (Young et al., 1987), GOSSYM (Cotton simulation and management model) (Baker et al., 1983) and DAISY (Hansen et al., 1991) have all been used in different applications for specific agricultural systems. The range of application of these models has been from individual soils within a field to large watersheds. Models are continually being developed and applied in attempts to represent agricultural systems. It is not our intent in this chapter to review each of these models and to compare among models, but instead we use the current state of information as a springboard into identifying knowledge gaps.

SOIL COMPONENTS

Soil components included in most models treat hydrologic, chemical and biological processes in a series of layers within the soil profile. These layers are defined either by soils information gleaned from soil survey information available in county soil survey reports or at the regional scale through the STATSGO database (USDA-SCS, 1992). Interrelationships among soil layers are treated through a variety of computational methods. One of the largest gaps that exist in the use of agricultural systems models is the characterization, quantification and simulation of the upper soil layer as affected by tillage and crop residue management. As an example, the NTRM (Nutrient, Tillage, Residue Management) model described by Shaffer and Larson (1987) was one of the first to try and simulate the interactions among tillage and residue management. Williams et al. (1984) has proposed a simple relationship between soil bulk density and tillage for use in EPIC as

$$\rho_t = \rho_{t-1} - \left[\left(\rho_{t-1} - 0.667 \rho_c \right) I_i \right] \tag{16.1}$$

where ρ_t is the bulk density after tillage (Mg m^{-3}), ρ_{t-1} is the bulk density before tillage, ρ_c is the consolidation bulk density at 33 kPa pressure (kg m^{-3}) and I_i is the tillage intensity value (0–1) and varies with type of equipment. This is an empirical model assembled from observations; however, the change in bulk density after tillage due to rainfall events or subsequent tillage is not well defined. We assume in many of the current models a very limited temporal interaction of management effects. One of the primary limitations to Eq. 16.1 is that these changes in bulk density are assumed to be uniform across the soil surface and therefore, they do not account for some of the modern tillage equipment, which till only a small portion of the soil surface. These types of tillage systems induce a spatial pattern onto the soil surface, which requires a three-dimensional treatment of the soil surface in order to be realistic. Such consideration of the dynamics of the soil surface has been based on adaptation of limited empirical observations; yet, it has beenapplied across an extremely wide range of soils and management systems.

The temporal dynamics of this upper surface layer are critical to linking agricultural management practices to the water flow, gas exchange, biological activity and carbon sequestration. Bresler (1991) outlined some approaches in a chapter on soil spatial variability and concluded that chemical reaction, chemical decay, adsorption and exclusion, soil structural changes and the influence of root uptake were viable research needs in perfecting models that simulate soil processes. Of these topics, the quantification of soil structural changes at the surface seems to be most elusive. There is little quantitative understanding of the management effects on soil properties and processes at the surface or throughout the soil profile that provides enough information to guide incorporation into existing models.

Soil water dynamics within models is one of the critical factors because of the role in the transport of nutrients and chemicals and supplying water for the transpiration requirements of the crop. Incorporation of water into models has been built around the Richards equation. Feddes et al. (1978) documented one of the earlier soil water balance models constructed as a multilayer model that could handle movement into field drain lines and downward movement through the soil profile. Lascano and van Bavel (1986) developed a soil water evaporation model that used multiple layers and solved for closure of the energy balance through surface temperature. Various forms of soil water models are being used in various simulation models and the results are fairly realistic. One critical aspect of model development is to compare the soil water change patterns over time, space and depth for different soils within fields to determine how realistic their performances are for complex terrain. One of the major limitations in soil water components has been the treatment of macropore flow and how to describe the continuity of these macropores through the soil profile. There is qualitative evidence for these processes but little quantitative evidence. Macropores may be a good example of how management practices might disrupt macropore continuity, but little is known about how these macropores change over time.

Lascano and Hatfield (1992) measured soil water evaporation from bare soil in transects across a field. In this study, they were able to measure the spatial variation of the soil properties and also the variance of the evaporation rates within the field. Evaporation rates varied across the field due to small differences in topography and surface roughness. When the CONSERVB model (Lascano and van Bavel, 1986) predictions were compared with measured values of soil water evaporation the predicted values were within one standard deviation of the measured mean when average soil hydraulic properties were used in the model. Comparisons of this type across larger transects within fields would help evaluate soil water models; however, these data are not often available for many sites and site variability is difficult to assess *a priori*.

Ahuja et al. (2000a) stated that theories of soil water movement and transport of heat and chemicals are understood for simple soil systems; however, for complex soil systems, several aspects are not treated mechanistically. They stated that research should address several gaps including:

1. Role of macropores on water and chemical flow from the surface to deeper layers
2. Role of surface aggregates on the water transport in the upper soil layers
3. Role of inter-aggregate, immobile pore space and chemical kinetics in the soil matrix
4. Changes in soil properties, continuity of macropores and aggregation bought about by tillage
5. Reconsolidation of soil surface after tillage due to drying and rewetting
6. Role of high water table and drainage systems on water and chemical transport

These soil structural aspects have not been considered in many of the models, even though they can profoundly affect most soil processes. These parameters, if available, would help in developing improved models. One area, in addition to these needs is to develop a quantitative understanding of the changes in surface roughness and surface detention of water across slopes or landscapes that could be used to model infiltration, runoff and chemical transport. Continual development of improved models of water dynamics at the soil surface will improve many soil transport models.

For some aspects, far more is known experimentally than has generally been incorporated into the models. For example, P and K concentrations are listed on every commercial fertilizer bag and farmers apply these nutrients to improve crop growth on soils where these elements are limiting, yet N is the main soil nutrient simulated by most of these models. Although, N, P and K are considered the dominant nutrients limiting plant growth, a major void in the soil component is the treatment of micronutrients in the soil profile, changes of pH due to liming, changes in cation exchange capacity (CEC) over time and movement of heavy metals or other organic compounds through the soil profile. All these areas will require more quantitative understanding before impacting the current models.

The recent concerns about global climate change and the possibility of trading carbon credits (whereby power companies pay farmers to sequester carbon in their soil organic matter) has highlighted gaps in knowledge with respect to predicting C storage in soils. These include:

1. How will possible changes in elevated CO_2 and climate affect plant residue quantity and quality inputs to the soil?
2. How will C sequestration be affected by different cropping systems across a range of soils and climates? Some evidence is available, but more is needed to help determine a wider range of expected changes in C due to agricultural practices.
3. How fast will various fractions of plant residue and soil organic matter change from one pool to another, classified according to residence time, as affected by crop species, environmental conditions, soil properties and tillage management practices?

These needs have not been considered in many of the models; however, they represent the conditions in which soils play a role in agricultural system simulation.

Concerns about global climate change have also stimulated interest in emission and consumption rates of soil gases, especially radiatively active (i.e., greenhouse) gases such as CO_2, N_2O and CH_4. Emissions from soils have been studied in different settings and management practices; however, an analysis of the factors that affect these emissions on a year-round basis is lacking. Conrad (1996) provided a review of the role of soil microorganisms as controllers of atmospheric trace gases. On the other hand, soil microbiological components in agricultural systems models are treated in relatively gross terms. For example, recent studies by Wagner-Riddle and Thurtell (1998) suggest that N_2O emissions from soils increase in the spring because of the build-up of the gas levels in the upper soil layers, so that when spring thaw occurs, the movement of gases into the atmosphere increases. They suggested that management practices that would alter the N pools would have a significant impact on N_2O emissions. Another another practice that affects N_2O emission is manure application. For example, Ferm et al. (1999) observed large fluxes immediately after application. Soil management effects, e.g., tillage coupled with either manure or fertilizer application, need to be understood if agricultural systems models that estimate global impacts are to be developed. Development of simulation models that could accurately assess the interactions of variables across a range of soils, management practices and weather scenarios would be an advance in the soil component.

Soil biological processes, particularly the distribution and activity of microbial and mesofaunal populations throughout the soil profile and over the course of time, is another aspect that is affected by changes in soil management, but sufficient knowledge is lacking to improve the soil component of models. Mycorrhyzal infections are known to improve phosphorous and water uptake by many plant species, but modeling this interaction is problematic. Recently, Rillig et al. (2001) reported that elevated CO_2 concentrations, such as expected near the middle of this century, doubled the length of hyphae of arbuscular mycorrhizal fungi and increased aggregate water stability 20 to 40% in soil samples taken from a sorghum field exposed to free-air CO_2 enrichment. Predicting the extent of such changes in soil structure and how they will affect future water permeability and runoff is beyond our current capability.

Another large knowledge gap in the soil component of agricultural systems models is the incorporation of the variation of soil properties in a field into the model. As an example, a description of a typical central Iowa field is shown in Figure 16.1. Examining soil property values as reported by USDA-SCS, reveal a variance associated with each property (Table 16.1). Proper accounting of such variable soil properties by the simulation models will be required in order to obtain any meaningful regionally averaged information. One of the primary questions to be addressed is whether it is best to simulate field variation by averaging across soil properties or by averaging after simulating for each soil.

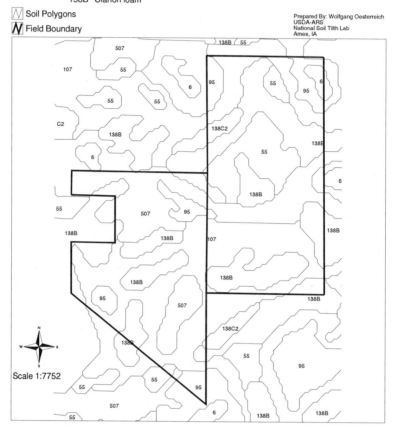

Figure 16.1 Spatial variation of soil types across an agricultural production field in central Iowa.

Table 16.1 Variation in Surface Soil Properties for Four Soils form Story County, Iowa

Soil	Percent Clay (%)	Permeability (Mm/hr)	Available Water Capacity (Mm/mm)	Soil pH	Organic Matter (%)
Clarion	18–24	15–50	0.2–0.22	5.6–7.3	1–3
Webster	26–36	15–50	0.19–0.21	6.6–7.3	6–7
Nicollet	24–35	15–50	0.17–0.22	5.6–7.3	5–6
Okoboji	35–42	5–15	0.21–0.23	6.6–7.8	9–18

Source: USDA-SCS, *Soil Survey of Story County,* Iowa, 1984.

PLANT COMPONENTS

Within agricultural systems, plants are critical factors because crop production is one of the endpoints simulated. The initial development of the CERES-type models was built around accurate simulation of phenological stages of plant development. The philosophies of the phenological models were described for corn (Kiniry, 1991), wheat (Ritchie, 1991) and soybeans (Jones et al.,

1991) to show how temperature had to be adjusted to predict the development of the different crops. Using the phenological approaches provided a basis for continued addition of other factors that affected growth, e.g., solar radiation, soil water, nutrients and carbon dioxide. Specific crop growth models utilize a basic foundation of development as influenced by temperature, light interception, photoperiod, carbon dioxide uptake and adjustment of these factors by soil water or nutrient supply. Partitioning of carbon allocations within the plant into roots and shoots and then further partitioning into reproductive organs, is controlled by the onset of phenological stages. Phenology and partitioning are generally handled empirically in the models, but a huge knowledge gap exists in understanding the fundamental controls within the plants for these important aspects. For example, water stress accelerates the rate of development in wheat, whereas it slows development in sorghum before anthesis, but no one knows why. Progress is being made by plant physiologists in understanding gene expression and hormonal signaling among plant organs (e.g., Beveridge, 2000), but we are far from being able to utilize such information for prediction. Nevertheless, the research community has had success in developing plant growth models that are able to estimate the development, total biomass and yield of several major crops.

The research community associated with plant-growth models appears fairly confident that current models are able to estimate the development of crops, total biomass and yield. A number of other knowledge gaps still present challenges for plant-growth modelers. These challenges can be expressed as a series of research needs.

1. *Incorporation of the genetic diversity into crop models in terms of phenology and partitioning, as mediated by environmental conditions* — Many of the current approaches used in changing phenological response use simple linear models to account for the effect of environmental stresses. For example, time within a phenological stage increases or decreases linearly with exposure to stress. Observations would suggest a more complex interaction with environmental conditions and even changes throughout the life cycle of a plant.
2. *Quantitative understanding of root growth and development due to genetic differences and environmental conditions* — Root-shoot partitioning in different genetic material and the interaction of this partitioning with environmental conditions. Often, these relationships are driven by soil temperature but do not incorporate water or nutrient interactions. These observations are difficult to collect under experimental studies and incorporation of these interactions into simulation models will require continued feedback to plant physiologists.
3. *Controls on root growth as affected by both plant and soil characteristics* — Simple descriptions are available regarding the effect of soil temperature on root growth. These need to be expanded to include the effects of water, nutrient and potentially even agricultural chemical responses.
4. *Estimation of the role of root exudates and root mortality on the soil biological component and nutrient cycling in the soil profile* — The feedbacks and interactions of root exudates and decay on soil biological populations remain large challenges for simulation models. Estimates of root turnover during the growing season and after harvest of annual crops will require understanding the temporal and spatial dynamics.
5. *How root mechanical restrictions, nutrient limitations, or conditions that slow translocation within a plant feed back on photosynthetic processes.*
6. *How humidity, as well as light, CO_2 concentration and root signaling, control stomatal conductance* (e.g., Jarvis and Davies, 1998).
7. *Quantify how stress conditions due to water, temperature, light, or nutrients affect vegetative growth, photosynthesis, C:N allocation among plant components, or root: shoot ratios across a range of growth stages and varieties within a species.*
8. *Interactions among plant growth and other biotic factors, e.g., insects diseases, weeds, etc.* — Weeds offer a unique competition with plant growth because early in the growing season they complete for light, soil water and nutrients. Late-season weeds, however, provide more of a competition for light because the plant is established and near maturity. The amount of total weed biomass may be the same but the effect on the crop many be entirely different. Likewise, insect populations will have a differential temporal effect on plant growth. Simulation models will have

to account for the differences among competitors in plant communities and their interactions with plants.

9. *Quantification of the competitive component to allow for better estimation of the effect of varying plant population on plant growth and partitioning or the competitive aspects of weed–crop mixtures.*

10. *Quantification of the effect of stresses and management on the quality of grain (protein, oil, starch) or forage* — Observations of grain or fiber quality have been somewhat empirical, although as interest increases in examining product quality, simulation models will have to begin to incorporate these components and characteristics.

Similar to the soil component, knowledge gaps also exist about how to account for variation of individual plants. Plant growth models predict phenology, biomass, or leaf area accumulation without an estimate of the variance of response among plants. Within this description of variation is the genetic diversity that exists within species and within hybrids, cultivars, or varieties of agricultural crops in responding to changes in temperature or the effect of stress (nutrient or water) on phenological stages, growth rates and partitioning. Plant sampling within the field from the same treatment will have a given amount of error associated with each measurement, which might simply reflect measurement errors or "real" plant differences caused by many factors. Comparison of these measured data with simulated results can be made by ensuring the simulated results fit within the variance and/or that the simulated variances match those found in the field. Most models, however, do not simulate the variances expected for various management scenarios.

ATMOSPHERIC CONSTITUENTS

Within agricultural systems models the major component that describes the atmosphere is the evapotranspiration (ET) models. The bridge among the soil–plant–atmosphere components is through the water loss or ET from the soil or plant surface as estimated through a variety of methods. Agricultural systems models vary in the method that is used to estimate ET from very simple monthly average models such as Blaney–Criddle (Blaney and Criddle, 1950) to the treatment of sparse canopies as described by Shuttleworth and Wallace (1985). Hatfield and Allen (1996) showed that many of the current ET models are well suited to estimation of potential ET (unlimited soil water supplies) but have problems when applied to deficit soil water conditions. Part of this difficulty is in quantifying how the plant canopy responds to deficit soil water conditions. Continual measurements of the flux rates of latent heat, sensible heat, or carbon dioxide fluxes over an agricultural surface have shown large amounts of variation within short time intervals. As an example, one of the major factors within the Shuttleworth–Wallace model is the inclusion of a canopy resistance term. It is assumed that this term is relatively constant; however, observations of the energy balance over crop canopies that permit the calculation of canopy resistance have revealed how this changes throughout the day and across days, due to crop growth and soil water conditions (Hatfield, 1985). Variations from 10 to 1000 m s^{-1} have been observed within a five-day period. Large variations in surface resistance were observed immediately after tillage operations and quickly changed as the soil surface dried. We have observed differences within fields in the drying rate due not only to varying water holding capacity of the soil but also to differing amounts of residue on the soil surface. We can describe these processes with simulation models, but accounting for the spatial and temporal responses to various conditions is not done very well within the current models.

Atmospheric components into agricultural systems models include air temperature, solar radiation and precipitation as the minimum data set, although the more sophisticated models also need humidity and wind speed. These data are often used from observational sites within an area and there have been extensive efforts to develop tools to estimate weather parameters. An example of this approach is described by Yu (2000) for CLIGEN. The values generated through this method represent the general conditions for a site and do not account for the small variations that may be induced by microtopographic effects on air temperature, soil temperature, windspeed, or solar radiation. Variations in air or soil temperature can be quite large over small distances, although little quantitative evidence has been assembled to guide the development of submodels that would estimate this variation. It is often assumed that in the energy balance that soil heat flux is relatively constant over small areas. This energy balance component varies greatly over due to microtopographic effects as shown by Kustas et al. (2000). They found that differences at midday on the order of 200 W m^{-2} were not uncommon in the sparse mesquite dune canopies. Hatfield and Prueger (2001) found that within a corn (*Zea mays* L.) canopy the variation among 20 soil heat flux measurements in a 1.5 by 4 m area was 100 W m^{-2} when the canopy was small and variation decreased as the canopy developed. Likewise, soil temperature in the upper 10 cm of the soil profile also showed a large variation with differences at midday of nearly 10°C. This example demonstrates the potential variation that exists in one component of the energy balance that would be critical for simulation of plant water use. We expect less difference in air temperature across this same distance because air is more fluid and has a greater degree of mixing. Most models would assume a single value of soil heat flux or predict a single value. This degree of variation would have a large impact on nutrient cycling, chemical reactions and biological activity.

Water budget methods have been combined with soil variation to estimate corn yields. Timlin et al. (2001) used a water budget from the PLANTGRO model to calculate grain yield as a function of soil water. They incorporated spatial variation of rooting depth into this model and then simulated grain yield. This effort was an expansion of earlier work by Timlin et al. (1986), which defined how soil water variation affected crop production. This type of approach integrates several different components of the soil–plant–atmosphere continuum into a yield assessment tool.

The atmospheric component of agricultural system models needs improvement in the following ways:

1. *Quantification of the variation in atmospheric parameters required in agricultural system models —* Variations exist in time and space. These variations are often assumed to be small; however, as with soil heat flux or air temperature across small distances, they could be quite large. One example of the variation expressed in most models is the effect of slope and azimuth on the solar radiation input.

2. *Incorporation of realistic atmospheric variation induced by microtopographic changes within the field —* Tillage induces change in the surface roughness and at the same time changes the microtopography of the soil surface. This affects the surface energy balance. Evidence for this affect is easily seen when the south facing sides of ridges have a large plant growth than the north side of the ridge even when the ridge is only 25 cm wide. Most models assume a level soil surface and do not account for the changes in surface roughness after tillage.

3. *Incorporation of realistic atmospheric parameters that represent changes within the field microclimate —* Often, meteorological data are used from nearby weather stations that are not representative of the microclimate induced by the developing crop or soil management treatment zones.

4. *Utilization of improved simulation modules that represent actual conditions rather than potential conditions —* Realistic simulations of ET deficit under less-than-full soil water conditions need to be developed. Several models can estimate potential ET (PET) and these work reasonably well for a variety of conditions. The main limitation is how we adjust PET to estimate ET through some type of canopy parameterization process.

5. *Incorporation of sufficient detail of the microclimate within and around agricultural systems that would link with other biotic models (insects, weeds, or diseases) —* This remains one of the largest challenges for models because of the number of interactions that are required to understand and quantify.

6. *Agronomic practices, such as row spacing and row direction, affect the microclimate —* Although some evidence exists regarding the degree of influence of these changes on the soil–plant–atmosphere interaction, there remains a lack of quantitative understanding of these changes throughout the growing season.

Atmospheric components in the models must address a number of challenges. For many of these needs, some qualitative evidence can be used to begin the development of process models, but little quantitative evidence has been collected over a range of environments or cropping systems. Incorporation of these factors into the current models would improve performance and reliability in estimating agricultural systems.

ISSUES IN SCALING

The largest challenge in the further refinement of agricultural systems models is the issue of scaling. As an example, RZWQM is a one-dimensional model, while SWAT is constructed to simulate watershed scale processes using lumped parameters. Each model has its own utility and purpose; however, scaling becomes an issue on agricultural system models because results from fields or portions of a field need to be extrapolated to another scale. This transfer of information across scales (time and space) may be the greatest research need for agricultural systems models. Bierkens et al. (2000) describes a philosophy and process for scaling that provides some useful insights for agricultural systems models. As an illustration, if we use precipitation in a surface runoff model for a watershed then we have to account for the spatial variation in precipitation. For most watersheds, this is unknown. In a study in central Iowa, Hatfield et al. (1999) showed that for individual storms there was a large variation among rain gauges across a 5400 ha watershed in nearly level terrain and during individual convective storms the variation was tenfold of the mean for the storm, but there were no significant differences in the annual totals among the 25 rain gauges. The method of accounting for or understanding spatial and temporal variation will be critical in moving among scales with models. A single rain gauge would not provide an adequate estimate of the variation across the watershed during the times of the year in which estimating the impact of surface runoff would be critical. Using lumped data into a model to estimate surface runoff or water distribution in a watershed is not appropriate.

The ability to move across scales is an assumption that most users make about models. Increasing aggregation size (time or space) can be accomplished by two methods: averaging the model inputs across the desired intervals (lumping) and using those averages as inputs for a single run of the model; or, running the model many times with the individual input data points (distributing) and then integrating the outputs over the desired aggregation scale. Using Figures 16.1 as a case study, we could either run a model with individual values for each soil or compute the weighted average across the field as inputs. Each approach has advantages and disadvantages. Averaging the inputs and running the model once is less time consuming, but nonlinearities in the responses to the inputs would likely reduce accuracy of the overall field estimate. Analyses need to be conducted for a number of models to determine the potential problems in scaling issues. A primary issue will be the method of handling variation within a soil type, as depicted in Table 16.1. One of the most important issues is to determine the scale appropriate parameters for different components of the model. Incorporation of a quantitative treatment of variability within the unit area of the simulation model needs to be evaluated and more importantly compared to observed results to determine if the model treatment of variability is adequate. For example, if a field average determination of leaching losses of a pesticide is required, what would be the difference between simulating a field average set of soil parameters compared with averaging the results for the individual soils within the field? Assessment of the impact of different methods of scaling relative to model inputs and outputs will require data sets collected specifically for this type of model validation study. Development of sampling methods for fields that address both spatial and temporal variation are being undertaken as continuation of the geostatistical approach described by Journel and Huijbrechts (1978).

Problems of this nature have been addressed in several of the models. Haskett et al. (1995) used an aggregation of the input parameters by county across Iowa as input into GYLSIM to estimate regional scale soybean yields. Predicted yield values at the regional and state scale agreed reasonably with reported yield from statewide surveys. To estimate surface and subsurface water quality, the SWAT model has been combined with HUMUS using GIS and regional databases to provide aggregate input into the SWAT model (Srinivasan et al., 1993). Development of GIS layers for different parameters has helped to provide a more realistic regional scale assessment. Spatial patterns are represented in this way for basin and regional scale estimates; however, the greatest challenge may be at the field scale in which crop growth and yield differences are responding to microscale variations. Another approach to scaling within agricultural systems models is to utilize remote sensing inputs as a representation of spatial scale and also feedback into the model prediction. Maas (1988 a, b) described an approach to incorporate remotely sensed observations into crop growth and yield models. Crop growth models were coupled with remotely sensed observations of leaf area index and plant biomass to provide spatial feedback into the model to accommodate periodic calibration of model performance. This technique has a large amount of potential utility for application to crop management models, although the value to the decision making process has yet to be demonstrated.

There remain a number of challenges with scaling. Properties have been estimated with geostatistical methods in order to provide estimates of soil parameters at different locations within a field or management unit. Development in interpolation methods continues to take observations into another representation. Methods such as jackknifing, bootstrapping and kriging have become routine words in many scientists' vocabularies as well as in statistical analyses (Journel and Huijbrechts, 1978). We would suggest that, in the next decade, refinement of these methods should continue as more spatial data are collected that permit rigorous evaluation of these techniques.

CONCLUSIONS

A number of challenges for the future of agricultural system models must be addressed by researchers, both experimentalists and modelers, over the next decade. The authors' goal has been to provide a spark of interest in areas that increasing the value of the current and the next generation of models. These can be examined through the components of current models that express the soil, plant, or atmospheric parameters. On examination, the changes that occurred in agricultural models over the past 10 years, resulted in large strides being made in the assessment of different scales. The primary questions are "Who is the intended user of the model output?" and "What type of decision will be made with the output?" It is unreasonable to assume that the current models will address all of the needs or expectations from a wide range of users. The scientific community has been developing models that accurately simulate single agricultural systems. Few models truly allow for diversity in crop rotations or feedbacks among changes induced by management and the long-term effect on varying cropping or livestock systems.

Future research will have to focus on quantifying the interactions among system components. Endless challenges exist in this area and increased maturity of our understanding of these interactions will help guide our process. The world of decision making and artificial intelligence will provide opportunities for us that we can not imagine today.

REFERENCES

Ahuja, L.R., K.E. Johnsen and K.W. Rojas. 2000a. Water and chemical transport in soil matrix and macropores. In *Root Zone Water Quality Model*, Ahuja, L.R. et al, Eds., Water Resources Publications LLC, Highlands Ranch, CO, 13–50.

Ahuja, L.R. et al., Eds., 2000b. *Root Zone Water Quality Model,* Water Resources Publications LLC, Highlands Ranch, CO.

Arnold, J.G., R. Srinivasan, R.S. Muttiah and J.R. Williams. 1998. Large area hydrologic modeling and assessment: I. Model development, *J. Amer. Water Res. Assoc.,* 34:73–89.

Ascough, J.C. II, G.S. McMaster, M.J. Shaffer, J.D. Hanson and L.R. Ahuja. 1998. Economic and environmental strategic planning for the whole farm and ranch: the GPFARM decision support system, *Proc. First Interagency Hydrologic Modeling Conf.,* U.S. Department of Agriculture/U.S. Geological Survey, U.S. Environmental Protection Agency, Las Vegas, Nevada, April 19–23, 1998.

Baker, D.N., J.R. Lambert and J.M. McKinion. 1983. GOSSYM: A simulator of cotton crop growth and yield, *South Carolina Agric. Exp. Stn. Tech. Bull. 1089.*

Beveridge, C. 2000. The ups and downs of signaling between root and shoot, *New Phytol.,* 147:413–416.

Bierkens, M.F.P., P.A. Finke and P. de Willigen. 2000. *Upscaling and Downscaling Methods for Environmental Research,* Kluwer Academic Publishers, Dordrecht, The Netherlands.

Blaney, H.F. and W.D. Criddle. 1950. *Determining Water Requirements in Irrigated Areas from Climatological and Irrigation Data,* SCS-TP 66, USDA-SCS, U.S. Government Printing Office, Washington, D.C.

Bresler, E. 1991. Soil spatial variability. In *Modeling Plant and Soil Systems, Agronomy 31,* Hanks, R.J. and J.T. Ritchie, Eds., American Society of Agronomy, Madison, WI, 145–180.

Conrad, R. 1996. Soil microorganisms as controllers of atmospheric trace gases (H_2, CO, CH_4, OCS, N_2O and NO), *Microbiol. Rev.,* 60:609–640.

Feddes, R.A., P.J. Kowalik and H. Zaradny. 1978. *Simulation of Field Water Use and Crop Yield,* Pudoc, Wageningen, The Netherlands.

Ferm, M., A. Kasimir-Klemedtsson, P. Weslien and L. Klemedtsson. 1999. Emission of NH_3 and N_2O after spreading of pig slurry by broadcasting or band spreading, *Soil Use Manage,* 15:27–33.

Foy, J.K., W.R. Teague and J.D. Hanson. 1999. Evaluation of the upgraded SPUR model (SPUR2.4), *Ecol. Modelling,* 118:149–165.

Godwin, D., J. Ritchie, U. Singh and L. Hunt. 1990. *A User's Guide to CERES-Wheat V2.10,* 2nd ed., International Fertilizer Development Center, Muscle Shoals, AL.

Grant, R.F. 1995. Dynamics of energy, water, carbon and nitrogen in agricultural ecosystems: simulation and experimental validation, *Ecol. Modelling,* 81:169–181.

Hanks, R.J. and J.T. Ritchie, Eds., 1991. *Modeling Plant and Soil Systems, Agronomy 31,* American Society of Agronomy, Madison, WI.

Hansen, S., H.E. Jensen, N.E. Neilsen and H. Svendsen. 1991. Simulation of nitrogen dynamics and biomass production in winter wheat using the Danish simulation model DAISY, *Fert. Res.,* 27:245–259.

Hanson, J.D. 2000. Generic crop production. In *Root Zone Water Quality Model,* Ahuja, L.R. et al., Eds., Water Resources Publ. LLC, Highlands Ranch, CO.

Haskett, J.D., Y.A. Pachepsky and B. Acock. 1995. Estimation of soybean yields at county and state level using GLYSIM: a case study in Iowa, *Agron. J.,* 87:926–931.

Hatfield, J.L. 1985. Wheat canopy resistance determined by energy balance techniques, *Agron. J.,* 77:279–283.

Hatfield, J.L. and R.G. Allen. 1996. Evapotranspiration estimates under deficient water supplies, *J. Irrig. Drain. Eng.,* 122:301–308.

Hatfield, J.L. and J.H. Prueger. 2001. Variation of soil heat flux under a corn canopy, *Agron. J.,* (in review).

Hatfield, J.L., J.H. Prueger and D.W. Meek. 1999. Spatial variation of rainfall over a large watershed in central Iowa, *Theor. Appl. Clim.,* 64:49–60.

Hill, H.S.J., D. Butler, S.W. Fuller, G. Hammer, D. Holzworth, H.A. Love, H. Meinke, J.W. Mjelde, J. Park and W. Rosenthal. 2000. Effects of seasonal climate variability and the use of climate forecasts on wheat supply in the United States, Australia and Canada. In *Impacts of El Niño and Climate Variability on Agriculture,* Rosenzweig, C.K. Boote et al., Eds., American Society of Agronomy Special Publ. No. 63. Madison, WI, 101–123.

Jarvis, A.J. and W.J. Davies. 1998. The coupled response of stomatal conductance to photosynthesis and transpiration, *J. Exp. Bot.,* 49:399–406.

Jones, C.A. and J.R. Kiniry, Eds., 1986. *CERES-Maize: A Simulation Model of Maize Growth and Development,* Texas A&M University Press, College Station, TX.

Jones, J.W. et al. 1991. Soybean development. In *Modeling Plant and Soil Systems, Agronomy 31,* Hanks, R.J. and J.T. Ritchie, Eds., American Society of Agronomy, Madison, WI, 71–90.

Journel, A.G. and Ch.J. Huijbrechts. 1978. *Mining Geostatistics,* Academic Press, London.

Kiniry, J.R. 1991. Maize phasic development. In *Modeling Plant and Soil Systems, Agronomy 31,* Hanks, R.J. and J.T. Ritchie, Eds., American Society of Agronomy, Madison, WI, 55–70.

Kiniry, J.R., J.R. Williams, P.W. Gassman and P. Debaeke. 1992. A general process-oriented model for two competing plant species, *Trans. ASAE,* 35:801–810.

Kustas, W.P., J.H. Prueger, J.L. Hatfield, K. Ramalingam and L.E. Hipps. 2000. Variability of soil heat flux from a mesquite dune site, *Agric. For. Meteorol.,* 103:249–264.

Lascano, R.J. and C.H.M. van Bavel. 1986. Simulation and measurement of evaporation from a bare soil, *Soil Sci. Soc. Am. J.,* 50:1127–1132.

Lascano, R.J. and J.L. Hatfield. 1992. Spatial variability of evaporation along two transects of a bare soil, *Soil Sci. Soc. Am. J.,* 56:341–346.

Leonard, R.A., W.G. Knisel and D.S. Still. 1987. GLEAMS: groundwater loading effects of agricultural management systems, *Trans. ASAE,* 30:1403–1418.

Maas, S.J. 1988. Use of remotely sensed information in agricultural crop growth models, *Ecol. Modelling,* 41:247–268.

Maas, S.J. 1988b. Using satellite data to improve model estimates of crop yield, *Agron. J.,* 80:655–662.

Meinke, H., G.L. Hammer, H. van Keulen and R. Rabbinge. 1998. Improving wheat capabilities in Australia from a cropping systems perspective. III. The integrated wheat model (I-Wheat), *Eur. J. Agron.,* 8:101–116.

Parton, W.J., D.S. Ojima, C.V. Cole and D.S. Schimel. 1994. A general model for soil organic matter dynamics: Sensitivity to litter chemistry, texture and management. In *Quantitative Modeling of Soil Forming Processes,* Bryant, R.B. and R.W. Arnold, Eds., Soil Science Society of America Special Publication No. 39, Madison, WI, 147–167.

Rillig, M.C., S.F. Wright, B.A. Kimball, P.J. Pinter, Jr., G.W. Wall, M.J. Ottman and S.W. Leavitt. 2001. Elevated carbon dioxide (free-air CO_2 enrichment, FACE) and irrigation effects on water stable aggregates in a sorghum field: a possible role for arbuscular mycorrhizal fungi, *Global Change Biol.,* 7:333–337.

Ritchie, J.T. 1991. Wheat phasic development. In *Modeling Plant and Soil Systems, Agronomy 31,* Hanks, R.J. and J.T. Ritchie, Eds., American Society of Agronomy, Madison, WI, 31–54.

Rosenthal, W.D., G.L. Hammer and D. Butler. 1998. Predicting regional grain sorghum production in Australia using spatial data and crop simulation modeling, *Agric. For. Meteorol.,* 91:263–274.

Rossi, P. 1968. *Francis Bacon: From Magic to Science,* University of Chicago Press, Chicago, IL.

Shaffer, M.J. and W.E. Larson, Eds., 1987. NTRM, A soil-crop simulation model for nitrogen, tillage and crop-residue management, *USDA-ARS Conservation Research Report 34-1,* National Technical Information Service, Springfield, VA.

Sharpley, A.N. and J.R. Williams. 1993. EPIC, erosion/productivity impact calculator: 1. Model documentation, *USDA-ARS Technical Bulletin. No. 1768,* Washington, D.C.

Shuttleworth, W.J. and J.S. Wallace. 1985. Evaporation from sparse crops — an energy combination theory, *Q. J. R. Meteorol. Soc.,* 11:839–855.

Srinivasan, R., J.G. Arnold, R.S. Muttiah, C. Walker and P.T. Dyke. 1993. Hydrologic unit model for the United States (HUMUS). In *Proc. of Advances in Hydro-Science and Engineering,* University of Mississippi, Oxford, MS, 451–456.

Thornton, P.K., H.W.G. Booltink and J.J. Stoorvogl. 1997. A computer program for geostatistical and spatial analysis of crop model outputs, *Agron. J.,* 89:620–627.

Timlin, D.J. et al. 1986. Modeling corn grain yields in relation to soil erosion using a water budget approach, *Soil Sci. Soc. Am. J.,* 50:718–723.

Timlin, D.J., Y. Pachepsky, V.A. Snyder and R.B. Bryant. 2001. Water budget approach to quantify corn grain yields under variable rooting depths, *Soil Sci. Soc. Am. J.,* 65:1219–1226.

Tsuji, G.Y., G. Hoogenboom and P.K. Thornton. 1998. *Understanding Options for Agricultural Production,* Kluwer Academic Press, New York, NY.

USDA-SCS. 1992. *States Soil Geographic Database (STATSGO) Data User's Guide,* Publ. No. 1492, U.S. Government Printing Office, Washington, D.C.

USDA-SCS. 1984. *Soil Survey of Story County, Iowa,* USDA-SCS, Washington, D.C.

Wagner-Riddle, C. and G.W. Thurtell. 1998. Nitrous oxide emissions form agricultural fields during winter and spring thaw as affected by management practices, *Nutrient Cycling in Agroecosys.,* 52:151–163.

Williams, R.R., C.A. Jones and P.T. Dyke. 1984. A modeling approach to determining the relationship between erosion and soil productivity, *Trans. ASAE,* 27:129–142.

Young, R.A., C. A. Onstad, D. D. Bosch and W. P. Anderson. 1987. AGNPS, agricultural non-point-source pollution model, *Conserv. Res. Rep. 35,* USDA-ARS, Morris, MN.

Yu, B. 2000. Improvement and evaluation of CLIGEN for storm generation, *Trans. ASAE,* 43:301–307.

Index